Algebra II
Student Workbook

Pathways 3rd Edition

Marilyn P. Carlson
Professor, Mathematics Education
Arizona State University

Alan OBryan
Arizona State University

Kacie Joyner
Arizona State University

Phoenix

Algebra II – Student Workbook, Pathways Third Edition

Published by Rational Reasoning, LLC., 477 N Mondel Dr, Gilbert, AZ 85233.
Copyright © 2014 by Rational Reasoning, LLC. All rights reserved. No part of this book may be reproduced or utilized in any form or by any means, electronic or mechanical, including photocopying, recording or by any information storage and retrieval system without permission in writing.

This book was typeset in 11/12 Times Roman

ISBN 978-0-9845795-8-7

Printed in the United States of America

10 9 8 7 6 5 4 3 2

www.rationalreasoning.net

Table of Contents

MODULE 1: Changes in Quantities, Constant Rate of Change, and Linear Functions............................. 1
Module 1 promotes a conceptually powerful understanding of constant rate of change and leverages this image to bring meaning to linear functions. Students learn about constant rate of change as a proportional relationship between the changes in two quantities, and thus understand how the change in one quantity can be used to find the corresponding change in another quantity. When students apply this reasoning to find new function values, and then generalize the process, they are able to develop meaningful formulas representing linear relationships. After students have spent time working with these first examples of functions, they learn the basics of function notation and domain and range. These ideas reappear frequently throughout the other modules.

MODULE 2: Sequences and Series... 79
Module 2 leverages student understanding of functions to develop meaning for sequences as function relationships between term values and term positions. Students explore the long-term behavior of a sequence to build an understanding of *limit*, and then apply their understanding of linear functions to make sense of arithmetic sequences. Students then consider patterns formed by repeated multiplication (geometric sequences) as an introduction to exponential functions. Students also learn how to build a sequence of partial sums, find the sum of a finite series, represent series with sigma notation, and look for a limit to the sequence of partial sums to find the sum of an infinite series.

MODULE 3: Exponential and Logarithmic Functions.. 131
In Module 3 students build on their understanding of discrete exponential growth first seen in geometric sequences. Students consider how to model continuous exponential growth, including how to determine partial growth factors or n-unit growth factors to calculate a greater variety of output values for an exponential function or to rewrite an exponential function in a different (but equivalent) form. Students then consider logarithmic functions as representing the same relationship as an exponential function but with the input and output quantities reversed. This leads to a general treatment of inverse functions and function composition at the end of the module.

MODULE 4: Quadratic Functions... 197
In Module 4 students learn about what makes quadratic functions unique (the average rates of change over successive intervals change at a constant rate), and learn how to find extrema, zeros, and lines of symmetry for quadratic functions and their graphs. Students learn and practice factoring techniques and apply these skills to help solve quadratic and polynomial equations, including systems of equations involving quadratic functions. Students conclude the module by exploring complex numbers and building meaning for and finding complex roots of quadratic functions.

MODULE 5: Polynomial Functions... 273
Module 5 formalizes and strengthens ideas developed throughout the previous modules. Students begin the module by exploring the commonly seen "Box Problem" as a way to understand polynomial functions in context and in a variety of representations. Students then work to understand and explain changing rates of change for a function and to generalize characteristics of polynomial functions such as end behavior and repeated roots. Finally, students begin to study function transformations and look at transformations as a way of describing the relationships between two pairs of co-varying quantities.

MODULE 6: Rational Functions... 329
Module 6 continues the focus on reasoning about how the values of two quantities change together. Students apply their previous work with the long-run behavior of polynomial functions to understand the long-run behavior of rational functions. Students also examine a function's behavior as it get closer and closer to some value(s) of the input variable that makes the function undefined in order to build a conceptual foundation for vertical asymptotes. Students end the module by learning how to solve rational equations by first finding common denominators for the expressions involved.

Table of Contents

MODULE 7: Radical Functions.. 367
In Module 7 students learn techniques for solving radical equations and how a solution process can generate extraneous solutions. Students explore the co-variational relationship indicated by radical functions and apply ideas about function transformations from Module 5 to this function family.

MODULE 8: Trigonometric Functions... 399
The investigations and homework for this module are available online at www.rationalreasoning.net. Module 8 introduces the idea of periodic behavior and how to model periodic phenomena. Students begin the module by building meaning for the concept of angle measure and learn how to measure angles in a variety of units. Students then study circular motion and develop the sine, cosine, and tangent functions as representing important relationships that emerge during circular motion. Students continue to explore ideas of function domain, range, composition, inverse, and transformations in the context of trigonometric functions, and they end the module by exploring and verifying the Pythagorean Identity.

Module 9: Probability... 401
In this module you will learn how to organize, count, and report the set of results that can occur when you repeat an activity with random outcomes and how to calculate and interpret the probability that a specific event from this activity occurs. You will also consider how to calculate the probability that more than one event occurs (such as drawing two specific cards one after another from a well-shuffled deck of cards) in a variety of contexts.

Module 10: Univariate Statistics.. 469
In this module you will learn how to calculate and interpret the mean and standard deviation for a set of numbers and explore what it means to have a distribution of results when an activity is repeated many times. You will study a special type of distribution called *normal distribution* with a focus on making predictions and answering questions about a population. You will also learn about sampling and how to use the results of simulations and samples to gain insights about a population.

Module 11: Bivariate Statistics.. 541
In this module you will explore patterns generated by comparing sets of measurements for two different quantities and attempt to describe observed trends with mathematical formulas. You will consider when this technique is appropriate and when it is not appropriate. You will learn how to mathematically compare the "fit" of different functions that attempt to describe trends and how to use technology to find the function with the best fit. Finally, you will use functions describing trends to make predictions and consider the benefits and dangers of this practice.

Algebra II | **Module 1: Investigation 1**
Reasoning with Quantities

> ### Quantity
> A *quantity* is some attribute of an object that you can imagine as being measured.
>
> *Examples:*
>   i) your height
>   ii) the number of people in this room
>   iii) the amount of money in your pocket
>   iv) the distance around a 400-meter track

1. Sarah decides to ride a Ferris wheel at a local fair. The animation shown models her ride around the Ferris wheel. Assume she is seated in the brown cart and starts at the bottom of the Ferris wheel. Watch the Ferris wheel animation as Sarah makes two full rotations and then respond to the following questions.

    a. What varying quantities can you identify in this situation?

    b. How does Sarah's *height above the ground* vary?

    c. How does the *time since the Ferris wheel began moving* vary?

    d. How do the quantities *Sarah's height above the ground* and *time since the Ferris wheel began moving* vary together?

Module 1: Investigation 1
© 2014 Carlson, O'Bryan & Joyner

Algebra II

Module 1: Investigation 1
Reasoning with Quantities

> **Points and Graphs**
>
> A ***point*** shows the corresponding values of two quantities that we are comparing. A ***graph*** of the relationship between the two quantities is a collection of points showing all pairs of corresponding values.

2. a. The given axes are labeled *time since the Ferris wheel began moving* and *Sarah's height above ground*. Your teacher will play the animation again while you represent how these quantities are changing using your fingers. Each time the video stops, plot a point on the axes that represents the corresponding values of *time since the Ferris wheel began moving* and *Sarah's height above ground*.

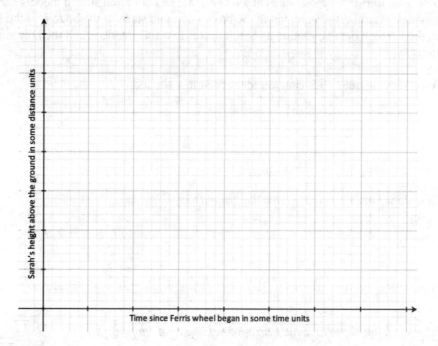

   b. On the axes in part (a), show all of the corresponding values of *Sarah's height above the ground* and *the amount of time since the Ferris wheel began moving* as Sarah travels around the Ferris wheel twice. (Your graph can be a rough sketch of how these two quantities change together – you do not need to worry about making your graph exact.)

   c. Choose a point on your graph. Describe the information that this point conveys about the situation.

Algebra II

Module 1: Investigation 1
Reasoning with Quantities

In Exercises #3-6, use the following context.
Sam rides a different Ferris wheel. He boards at the bottom and rides around a few times before getting off. The following graph represents Sam's height above the ground (in feet) with respect to the amount of time since the Ferris wheel began moving (in minutes) for one complete rotation around the wheel.

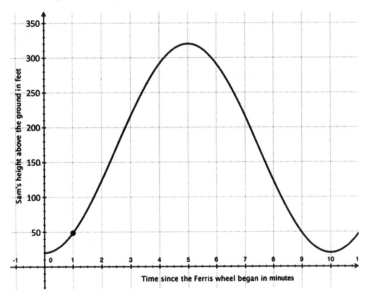

3. The point (1, 50) is plotted on the graph. Explain what information this point conveys in the context of the problem.

4. a. How much did the *time since the Ferris wheel began moving* **change** from 1 to 4 minutes?

   b. Represent this change on the graph.

   c. Represent the corresponding **change** in *Sam's height above the ground* from 1 to 4 minutes since the Ferris wheel began moving.

   d. Approximately how much did *Sam's height above the ground* **change** over this time interval? Explain how you determined the change.

   e. Estimate Sam's height above the ground 4 minutes since the Ferris wheel began moving and plot the point representing this information.

Module 1: Investigation 1
© 2014 Carlson, O'Bryan & Joyner

Algebra II

Module 1: Investigation 1
Reasoning with Quantities

5. Plot the point (5, 320) on the graph and explain its meaning in the context of the problem.

6. a. How much did the *time since the Ferris wheel began moving* **change** from 5 to 7 minutes?

   b. Represent this change on your graph.

   c. Represent the corresponding change in *Sam's height above the ground* from 5 to 7 minutes since the Ferris wheel began moving?

   d. Approximately what was the **change** in *Sam's height above ground* over this time interval? Explain how you determined the value.

   e. Estimate *Sam's height above the ground* 7 minutes since the Ferris wheel began moving and plot the point representing this information.

Algebra II

Module 1: Investigation 2
Changes in Quantities' Values

1. In high school Maria grew from 62 inches tall to 74 inches tall. What was Maria's change in height during high school?

2. Matt started a new diet two months ago. His current weight is 181.2 pounds and his weight before starting the diet was 196.7 pounds. What was the change in Matt's weight?

3. When Cyndi woke up this morning (6:00 am) the temperature outside was $-6°F$. When she got home from school (4:00 pm) the temperature outside was $28°F$. What was the change in temperature from 6:00 am to 4:00 pm?

Variable

A ***variable*** is a letter used to represent the possible values for the measurement of a varying quantity.

Ex: A bathtub, with a maximum volume of 42 gallons, starts off full and then drains to empty. Let v be the volume of water remaining in the bathtub (in gallons). Then v can represent any possible number from 0 to 42 as the water drains, such as $v = 38$ or $v = 5.7$.

---

4. a. Define variables to represent the quantities' values in Exercises #1-3.

   b. What values might each variable take on?

5. In January 2012, Caitlyn began saving money to make a down payment on a house. Let m be the number of months she's been saving and let v represent the amount of money she's saved (in dollars).
   a. What is the change in v from $v = 940$ to $v = 2,360$? What does this represent?

Algebra II
Module 1: Investigation 2
Changes in Quantities' Values

b. The change in v described in part (a) occurred as m changed from $m=3$ to $m=7$? What is the change in m and what does it represent?

6. Let x, y, and z represent the values of three different quantities.
   a. If the value of x goes from $x=1$ to $x=10$, what is the change in x?

   b. If the value of x changes to $x=-5$ from $x=3$, what is the change in x?

   c. If the value of z changes to $z=18.07$ from $z=3.15$, what is the change in z?

   d. If the value of y goes from $y=6$ to $y=p$, what is the change in y?

Changes in Quantities' Values

Suppose x is the value of *Quantity A* and y is the value of *Quantity B*. Then Δx represents a **change** in the value of *Quantity A* and Δy represents a **change** in the value of *Quantity B*.

Ex 1: If we let x change from $x=4$ to $x=11$, then $\Delta x = 7$ which we calculate with the expression $11-4$.

Ex 2: If we let y change from $y=8$ to $y=5$, then $\Delta y = -3$ which we calculate with the expression $5-8$.

---

7. Suppose Gerald bought a tree and planted it in his yard last summer. At the time of planting it was 4.5 feet tall. Let h represent the height of the tree in feet.
   a. What is the difference between saying $h=5$ and $\Delta h = 5$?

   b. If we substitute $h=6.8$ into the expression $h-4.5$, what are we calculating?

c. If we substitute $h = 10.7$ into the expression $h - 6.8$, what are we calculating?

8. Fill in the following tables showing the appropriate changes in the variable values for consecutive entries in each table.

a.

x	Δx
2	
5	
21	
30	

b.

y	Δy
10	
7.4	
2.1	
−1.3	

c.

w	Δw
−7	
−2	
2	
−3	

9. Each of the following graphs shows two ordered pairs (x, y). Find Δx and Δy from the point on the left to the point on the right.

a.
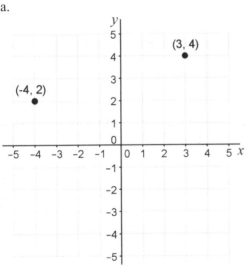

b.
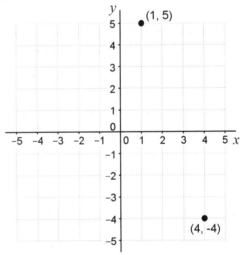

10. Show how we can illustrate the values of Δx and Δy in Exercise #7 on the graphs.

11. Do your answers to Exercise #9 change if we instead want to find Δx and Δy from the point on the right to the point on the left? Explain.

Algebra II

Module 1: Investigation 2
Changes in Quantities' Values

12. Suppose we start with a value of $x=-3$.
    a. What expression represents the change in value for x from $x=-3$ to any possible value of x?

    b. Repeat part (a) if our initial reference value is $x=-7$ instead of $x=-3$.

    c. Repeat part (a) if our initial reference value is $x=1.2$ instead of $x=-3$.

13. Think about the idea of the change in a variable from one value to another. What do the following expressions represent when we substitute a value for the variable?
    a. $x-34$
    b. $y-(-2)$ [or $y+2$]

    c. $r-4.7$
    d. $p-(-13.8)$ [or $p+13.8$]

Algebra II

Module 1: Investigation 3
Constant Rate of Change

You've probably heard someone use the phrase "constant speed" before. For example, you might hear "He was driving at a constant speed of 45 miles per hour." In this investigation we will explore the idea of **constant speed** (and more generally **constant rate of change**) and link this back to your work with changes in quantities from Investigation 2.

For Exercises #1-6 you may use the "Jane Walking" applet to help you visualize the situation.
Jane is walking from her home to work. When she passes her mailbox she is 25 feet from her house. Between the mailbox and a tree she walks at a constant speed, covering 40 feet in 8 seconds.

1. After Jane passes the mailbox, how long does it take her to travel 20 feet? Explain your thinking.

2. After Jane passes the mailbox, how far does she travel in 2 seconds? 6 seconds? Explain your thinking.

3. Let's draw line segments to represent the total distance traveled from the mailbox to the tree and the amount of elapsed time it takes to cover this distance.

    change in distance traveled of 40 feet
    _____

    change in time elapsed of 8 seconds
    _____

    Demonstrate how we can use these line segments to represent the reasoning in Exercises #1-2.

4. Use the line segments in Exercise #3 and the applet to help you answer the following questions.
    a. How far will she travel during any 1-second time interval between the mailbox and the tree?

    b. How long will it take her to travel any 1-foot distance between the mailbox and tree?

Algebra II

Module 1: Investigation 3
Constant Rate of Change

5. How do your answers to Exercise #4 help you find
   a. how far Jane travels for *any* change in the time elapsed since she passed the mailbox? *(For example, how far does Jane travel for a change in time elapsed of 2.8 seconds?)*

   b. how long it takes Jane to travel *any* distance between the mailbox and tree? *(For example, how long does it take Jane to travel 33 feet?)*

6. Think about the work you did in Exercises #1-5. What does it mean for an object to move at a constant speed? *(Note: Please say something more than "The speed doesn't change" – be descriptive and reference specific quantities.)*

7. Felipe was walking to school today. Assume that he walked at a constant speed during the entire trip, and also suppose that during one part of the trip he walked 70 feet in 16 seconds.
   a. Provide at least four conclusions we can draw from the given information.

Algebra II

b. During this interval, how far did Felipe travel in 3 seconds?

c. Does your answer to part (b) depend on which 3-second interval we're talking about? Explain.

d. How long did it take Felipe to travel any 10-foot distance during this part of his trip?

8. Assume that a baseball and a tennis ball are both traveling at constant speeds (but not necessarily the same speed).
   a. What does it mean to say that the tennis ball is traveling *faster* than the baseball? *Be descriptive, and reference the specific quantities involved.*

   b. What does it mean if the baseball and tennis ball have the same constant speed? *Be descriptive, and reference the specific quantities involved.*

Algebra II

Module 1: Investigation 3
Constant Rate of Change

9. Suppose we want to rank the following objects in order from fastest to slowest. Assume each object is traveling at a constant speed (but not necessarily the same constant speed).

   Bike: 11 feet in 0.5 second
   Car: 212 feet in 10 seconds
   Bird: 4.24 feet in 0.2 second

   Runner: 112 feet in 7 seconds
   Football: 52.5 feet in 3 seconds
   Water Balloon: 24 feet in 1.5 seconds

   a. Why is it difficult to compare the objects' speeds based on the information given?

   b. What strategy can we use to help us complete the task?

   c. Rank the objects in order from fastest to slowest.

10. What are the benefits of stating speeds in terms of distance traveled in one unit of time, such as 20 miles traveled in one hour ("20 mph""), 9.8 meters traveled in one second ("9.8 m/s"), and so on? Brainstorm some ideas with a classmate and summarize your thinking.

Algebra II

Module 1: Investigation 3
Constant Rate of Change

11. Suppose that Joe is jogging at a constant speed of 0.12 miles per minute. Let *d* represent the total distance (in miles) Joe has traveled during his jog and let *t* represent the total number of minutes he's been jogging.
    a. Use Δ notation to represent a 4-minute time period during Joe's jog.

    b. How far does Joe travel during any 4-minute time period? Represent your answer using Δ notation.

    c. Repeat parts (a) and (b) if instead of a 4-minute time period we look at a
       i) 7.8-minute time period

       ii) 0.3-minute time period

       iii) any *x*-minute time period

    d. For any change in *t*, what happens to the change in *d*? Explain, then represent your thinking using Δ notation.

So far we've focused on the idea of *constant speed*. However, this is just a specific example of the more general concept of *constant rate of change*.

Constant Rate of Change

A ***constant rate of change*** of one quantity with respect to another quantity exists when, for **any** uniform change in one quantity's value, the other quantity's value changes by equal amounts.

This is equivalent to saying that the change in one quantity's value is always a constant multiple of the change in the other quantity's value.

Algebra II

Module 1: Investigation 3
Constant Rate of Change

12. If we know that the total cost of purchasing a bag of walnuts increases at a rate of $7 per pound, then it's easy to determine the change in total cost if we change the amount of walnuts we purchase. Let c represent the total cost of purchasing walnuts (in dollars) and let w represent the total weight of walnuts purchased (in pounds).

    a. Suppose we have some walnuts in a bag and we add 3 more pounds of walnuts to the bag. Represent the change in the number of pounds of walnuts using Δ notation.

    b. How much does the total cost of the walnuts we purchase change? Represent your answer using Δ notation.

    c. Repeat parts (a) and (b) if instead of adding 3 more pounds of walnuts we
        i) add 2.8 pounds of walnuts to the bag.

        ii) add 0.65 pounds of walnuts to the bag.

        iii) remove 1.2 pounds of walnuts from the bag.

    d. For any change in w, what happens to the change in c? Explain, then represent your thinking using Δ notation.

    e. If your friend gives you $10 to purchase more walnuts, how many additional pounds of walnuts could you add to your bag?

13. Suppose we want to add some sand to a child's sandbox, and that we know an 8-liter bag of sand weighs 25.6 pounds. Define variables to represent the quantities in this context and then represent the relationship described using Δ notation.

Algebra II

Module 1: Investigation 5
Applying Constant Rate of Change to Determine New Values

> Investigation 4: *Practice with Constant Rate of Change* and associated homework is available online at www.rationalreasoning.net.

1. Suppose we know that $\Delta y = m \cdot \Delta x$ for some constant m, and we are given the information in the following table. What is the value of m?

x	y
−7	15
−4	9
1	−1
9	−17

For Exercises #2-4, suppose we know that the changes in the values of two variables are related according to $\Delta y = 4 \cdot \Delta x$.

2. a. If we start off at $x = 7$ and let x change to be $x = 10$, what is the change in x?

   b. By how much does y change for the change in x you found in part (a)?

   c. Suppose we know that $y = 15$ when $x = 7$. What is the value of y when $x = 10$? How did you find this?

3. a. If we end at $x = 12$ after starting at $x = 7$, what is the change in x?

   b. By how much does y change for the change in the value of x you found in part (a)?

   c. Suppose we know that $y = 15$ when $x = 7$. What is the value of y when $x = 12$?

4. a. If we start off at $x = 7$ and let x change to be $x = 4$, what is the change in x?

   b. By how much does y change for the change in the value of x you found in part (a)?

   c. Suppose we know that $y = 15$ when $x = 7$. What is the value of y when $x = 4$?

Algebra II

Module 1: Investigation 5
Applying Constant Rate of Change to Determine New Values

5. The given graph displays the point $(7, 15)$. Show how we can represent the reasoning from Exercises #2-4 visually on the graph.
   Hint: Use horizontal line segments to represent changes in the value of x and vertical line segments to represent changes in the value of y.

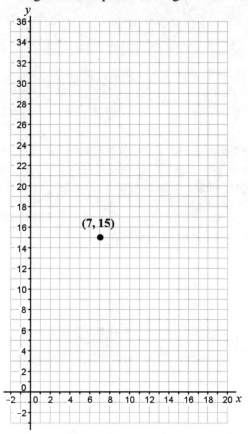

6. Apply your understanding of constant rate of change to complete the following table of values given that $\Delta y = 4 \cdot \Delta x$. *Hint: Your first step should be to determine the change in x from one value to another value.*

x	y
−2	
5	
7	15
9.5	
10.25	

Algebra II

Module 1: Investigation 5
Applying Constant Rate of Change to Determine New Values

In Exercises #2-6 we thought about how to find a new value for y by applying our understanding of constant rate of change. We started by finding a change in the value of x and then used our constant rate of change to determine the corresponding change in the value of y. Finally we used this change in y to find a final (new) value for y. Use this technique and reasoning to complete the remaining tasks in this investigation.

7. Suppose you are considering shipping a package with a certain shipping company. Let w represent the weight of the package (in pounds) and let c represent the cost of shipping the package (in dollars). Furthermore, suppose $c = 15$ when $w = 5.5$ and that $\Delta c = 1.20 \cdot \Delta w$.

   a. What is the value of c when $w = 7.5$? What does this tell us?

   b. What is the value of c when $w = 0$? What does this tell us?

8. Given that $\Delta r = -1.25 \cdot \Delta h$, complete the following table of values.

h	r
–7	
–4	
4	6
6	
18	

Linear Function

If there is a constant rate of change of one quantity with respect to another quantity, then we say that the relationship is a ***linear function***.

Ex 1: Suppose $\Delta y = 2 \cdot \Delta x$. Then we say that "$y$ is a linear function of x."

Ex 2: Suppose a car is traveling at 50 mph. Let d represent the distance the car has traveled from some location (in miles) and t represent the corresponding elapsed time (in hours) since leaving the location, then $\Delta d = 50 \cdot \Delta t$. Then we say that the distance traveled is a linear function of the time elapsed.

Algebra II

Module 1: Investigation 5
Applying Constant Rate of Change to Determine New Values

9. Use the given graph to answer the questions that follow given that $\Delta y = \Delta x$.

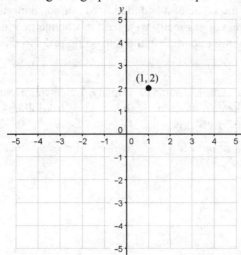

   a. What does it mean to say $\Delta y = \Delta x$?

   b. What is the value of y when $x = 3.5$? Represent your reasoning on the graph.

   c. What is the value of y when $x = -4$? Represent your reasoning on the graph.

10. The length of a burning candle decreases 1.6 inches every hour at a constant rate.

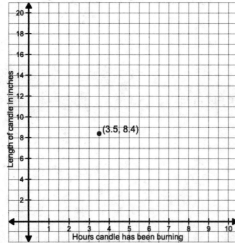

   a. What information does the point (3.5, 8.4) represent in this situation?

   b. If L represents the length of candle remaining (in inches) and t represents the number of hours the candle has been burning, write a statement showing the relationship between the changes in these variables.

   _____ = _____

   c. What is the length of the candle 5.5 hours since it began burning? Find the answer and demonstrate your reasoning on the graph.

   d. What is the length of the candle 1.7 hours since it began burning? Find the answer and demonstrate your reasoning on the graph.

   e. What was the original length of the candle (before burning)?

Algebra II

Module 1: Investigation 6
Developing a Formula for Linear Functions

1. Let $\Delta y = 3 \cdot \Delta x$.

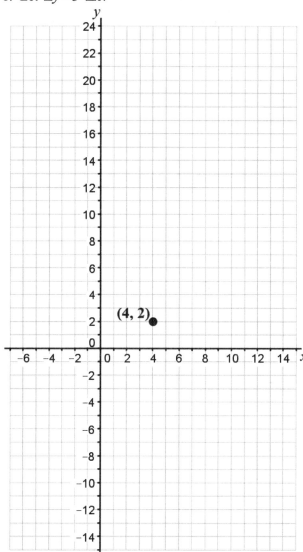

For parts (a) through (c) you are given a value for x. For each, do the following.

i) Determine the change in x from $x = 4$ to the given x-value. Give the answer and write the expression that calculates this answer.

ii) Find the corresponding change in y for the change in x from part (i). Give the answer and write the expression that calculates this answer.

iii) Determine the new value of y for the given value of x. Give the answer and write the expression that calculates this answer.

iv) Show how you can represent your thinking in parts (i)-(iii) on the graph.

a. $x = 11$

b. $x = 5.75$ c. $x = -1$

d. How can we determine the value of y for some unknown x value, such as $x = n$?

Algebra II

Module 1: Investigation 6
Developing a Formula for Linear Functions

In Exercises #2-4 we will continue to use the linear relationship from Exercise #1.

2. Let $x = 10$. The process to find the corresponding value of y is shown. Use this to answer the given questions.

   $y = 3(x-4) + 2$
   $y = 3(10-4) + 2$ a. What does $10-4$ represent?
   $y = 3(6) + 2$ b. What does $3(6)$ represent?
   $y = 18 + 2$ c. What does $18 + 2$ represent?
   $y = 20$ d. What does 20 represent?

3. Let $x = 1.5$. The process to find the corresponding value of y is shown. Use this to answer the given questions.

   $y = 3(x-4) + 2$
   $y = 3(1.5-4) + 2$ a. What does $1.5 - 4$ represent?
   $y = 3(-2.5) + 2$ b. What does $3(-2.5)$ represent?
   $y = -7.5 + 2$ c. What does $-7.5 + 2$ represent?
   $y = -5.5$ d. What does -5.5 represent?

4. The linear function formula allows us to find the value of y for *any* value of x.

   $y = 3(x-4) + 2$ a. What does $x-4$ represent?

                          b. What does $3(x-4)$ represent?

                          c. What does $3(x-4) + 2$ represent?

Algebra II
Module 1: Investigation 6
Developing a Formula for Linear Functions

> **Linear Function Formula**
>
> If y is a linear function of x with a constant rate of change of m and if (x_1, y_1) is an ordered pair solution of the function, then the following is a formula for the linear relationship.
>
> *Ex:* If the constant rate of change of y with respect to x is -3, and if $(-1, 7)$ is a point on the graph, then the formula for the linear relationship is

5. The constant rate of change of y with respect to x is 3.5, and $(6, -2)$ is a point on the graph.
   a. Write the formula for the linear function that calculates y given any value of x.

   b. Find the value of y when $x = 14$.

6. The constant rate of change of p with respect to t is -4, and $(t, p) = (-3, 5)$ is a point on the graph.
   a. Write the formula for the linear function that calculates p given any value of t.

   b. Find the value of p when $t = -10$.

Algebra II

Module 1: Investigation 6
Developing a Formula for Linear Functions

7. Write the formula that defines the linear relationship represented in each of the following graphs.
   a. 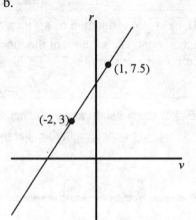   b.

8. Write the formula that defines the linear relationship given in each of the following tables.
   a.

x	y
−3	−23
−1	−13
4	12
9	37

   b.

w	d
−7	−2
−4	−5
2	−11
20	−29

Algebra II
Module 1: Investigation 6
Developing a Formula for Linear Functions

9. Recall the burning candle context from Investigation 5. The length of a burning candle changes at a constant rate of −1.6 inches per hour. When the candle had been burning for 3.4 hours it was 8.4 inches long. Let L represent the length of candle remaining (in inches) and let t represent the number of hours the candle has been burning.

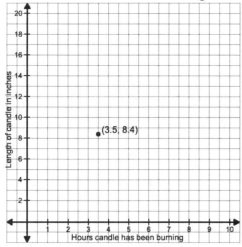

a. Write the formula that will calculate the length of candle remaining (in inches) given the number of hours the candle has been burning.

b. Describe what the different parts of the formula represent.

c. Use your formula to find the length of candle remaining for each of the following number of hours burned.
   i. 6 hours
   ii. 3 hours
   iii. 0 hours (prior to burning)

d. Graph the function. What does the graph represent?

In this investigation we developed the general linear formula $\boxed{y = m(x - x_1) + y_1}$. This is one version of *the point-slope formula for a linear function* [named this because we can write the formula if we know the constant rate of change (which is sometimes called the slope) and one point]. Most textbooks give the point-slope formula as $\boxed{y - y_1 = m(x - x_1)}$. The two formulas provide the same information but one is solved for y.

Algebra II

Module 1: Investigation 6
Developing a Formula for Linear Functions

10. Explain why $y - y_1 = m(x - x_1)$ is just another way of writing $\Delta y = m \cdot \Delta x$.

11. Given the formula $y - 1 = 4(x - 9)$, answer the following questions.
    a. What is the constant rate of change of *y* with respect to *x*?

    b. Based on the formula, what point do we know for sure must be on the graph of the function?

    c. For some ordered pair solution (x, y), explain what each of the following represents.
        i. $x - 9$ ii. $4(x - 9)$ iii. $y - 1$

    d. Solve for *y* to rewrite the formula in the form we used earlier in this investigation.

12. In the candle burning context we could have written $L - 8.4 = -1.6(t - 3.5)$. Explain what each of the following represents.
    a. $t - 3.5$ b. $-1.6(t - 3.5)$ c. $L - 8.4$

Algebra II
Module 1: Investigation 8
Slope-Intercept Form of a Linear Function

> Investigation 7: *Extra Practice with Linear Functions* and associated homework is available online at www.rationalreasoning.net.

In previous investigations we learned how to write the formula for a linear relationship if we know the constant rate of change of one quantity with respect to the other quantity (m) and one ordered pair for the relationship (x_1, y_1): $\boxed{y = m(x - x_1) + y_1}$, or alternatively $\boxed{y - y_1 = m(x - x_1)}$. In this investigation we will explore the special case where $x_1 = 0$ (we are using the vertical intercept as our reference point).

1. Suppose y changes at a constant rate of -2 with respect to x.

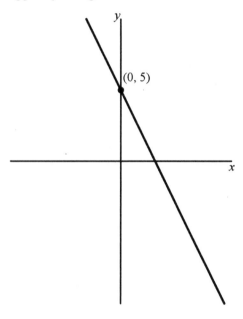

   a. What is the change in x from the given reference point to $x = 7$? Represent this on the graph.

   b. What is the value of y when $x = 7$? Represent this on the graph.

   c. What is the change in x from the given reference point to $x = -1$? Represent this on the graph.

   d. What is the value of y when $x = -1$? Represent this on the graph.

   e. Write a formula that calculates the value of y for any possible value of x.

   f. Is there a benefit to using the vertical intercept as our reference point? Explain.

Algebra II

Module 1: Investigation 8
Slope-Intercept Form of a Linear Function

Slope-Intercept Form of a Linear Function

If the constant rate of change of *y* with respect to *x* is *m* and if we know that (0, *b*) is a point on the graph, then the following is a useful formula for the relationship.

[   ]

In this form we rely on the fact that the value of *x* is also the change in *x* away from *x* = 0. **Note: This is just a special case of** $y = m(x - x_1) + y_1$ **– it's not a "new" formula to memorize!**

Ex: If *y* changes at a constant rate of 4.2 with respect to *x*, and the graph of the function passes through the point (0, −6), then [   ].

2. Consider the formula $y = 3.5x - 9$. Suppose we want to find the value of *y* when $x = 6.5$. Explain how the formula determines the corresponding value of *y* using the meaning of constant rate of change. (*You may sketch a graph or draw a diagram if it helps you explain.*)

3. Savanna and Zoe met for their daily run. Savanna arrived late so Zoe started running before Savanna arrived. When Savanna started running, Zoe had already run 3 laps. Savanna joined Zoe at the beginning of the 4th lap and they ran together for 5 more laps before they both stopped.
   a. Complete the table that represents the relationship between the number of laps run by Savanna *x* and the number of laps run by Zoe *y*.

Number of Laps Run by Savanna, *x*	Number of Laps Run by Zoe, *y*
0	
1	
2	
3	
4	
5	

   b. What is the constant rate of change in this context? What does it tell us?

c. Write a formula to express the number of laps run by Zoe y in terms of the number of laps run by Savanna x.

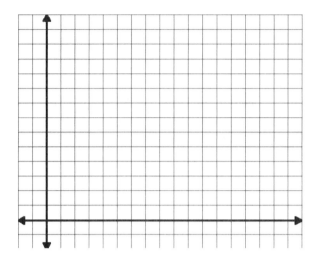

d. Construct a graph that represents the number of laps run by Zoe in terms of the number of laps run by Savanna. Be sure to label your axes.

e. Illustrate on your graph in part (d) a change from 2 to 5 laps run by Savanna and the corresponding change in the number of laps run by Zoe.

4. After pulling the plug from a bathtub, the water started draining. In the first 20 seconds 5 gallons drained from the tub. Suppose that the tub originally contained 56 gallons of water.
   a. Assuming the water is draining at a constant rate, what is the change in the volume of water remaining after 10 seconds? After 1 second? How much water remains in each case?

   b. Write the formula that calculates the number of gallons of water *remaining* in the tub in terms of the number of seconds that have elapsed since pulling the plug. Be sure to define any necessary variables.

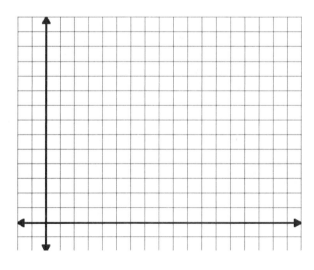

c. Sketch a graph that relates the number of gallons of water remaining in the tub to the amount of time elapsed since the tub began to drain. Be sure to label your axes.

d. State the vertical intercept of the graph and explain what it represents in this situation.

e. The *horizontal intercept* in this context is the number of seconds elapsed since pulling the plug where the graph intersects the horizontal axis. Find the horizontal intercept and explain what it represents in this situation.

5. Write the formula for each linear relationship described or represented in the following exercises.
   a. The constant rate of change of *y* with respect to *x* is 5 and the graph passes through the point (0, 3.6).

   b. Carlos left his house and started driving. When Carlos is 12 miles from his house he passes Exit 201. If he continues to drive at a constant speed of 58 miles per hour, write a formula that calculates the total distance he's traveled since leaving his house as a function of the number of hours since passing Exit 201.

   c.

x	y
−2	−9
0	4
6	43
9	62.5

6. Sometimes we want to write a linear relationship using the slope-intercept form ($y = mx + b$) but we don't know the vertical intercept.

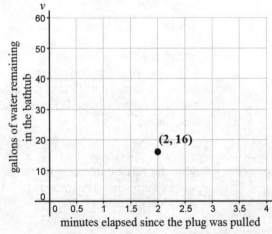

Let's revisit the bathtub draining context. Suppose we pull the plug on a different bathtub and allow it to drain. The volume is changing at a constant rate of −18 gallons per minute, and after 2 minutes there are 16 gallons of water remaining in the tub.

Let *v* be the number of gallons of water remaining in the tub and let *t* be the number of minutes that have elapsed since pulling the plug.

   a. How many gallons were in the tub before it started draining?

b. Write the relationship between v and t using the slope-intercept form ($v = mt + b$).

c. In a previous investigation we would have written the formula as $v = -18(t-2) + 16$. Use the distributive property and combine like terms to simplify $v = -18(t-2) + 16$. What did you find?

d. (*optional extension*) How do the calculations you performed in part (c) relate to the work you did in part (a)?

7. Write a formula for each of the following linear relationships in
    i) the form $y = m(x - x_1) + y_1$.
    ii) the form $y = mx + b$.

   a. y changes at a constant rate of 4 with respect to x, and (–6, 1) is a point on the line.

   b.
   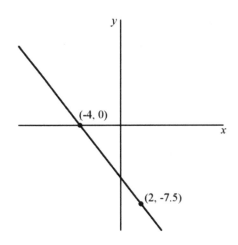

c.

x	y
−4.9	−8.2
−2.5	−4
1.7	3.35

8. Sketch the graph for each of the following functions.
   a. $y = -2x + 7$

   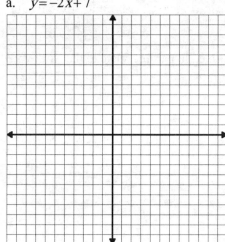

   b. $y = 1.5x$

   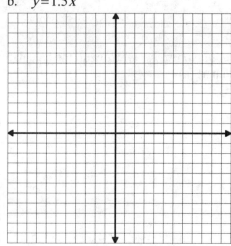

Algebra II

Module 1: Investigation 9
Proportionality

For Exercises #1-4, use the following context.
In computer word processing or layout programs you have the option to insert images. Then you can change the size of this image by clicking on it and dragging the edges. Suppose you insert a photo into a program that is a rectangle 5 inches wide by 4 inches high.

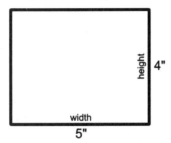

1. a. Suppose you know how wide the photo needs to be. What must you pay attention to so that the photo is not distorted from the original?

   b. The first step in correctly resizing the photo is to determine the height of the photo based on the width that you need. If you need the photo to be 13 inches wide, what should you make its height? Explain how you are thinking about this problem to come up with your answer.

   c. Complete the table that relates the heights h of the photo (in inches) to the widths w (in inches). How do we know that these measurements will produce a photo that is not distorted?

Width of photo, in inches (w)	Height of photo, in inches (h)
1	
2	
6	
13	

   d. Complete the following statement: As the width of the photo changes, the height of the photo changes so that …

   e. Write a formula that determines the height of the photo given its width such that the photo will not be distorted from the original.

Module 1: Investigation 9

© 2014 Carlson, O'Bryan & Joyner

2. a. Draw a graph of the height vs. the width for undistorted photos with a width up to 14 inches. Be sure to label your axes with the appropriate quantities and units of measure. Represent the photo widths on the horizontal axis.

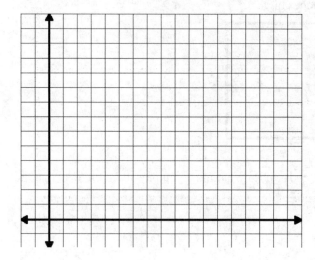

b. Explain what a point on your graph represents.

c. What is the constant rate of change for the relationship you graphed? What does this tell us?

d. Where does your graph cross the vertical axis? What does this point represent?

e. Write a formula that determines the width of the photo given its height such that the photo will not be distorted from the original.

3. a. What is the ratio $\frac{h}{w}$ equal to? What does this tell us?

b. Plot the points (0, 0) and (5, 4) on the graph in Exercise #2, then calculate Δw and Δh from the first point to the second point and represent these changes on the graph.

c. Talk about the link between $\frac{h}{w}$ and $\frac{\Delta h}{\Delta w}$ when we change away from (0, 0). Why did this happen?

d. Is the result you found in part (c) true for all linear functions? If so, why does this happen. If not, what makes this context different? *You may want to look back at exercises in previous investigations.*

Algebra II **Module 1: Investigation 9**
Proportionality

> **Proportionality**
>
> If *y* is a linear function of *x*, and if $y = 0$ when $x = 0$, then we say ***y is proportional to x*** and
>
> [ ]
>
> where *m* is the constant rate of change of *y* with respect to *x*.
>
> ---
>
> **Note 1:** When *y* is proportional to *x* the constant rate of change *m* is sometimes called the **constant of proportionality** because it tells us that *y* is always *m* times as large as *x*. You might also see *m* replaced with another letter, like *k*.
>
> **Note 2:** The relationship $y = mx$ is also sometimes called **direct variation** and we might say "***y*** **varies directly with** ***x***."

4. In earlier investigations we used the general formula $y = m(x - x_1) + y_1$ to represent a linear function.
   a. If we know *y* is proportional to *x*, what point must be on the graph of the relationship?

   b. Use your answer from part (a) as the reference point (x_1, y_1) to show that $y = mx$ is the correct formula when *y* is proportional to *x*.

---

*In the photograph context (Exercises #1-4), we say that **the height of the photo is proportional to the width of the photo** and the constant of proportionality is $\frac{4}{5}$. This creates some interesting properties. As the two quantities change together,*
   I) *the height of the photo will always be $\frac{4}{5}$ times as large as the width of the photo.*
   II) *the ratio of the height of the photo to the width of the photo will always be $\frac{4}{5}$.*

However, it's also true that
   III) *if one quantity is scaled by some factor m, then the other quantity will also be scaled by the same factor m.*

In Exercises #5-7, use the following context.
When making pancakes from a powdered mix, a recipe is given for the amounts of each ingredient to be used. One popular brand calls for 20 tablespoons of mix for every 6 ounces of water.

5. a. Suppose you want to use 10 tablespoons of powdered mix. How much water do you need? Explain how you are thinking about this problem to come up with your answer.

   b. Suppose you want to use 5 tablespoons of powdered mix. How much water do you need? What about for 15 tablespoons of mix?

c. How many tablespoons of powdered mix should be used with 12 ounces of water? 15 ounces of water? 29 ounces of water?

d. How many ounces of water should be used if you only have 8 tablespoons of powdered mix left in the box?

e. How much water should be used with one tablespoon of powdered mix? How much powdered mix should be used with one ounce of water?

6. Some people vary the ratio of powder to water based on how thick they like their pancakes. Some people like their batter "thicker" while others prefer "thinner" batter.
   a. In terms of the recipe, what does it mean for the batter to be thicker or thinner than the recommended mixture?

   b. If a person combines 3 ounces of water using 8 tablespoons of powdered mix, is her batter thicker or thinner than the recommended mixture? Explain your reasoning.

7. a. Construct a graph of the number of tablespoons of mix compared to the number of ounces of water based on the recommendations on the given recipe. Be sure to label your axes with the appropriate quantities and units, with the amount of water represented on the horizontal axis.

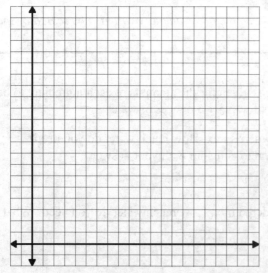

   b. Plot the point (4, 10) on your graph and explain what this point represents.

   c. Is the point you plotted in part (b) on the line? What does this indicate about the batter?

   d. Does this ordered pair represent batter that is thicker or thinner than batter made according to the recipe?

Algebra II

Module 1: Investigation 11
Introduction to Functions and Function Notation

> Investigation 10: *Extra Practice with Proportionality* and associated homework is available online at www.rationalreasoning.net.

In this module you've used formulas to describe how the values of two quantities are related. If every input value generates exactly one output value then we say that the relationship is a ***function***.

Function

If two quantities are related such that every input value produces exactly one output value, then the relationship is a ***function***.

Ex 1: Let P be the perimeter of a square (in inches) and let s be the length of one side (in inches). Then $P = 4s$. Since every value of s will only produce one value of P, then we say *"The perimeter of a square is a function of the length of one of its sides."*

Ex 2: Let $y = \pm\sqrt{x}$. Then y is not a function of x because there are values of x that produce more than one value of y. For example, if $x = 4$, then $y = 2$ and $y = -2$.

One way to visualize a function is as a "machine" that converts each input value into exactly one output value using some rule. In the given diagram we are representing P as a function of s.

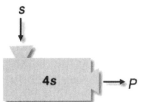

When we choose a value of s, the function applies the rule "$P = 4s$" to determine the output value P that goes with the input value we chose.

1. The formula $F = \frac{9}{5}C + 32$ converts temperature measurements in degrees Celsius (C) to the equivalent measurements in degrees Fahrenheit (F).
   a. What does it mean to say that *degrees Fahrenheit F* is a function of *degrees Celsius C*?

   b. The input variable is also called the *independent variable* and the output variable is also called the *dependent variable* (because its value *depends* on the value of the input). In the given formula, which variable is the independent variable and which is the dependent variable?

   c. Can we also think about the quantity *degrees Celsius C* as a function of the quantity *degrees Fahrenheit F*? Explain why or why not.

Algebra II

Module 1: Investigation 11
Introduction to Functions and Function Notation

d. Rewrite the formula so that it expresses a temperature measurement in degrees Celsius C as a function of the measurement in degrees Fahrenheit F.

2. A circle's circumference and radius are related by the formula $C = 2\pi r$.
   a. Determine the circumferences for radius lengths of 2 inches, 3.5 inches, and 4 inches.

   b. Does every radius length correspond with a single circumference length? Based on your answer, what can we then conclude?

   c. Suppose you are asked to calculate the circumferences for five different radius lengths. You determine the values, $C = 7.54$, $C = 18.85$, $C = 30.79$, $C = 43.98$, and $C = 50.27$. When asked what radius length gave you each circumference, you can't remember. How might you keep track of which radius length was used to calculate a circumference when you are doing several calculations?

In Exercise #2 we found that it would be useful to have notation that shows output values and links them back to the input value used to find them. In fact, it's useful to have this kind of notation to help represent function relationships in general. Thus, mathematicians developed ***function notation*** that assigns a letter to name the relationship between two (or more) specific quantities.

Algebra II

Module 1: Investigation 11
Introduction to Functions and Function Notation

Function Notation

Suppose y is a function of x, and call this function f. Then $y = f(x)$ and

$$f(x) = \langle \text{an expression representing the value of } y \text{ given } x \rangle$$

Some important notes:
- f is the name of the function
- $f(x)$ is NOT the name of the function, it's the output of the function for an input of x and it gives you a value of y [since $y = f(x)$]
- coordinate points for the relationship are of the form $(x, f(x)) = (x, y)$

---

3. Let C be a circle's circumference (in inches) and r its radius length (in inches). Then $f(r) = 2\pi r$ where $C = f(r)$.

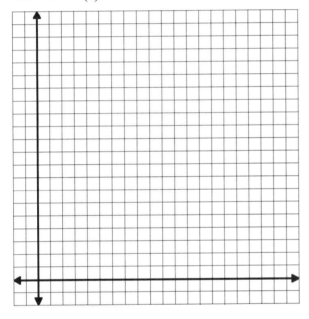

   a. What does $f(5.2)$ represent?

   b. What is the value of $f(5.2)$?

   c. What does the statement $f(3) = 18.85$ tell us?

   d. Create a graph of f and then plot and label the points $(0, f(0))$, $(1, f(1))$, and $(3, f(3))$.

   e. What does the graph of f represent?

4. Recall that the formula for determining the area of a circle with respect to the radius length is $A = \pi r^2$. Let g be the name of the function that inputs the radius length of the circle (in inches) and outputs the associated area of the circle (in square inches). In other words, $g(r) = \pi r^2$ where $A = g(r)$.

   a. Use function notation to represent (*not* calculate) the area of a circle whose radius is 3.5 inches, 18.2 inches, and 26.92 inches.

Algebra II

Module 1: Investigation 11
Introduction to Functions and Function Notation

b. Explain in your own words what $g(4)$ represents, then determine the value of $g(4)$.

c. Interpret what $g(4.9) = 75.43$ means in the context of the problem.

d. What does it mean to solve $g(r) = 141.026$ for r?

e. Determine the value of r such that $g(r) = 141.026$.

f. What's the difference between saying "g" and "$g(r)$"?

5. The ***domain*** of a function is the set of values that can be used as inputs to the function. A function's domain may be restricted because some values don't make sense for a certain quantity or in a specific function. The ***range*** of a function is the set of all of the output values you get when you use all of the values in your domain.
   a. What is the domain of g from Exercise #4 and what does it represent?

   b. What is the range of g from Exercise #4 and what does it represent?

   c. How do the domain and range change if we want to make sure the circle's radius is at least 2 inches long, but we also want to be able to draw the circle on an 8.5" x 11" piece of paper?

Algebra II

Module 1: Investigation 11
Introduction to Functions and Function Notation

d. How do the domain and range change if we remove the context? (*That is, we consider the general function $g(x) = \pi x^2$ where the input and output don't have to represent real world quantities.*)

6. Kim is riding her bike in a race. She rides at varying speeds for the first 5 miles of the race until she reaches the first checkpoint. For the next 14 minutes she travels 6 miles at a constant speed. Assume she continues traveling at this constant speed for the remainder of the race.
   a. After she passed the checkpoint, what was her constant speed in miles per minute?

   b. Define a function g that represents Kim's distance d (in miles) from the starting line in terms of the number of minutes t that have elapsed since she passed the first checkpoint.

   c. Determine the value of $g(12)$ and explain what it represents in the context of the problem.

   d. If a problem says "Solve $g(t) = 6$ for t," what are you being asked to find in this context?

7. Function formulas are also sometimes just defined algebraically and have no connection to a real world context. As an example we could define a function *k* as a mapping between any real number *x* to another real number using the rule $k(x) = \frac{4x+17}{2}$.

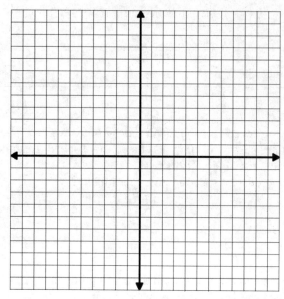

   a. Determine the values of $k(-2)$, $k(0)$, $k(3)$ and $k(5)$.

   b. Plot the points $(-2, k(-2))$, $(0, k(0))$, $(3, k(3))$ and $(5, k(5))$.

   c. Function *k* is a linear function. What is its vertical intercept and constant rate of change?

8. Evaluate each of the following functions for the specified value.
   a. Given $f(x) = \frac{3}{2}x - 11$, evaluate $f(-5)$.
   b. Given that $g(x) = \frac{5x+7}{10}$, evaluate $g(2)$.

   c. Given $h(x) = -7 - 11x$, evaluate $h(-4)$.
   d. Given $k(x) = 4x + 9$, what does it mean to evaluate $k(a+2)$?

9. Why is function notation useful?

Algebra II

Module 1: Investigation 12
Introduction to Piecewise Defined Functions

For Exercises #1-2, use the following context.
A snowstorm begins at 5:00 p.m. and lasts 3.5 hours before it stops snowing for the day. During the storm the snow falls steadily at a rate of 2 inches per hour. There is no snow on the ground before the storm begins. Suppose we want to model the depth of the snow on the ground (in inches) as a function of the number of hours elapsed since noon (12:00 p.m.). The graph of this relationship from noon to midnight is given.

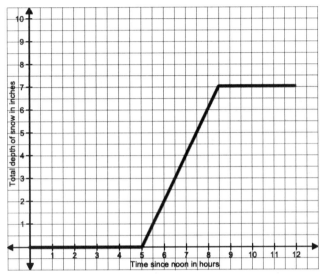

1. a. Can a single linear formula model this relationship? Explain.

We observe that there are three distinct parts of this graph.

   b. The first part represents the depth of snow in the 5 hours before the storm begins. Give the coordinate points at the beginning and the end of this part of the graph.

   c. The second part represents the depth of snow as it accumulates during the storm at a rate of 2 inches per hour. The storm lasts from 5:00 pm to 8:30 pm, a length of 3.5 hours (causing 7 total inches of snow to fall). Give the coordinate points at the beginning and the end of this part of the graph.

   d. The third part represents the depth of snow (*assuming no melting*) from 8:30 pm to midnight. Notice that the depth remains at 7 inches because no new snow has fallen. Give the coordinate points at the beginning and the end of this part of the graph.

Algebra II

Module 1: Investigation 12
Introduction to Piecewise Defined Functions

While we can't write a single formula to relate all values for total depth of snow (in inches) and hours since noon, we can write represent the relationship using three different formulas. Let t represent the number of hours since noon and $d(t)$ represent the total depth of snow on the ground (in inches).

2. a. Complete the following statements by filling in the blanks.
   i. During the first five hours we can model the depth of snow on the ground by
   $d(t) = $ _____ . However, this formula only works for values of t such that
   _____ $\leq t \leq$ _____ .

   ii. From 5:00 p.m. to 8:30 p.m. we can model the depth of snow on the ground by
   $d(t) = $ _____ . However, this formula only works for values of t such that
   _____ $< t \leq$ _____ .

   iii. From 8:30 p.m. to midnight we can model the depth of snow on the ground by
   $d(t) = $ _____ . However, this formula only works for values of t such that
   _____ $< t \leq$ _____ .

   b. Let's summarize:

   $$d(t) = \begin{cases} \underline{\hspace{1cm}} & \text{if } \underline{\hspace{0.5cm}} \leq t \leq \underline{\hspace{0.5cm}} \\ \underline{\hspace{1cm}} & \text{if } \underline{\hspace{0.5cm}} < t \leq \underline{\hspace{0.5cm}} \\ \underline{\hspace{1cm}} & \text{if } \underline{\hspace{0.5cm}} < t \leq \underline{\hspace{0.5cm}} \end{cases}$$

   c. Evaluate each of the following.
   i. $d(2.3)$ ii. $d(6)$ iii. $d(7.8)$ iv. $d(10.72)$

Exercises #1-2 demonstrated the reasoning and basic conventions of ***piecewise functions***. We use piecewise functions whenever we can't use a single formula to describe the relationship between two quantities. Instead, we have to define the relationship using several formulas, each one defining a *piece* of the overall relationship.

For Exercises #3-8, use the following context.
For the first 25 seconds of its run, a roller coaster is pulled up a ramp to its highest point at a constant speed of 12 mph. Let $s(t)$ represent the speed of the roller coaster (in miles per hour) based on t, the number of seconds since the roller coaster began its run.

3. *Complete the following statement*. For the first 25 seconds of its run we can model the roller coaster's speed by $s(t) = $ _____ . However, this formula only works for values of t such that
_____ $\leq t \leq$ _____ .

Algebra II

Module 1: Investigation 12
Introduction to Piecewise Defined Functions

4. At the top of the ramp the roller coaster starts its first downward plunge and its speed increases by 11 mph every second for the next 3 seconds until it reaches the bottom.
   a. How many seconds have passed since the roller coaster began its run when it starts the downward plunge? When it reaches the bottom of the downward plunge?

   b. How fast is the roller coaster moving at the moment it begins the plunge? When it ends the plunge?

   c. *Complete the following statement:* During this plunge we can model the roller coaster's speed by $s(t) =$ _____ . However, this formula only works for values of t such that _____ $< t \leq$ _____ .

   d. Evaluate $s(9.2)$ and $s(26)$ then explain what these values represent.

5. After the first plunge described previously, the roller coaster begins a 2.5-second climb. During this climb, gravity and friction cause the roller coaster's speed to decrease by 6 mph every second.
   a. How many seconds have passed since the roller coaster began its run when it starts the climb? When it reaches the top of the climb?

   b. How fast is the roller coaster moving at the moment it begins its climb? When it ends the climb?

   c. *Complete the following statement:* During this climb we can model the roller coaster's speed by $s(t) =$ _____ . However, this formula only works for values of t such that _____ $< t \leq$ _____ .

   d. Evaluate $s(28.5)$ and $s(29.5)$ then explain what these values represent.

6. *Let's summarize*:

$$s(t) = \begin{cases} \underline{\hspace{2cm}} & \text{if } \underline{\hspace{1cm}} \leq t \leq \underline{\hspace{1cm}} \\ \underline{\hspace{2cm}} & \text{if } \underline{\hspace{1cm}} < t \leq \underline{\hspace{1cm}} \\ \underline{\hspace{2cm}} & \text{if } \underline{\hspace{1cm}} < t \leq \underline{\hspace{1cm}} \end{cases}$$

7. Evaluate $s(28)$ and explain what this value represents.

8. Draw a graph of s.

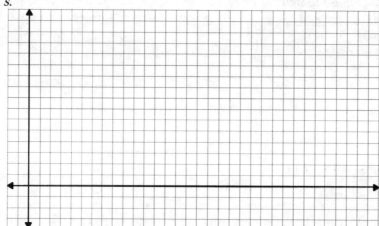

9. Most income tax systems (both state and federal) use a graduated income tax. Generally speaking, this means that the more money you earn per year the greater percentage of your income you owe in taxes. An individual's state income tax in a certain state is calculated using the formula below, where $T(i)$ is the amount an individual owes in state income tax (in dollars) and i is the person's annual income (in dollars).

$$T(i) = \begin{cases} 0.02i & \text{if } 0 \leq i \leq 55{,}000 \\ 0.10(i - 55{,}000) + 1{,}100 & \text{if } 55{,}000 < i \leq 100{,}000 \\ 0.14(i - 100{,}000) + 5{,}600 & \text{if } i > 100{,}000 \end{cases}$$

a. What percent of a person's income is owed in state taxes if he/she makes $13,000 per year? How is this represented in the function?

b. Evaluate each of the following, then explain what the value represents.
   i. $T(22{,}300)$ ii. $T(139{,}950)$ iii. $T(72{,}600)$

Algebra II

Module 1: Investigation 12
Introduction to Piecewise Defined Functions

c. In the second part of the formula, what does the expression $i-55{,}000$ represent?

d. In the second part of the formula, what does the expression $0.10(i-55{,}000)$ represent?

e. In the second part of the formula, what does 1,100 represent? Why is it needed?

f. Determine if you agree with the following statement, then explain your thinking: People who make over $100,000 per year have to pay 14% of their income in state taxes.

Algebra II **Module 1: Investigation 14**
Absolute Value

> Investigation 13: *Piecewise Defined Functions – Extra Practice* and associated homework is available online at www.rationalreasoning.net.

1. A carpenter needs to cut a wooden board to a length of 7.6 inches with a ***tolerance*** of 0.1 inch, meaning the actual length after the cut can be within 0.1 inch of 7.6 inches and still be usable.
   a. Let x represent the actual length of the board after cutting. List some possible values for x that represent usable board lengths, and then represent all possible values on the number line.

   b. Write a mathematical statement (inequality) to represent the possible usable board lengths.

   c. We will call the difference between the actual board length after the cut (x) and the desired board length the ***error***. Write an expression to represent the error.

   d. Write a mathematical statement (inequality) that represents errors within our tolerance of 0.1 inch.

   e. Write a mathematical statement (inequality) that represents actual board lengths x that are not usable, then represent these values on a number line.

2. Suppose we want to choose an x value within 4.5 units of 16.
   a. List some values of x that meet this constraint.

   b. Write a mathematical statement (inequality) to describe the possible changes in x away from 16.

   c. Write a mathematical statement (inequality) to describe the possible values of x.

   d. Represent the possible values of x on the number line.

Algebra II

Module 1: Investigation 14
Absolute Value

3. Suppose we want the change in *x* away from 5 to be positive but no more than 3.
   a. List some values of *x* that meet this constraint.

   b. Write a mathematical statement (inequality) to describe the possible changes in *x* away from 5.

   c. Write a mathematical statement (inequality) to describe the possible values of *x*.

   d. Represent the possible values of *x* on the number line.

4. Suppose we want to choose a value of *x* that is at least 4 units away from $x = -3$.
   a. List some values of *x* that are at least 4 units away from $x = -3$, then represent all possible values on the number line.

   b. Write a mathematical statement (inequality) that describes values of *x* at least 4 units away from $x = -3$. Try to think of at least two different ways to write this.

Definition of Absolute Value

Given a real number $c > 0$,

- [ ] represents all values of *x* that are exactly *c* units away from *h* on the real number line.

- [ ] represents all values of *x* that are less than *c* units away from *h* on the real number line.

- [ ] represents all values of *x* that are greater than *c* units away from *h* on the real number line.

The symbol $\geq$ ("greater than or equal to") is a combination of the signs $>$ and $=$. For example, $x \geq 2$ means that $x > 2$ OR $x = 2$. A similar idea describes $\leq$.

Algebra II

Module 1: Investigation 14
Absolute Value

5. Let's revisit your answers to the previous exercises. If you didn't already, use absolute value to rewrite the solutions to Exercises #1c, #2b, and #4b.

6. Jamie is assembling a model airplane that flies. The instructions say that the best final weight is 28.5 ounces. However, the actual weight can vary from this by up to 0.6 ounces without affecting how well the model flies.
   a. Represent the weights that are "acceptable" on the number line and describe them using an inequality.

   b. Write a mathematical statement (inequality) that represents the acceptable variation in the model's weight
      i) using absolute value.
      ii) without using absolute value.

7. a. What is the meaning of the statement "$|x-5|<2$"?

   b. Represent all values of x that make the statement true on the number line.

   c. Rewrite the statement in part (a) without using absolute value.

8. a. What is the meaning of the statement "$(w-1)\geq 5$ or $(w-1)\leq -5$"?

   b. Represent all values of w that make the statement true on the number line.

   c. Rewrite the statement in part (a) using absolute value.

Algebra II

Module 1: Investigation 14
Absolute Value

9. Use absolute value notation to represent the following.
   a. All numbers x whose distance from 6 is no more than 7.5 units.
   b. All numbers r whose distance from -2 is more than 1 unit.

10. For each statement, do the following.
    i) Describe the meaning of the statement.
    ii) Rewrite the statement without using absolute value.
    iii) Determine the values of the variable that make each statement true.
    iv) Represent the solutions on a number line.
    a. $|x-11|<6$
    b. $|y+7|\geq 1$

   c. $|a-3.3|>0.8$
   d. $|p+2|\leq 5$

11. a. Use the definition of absolute value to describe the solutions for each of the following statements, then give the values of x that make the statement true.
       i) $|x-0|=6$

       ii) $|x-0|<3$

    b. Based on your work in part (a), what does each of the following represent for a real positive number c?
       i) $|x|=c$

       ii) $|x|\geq c$

Algebra II

Module 1: Investigation 14
Absolute Value

12. For each statement, do the following.
    i) Describe the meaning of the statement.
    ii) Write the statement without using absolute value.
    iii) Represent the solutions on the number line.

   a. $|x| = 4$

   b. $|x| \leq 5.5$

   c. $|x| > 3$

Algebra II

Module 1: Homework
Linear Functions

INVESTIGATION 1: REASONING WITH QUANTITIES

Recall that a quantity is some attribute of an object that you can imagine measuring. The distance Josh has walked since leaving his home is an example of a quantity. Notice that when describing a varying quantity you need to specify the reference point for the measurement.

1. Imagine watching a racer running around a quarter mile oval track.

   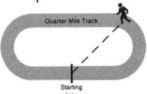

   a. Name 3 quantities that are varying in this situation (Your description should make clear exactly what you're imagining measuring). *Example of a varying quantity:* The direct distance (measured in yards) of the runner from the starting line as she moves around the track.
   b. Name at least 1 quantity in this situation that is not varying.
   c. How does *the distance in feet the runner has traveled around the track* vary?
   d. How does *the runner's direct distance in feet from the starting line* vary?
   e. How do the quantities *distance in feet the runner has traveled around the track* and *the runner's direct distance in feet from the starting line* vary together?
   f. Sketch a graph of the direct distance in feet of the racer from the starting line in terms of the distance in feet the runner has traveled around the track.

2. A basketball player runs a wind sprint from the west end (also called the baseline) of a basketball court.

   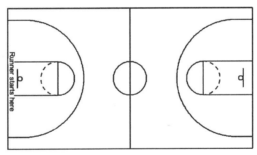

   The basketball player runs a wind sprint by:
   i. running from the west baseline of the court a distance of 25 feet toward midcourt
   ii. stopping and changing directions and returning back to the west baseline
   iii. stopping and changing directions and running 42 feet to midcourt
   iv. stopping and changing directions and returning to the west baseline
   v. stopping and changing directions and running 60 feet toward the east baseline
   vi. stopping and changing directions and returning to the west baseline
   vii. stopping and changing directions and running 84 feet to the east baseline
   viii. stopping and changing directions again and running back to the west baseline.

   a. Construct a graph of the distance (in feet) of the basketball player from the court's west baseline in terms of the total distance run by the basketball player since starting to run the wind sprint (label your axes).
   b. Pick three points on the graph and describe what each point represents in the context of this problem.
   c. Indicate all points on the graph that correspond to the runner being 15 feet from the west baseline.

Algebra II
Module 1: Homework
Linear Functions

3. John rides a Ferris wheel. He boards at the bottom and rides around a few times before getting off. The following graph represents John's height above the ground (in feet) with respect to the total distance John traveled in feet around the Ferris wheel for one complete rotation around the wheel.

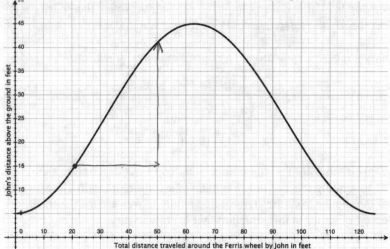

 a. The point (20.94, 15) is plotted on the graph. Explain what information this point conveys in the context of the problem.
 b. How much did John's distance traveled around the Ferris wheel **change** from 20.94 to 50 feet?
 c. How can you represent this change on the graph? Do this.
 d. Represent the corresponding **change** in *John's distance above the ground* from 20.94 feet traveled to 50 feet traveled.
 e. Approximately how much did *Jon's distance above the ground* **change** over this total distance traveled interval? Explain how you determined the change.
 f. Use the graph to estimate John's distance above the ground after traveling 50 feet around the Ferris wheel and plot the point representing this information.

4. Katie leaves a friend's house to jog home. The following graph represents Katie's distance in miles from home with respect to the amount of time elapsed in minutes since she began running.

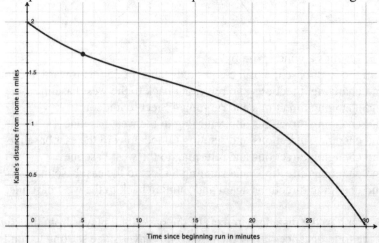

 a. The point (5, 1.6875) is plotted on the graph. Explain what information this point conveys in the context of the problem.
 b. How much did Katie's time elapsed **change** from 5 to 20 minutes since starting to jog?
 c. How can you represent this change on the graph? Do this.

(continues...)

Algebra II

Module 1: Homework
Linear Functions

d. Represent the corresponding **change** in *Katie's distance home* from 5 minutes elapsed to 20 minutes elapsed since starting to jog.
e. Approximately how much did *Katie's distance from home* **change** over this time interval? Explain how you determined the change.
f. Use the graph to estimate Katie's distance from home at 20 minutes elapsed and plot the point representing this information.

INVESTIGATION 2: CHANGES IN QUANTITIES' VALUES

5. In high school Matt grew from weighing 95 pounds to 162 pounds. What was Matt's change in weight during high school?

6. Sue did poorly on a History test causing her grade to go from 84.1% to 78.3%. What was the change in Sue's grade?

7. In winter, the temperatures in northern Wisconsin can be quite cold. When Lauren woke up this morning (6:00 am) the temperature outside was $-18°F$. When she got home from school (4:00 pm) the temperature outside was $4°F$. What was the change in temperature from 6:00 am to 4:00 pm?

8. Let x and y represent the values of two different quantities.
    a. If the value of x changes from $x=2$ to $x=17$, what is the change in x?
    b. If the value of y changes from $y=5$ to $y=-22$, what is the change in y?
    c. If the value of x changes to $x=-13$ from $x=-31$, what is the change in x?
    d. If the value of y changes from $y=9$ to $y=c$, what is the change in y?

9. Let x and y represent the values of two different quantities.
    a. If the value of x changes from $x=12$ to $x=1$, what is the change in x?
    b. If the value of y changes from $y=-8$ to $y=15$, what is the change in y?
    c. If the value of x changes to $x=-19$ from $x=4$, what is the change in x?
    d. If the value of y changes from $y=-3$ to $y=b$, what is the change in y?

10. Fill in the following tables showing the appropriate changes in the variable values.

a.
x	Δx
2	
8	
17	
36	

b.
y	Δy
10	
6.4	
1.8	
−4.6	

c.
w	Δw
−13	
−5	
7	
−4	

11. Fill in the following tables showing the appropriate changes in the variable values.

a.
x	Δx
4	
12	
31	
40	

b.
y	Δy
17	
12.2	
1.1	
−5.6	

c.
w	Δw
−9	
−1	
12	
4	

Algebra II

Module 1: Homework
Linear Functions

12. Each of the following graphs shows two ordered pairs (x, y). Find Δx and Δy from the point on the left to the point on the right.
    a.
    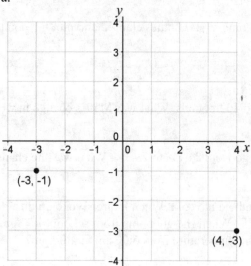
    b.
    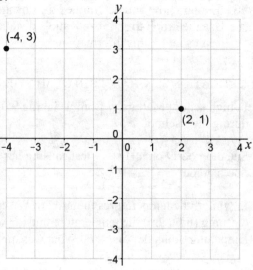

13. Each of the following graphs shows two ordered pairs (x, y). Find Δx and Δy from the point on the left to the point on the right.
    a.
    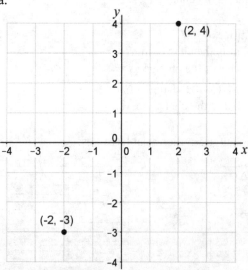
    b.
    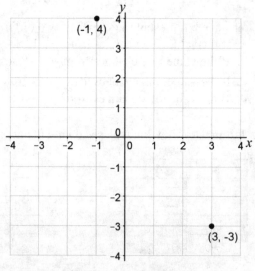

14. Suppose we start with a value of $x = 7$.
    a. What expression represents the change in x from $x = 7$ to any possible value of x?
    b. Repeat part (a) if our initial reference value is $x = 12$ instead of $x = 7$.

15. Suppose we start with a value of $x = -15$.
    a. What expression represents the change in x from $x = -15$ to any possible value of x?
    b. Repeat part (a) if our initial reference value is $x = -22$ instead of $x = -15$.

Algebra II   **Module 1: Homework**
Linear Functions

16. Suppose Lauren bought a puppy. At the time she bought the puppy it weighed 8 pounds. Let w represent the weight of the puppy in pounds.
    a. What is the difference between saying $w = 12$ and $\Delta w = 12$?
    b. If we substitute $w = 22.4$ into the expression $w - 8$, what are we calculating?
    c. If we substitute $w = 31$ into the expression $w - 17$, what are we calculating?

17. Think about the idea of the change in a variable from one value to another. What do the following expressions represent when we substitute a value for the variable?
    a. $x - 1$
    b. $y - (-7)$ [or $y + 7$]
    c. $r - 18.1$
    d. $p - (-4.6)$ [or $p + 4.6$]

18. Think about the idea of the change in a variable from one value to another. What do the following expressions represent when we substitute a value for the variable?
    a. $x - 23$
    b. $y - (-6)$ [or $y + 6$]
    c. $r - 41.2$
    d. $p - (-33.5)$ [or $p + 33.4$]

INVESTIGATION 3: CONSTANT RATE OF CHANGE

19. What does it mean for an object to move at a constant speed? (*Note: Please say something more than "The speed doesn't change" – be descriptive and reference specific quantities.*)

20. Jenny was riding her bike along a path. Assume that she rode at a constant speed during the entire ride, and also suppose that during one part of the trip she rode 88 feet in 6 seconds.
    a. Provide at least four conclusions we can draw from the given information.
    b. During this interval, how far did Jenny travel in 5 seconds?
    c. Does your answer to part (b) depend on which 5-second interval we're talking about? Explain.
    d. How long did it take Jenny to travel any 10-foot distance during this part of her ride?

21. Paul was walking in a park. Assume that he walked at a constant speed during the entire trip, and also suppose that during one part of the trip he walked 52.8 feet in 8 seconds.
    a. Provide at least four conclusions we can draw from the given information.
    b. During this interval, how far did Paul travel in 14 seconds?
    c. Does your answer to part (b) depend on which 14-second interval we're talking about? Explain.
    d. How long did it take Paul to travel any 20-foot distance during this part of his ride?

22. Assume that a football and a soccer ball are both traveling at constant speeds (but not necessarily the same constant speed as each other).
    a. What does it mean to say that the soccer ball is traveling *faster* than the football? *Be descriptive, and reference the specific quantities involved.*
    b. What does it mean if the football and soccer ball have the same constant speed? *Be descriptive, and reference the specific quantities involved.*

23. Suppose we want to rank the following objects in order from fastest to slowest. Assume each object is traveling at a constant speed (but not necessarily the same constant speed).
    Dog: 22 feet in 2 seconds   *Bird: 3.3 feet in 0.5 seconds*
    Baseball: 61 feet in 7 seconds   *Motorcycle: 190 feet in 10 seconds*
    Soccer Ball: 2.64 feet in 0.2 second
    a. Why is it difficult to compare the objects' speeds based on the information given?
    b. What strategy can we use to help us complete the task?
    c. Rank the objects in order from fastest to slowest.

Algebra II

Module 1: Homework
Linear Functions

24. Suppose we want to rank the following objects in order from fastest to slowest. Assume each object is traveling at a constant speed (but not necessarily the same constant speed).
    Bike: 88 feet in 6 seconds   Car: 70.4 feet in 4 seconds   Runner: 109 feet in 15 seconds
    Duck: 7.2 feet in 0.6 seconds   Cat: 13.2 feet in 1.5 second
    a. Why is it difficult to compare the objects' speeds based on the information given?
    b. What strategy can we use to help us complete the task?
    c. Rank the objects in order from fastest to slowest.

25. What are the benefits of stating speeds in terms of distance traveled in one unit of time, such as 34 mph (34 miles traveled in one hour), 14 m/s (14 meters traveled in one second), and so on?

26. Suppose that Mary is jogging at a constant speed of 0.10 miles per minute. Let d represent the total distance (in miles) Mary has traveled during her jog and let t represent the total number of minutes she's been jogging.
    a. Use Δ notation to represent a 6-minute time period during Mary's jog.
    b. How far does Mary travel during this 6-minute time period? Represent your answer using Δ notation.
    c. Repeat parts (a) and (b) if instead of a 6-minute time period we look at a
        i) 8.2-minute time period
        ii) 1.6-minute time period
        iii) any x-minute time period
    d. For any change in t, what happens to the change in d? Explain, then represent your thinking using Δ notation.

27. Suppose that Will is driving at a constant speed of 48 miles per hour. Let d represent the total distance (in miles) Will has traveled during his drive and let t represent the total number of hours he's been driving.
    a. Use Δ notation to represent a 2-hour time period during Will's drive.
    b. How far does Will travel during this 2-hour time period? Represent your answer using Δ notation.
    c. Repeat parts (a) and (b) if instead of a 2-hour time period we look at a
        i) 3.1-hour time period
        ii) 0.3-hour time period
        iii) any x-hour time period
    d. For any change in t, what happens to the change in d? Explain, then represent your thinking using Δ notation.

28. If we know that the total cost of purchasing a bag of candy increases at a rate of $9 per pound, then it's easy to determine the change in total cost if we change the amount of candy we purchase. Let c represent the total cost of purchasing candy (in dollars) and let w represent the weight of candy purchased (in pounds).
    a. Suppose we have some candy in a bag and we add 2 more pounds of candy to the bag. Represent the change in the number of pounds of candy using Δ notation.
    b. How much does the total cost of the candy we purchase change? Represent your answer using Δ notation.
    c. Repeat parts (a) and (b) if instead of adding 2 more pounds of candy we
        i) add 1.7 pounds of candy to the bag.
        ii) add 0.45 pounds of candy to the bag.
        iii) remove 1.2 pounds of candy from the bag.
    d. For any change in w, what happens to the change in c? Explain, then represent your thinking using Δ notation.
    e. If your friend gives you $13 to purchase more candy, how many additional pounds of candy would you add to your bag?

Algebra II

Module 1: Homework
Linear Functions

29. If we know that the total cost of purchasing a trail mix increases at a rate of $6 per pound, then it's easy to determine the change in total cost if we change the amount of trail mix we purchase. Let c represent the total cost of purchasing trail mix (in dollars) and let w represent the weight of trail mix purchased (in pounds).
    a. Suppose we have some trail mix in a bag and we add 4 more pounds of trail mix to the bag. Represent the change in the number of pounds of trail mix using Δ notation.
    b. How much does the total cost of the trail mix we purchase change? Represent your answer using Δ notation.
    c. Repeat parts (a) and (b) if instead of adding 4 more pounds of trail mix we
        i) add 4.2 pounds of trail mix to the bag.
        ii) remove 2.9 pounds of trail mix from the bag.
    d. For any change in w, what happens to the change in c? Explain, then represent your thinking using Δ notation.
    e. If your friend gives you $15 to purchase more trail mix, how many additional pounds of trail mix would you add to your bag?

30. Suppose we have a partially filled pitcher of water and that we want to add more water to the pitcher. We know that adding 60 ounces of water to the pitcher will increase the height of water in the pitcher by 7.8 inches, and that these two quantities are related by a constant rate of change. Define variables to represent the quantities in this context and then represent the relationship described using Δ notation.

31. Suppose we want to add some water to a bathtub, and that we know 5-gallons of water weighs 41.7 pounds. Define variables to represent the quantities in this context and then represent the relationship described using Δ notation.

INVESTIGATION 4: PRACTICE WITH CONSTANT RATE OF CHANGE

> Investigation 4: *Practice with Constant Rate of Change* and associated homework is available online at www.rationalreasoning.net.

INVESTIGATION 5: APPLYING CONSTANT RATE OF CHANGE TO DETERMINE NEW VALUES

48. Suppose we know that the changes in the values of two variables are related according to $\Delta y = 3 \cdot \Delta x$.
    a. If we start off at $x=5$ and let x change to be $x=12$,
        i. What is the change in x?
        ii. By how much does y change for the change in x you found in part (i)?
        iii. Suppose we know that $y=-2$ when $x=5$. What is the value of y when $x=12$? How did you find this?
    b. If we start off at $x=7$ and let x change to be $x=-3$,
        i. What is the change in x?
        ii. By how much does y change for the change in x you found in part (i)?
        iii. Suppose we know that $y=8$ when $x=7$. What is the value of y when $x=-3$? How did you find this?

Algebra II

Module 1: Homework
Linear Functions

49. Suppose we know that the changes in the values of two variables are related according to $\Delta y = -4.4 \cdot \Delta x$.
    a. If we start off at $x=2$ and let x change to be $x=-5$,
        i. What is the change in x?
        ii. By how much does y change for the change in x you found in part (i)?
        iii. Suppose we know that $y=-12$ when $x=2$. What is the value of y when $x=-5$? How did you find this?

    b. If we start off at $x=-5$ and let x change to be $x=16$,
        i. What is the change in x?
        ii. By how much does y change for the change in x you found in part (i)?
        iii. Suppose we know that $y=33$ when $x=-5$. What is the value of y when $x=16$? How did you find this?

50. Suppose you have a cell phone plan whose cost is based on the number of minutes you talk. Let n represent the number of minutes talked in a month and let c represent the monthly cost of using your phone (in dollars). Furthermore, suppose $c=45.70$ when $n=95$ and that $\Delta c = 0.06 \cdot \Delta n$.
    a. What is the value of c when $n=325$? What does this tell us?
    b. What is the value of c when $n=0$? What does this tell us?

51. Given that $\Delta b = 1.2 \cdot \Delta a$, complete the following table of values.

a	b
–2	
1	3
4	
9	

52. Given that $\Delta y = -3.4 \cdot \Delta x$, complete the following table of values.

x	y
–4	
4	2
6	
18	

53. In the following tables, you are given the values of x and y for two different situations in which the rate of change of y with respect to x is constant. The rate of change of y with respect to x for the values in the table on the left is 2.4, or $\Delta y = 2.4 \cdot \Delta x$. The rate of change of y with respect to x for the values in the table on the right is –1.7, or $\Delta y = -1.7 \cdot \Delta x$.

    Determine the missing values in each table *using your knowledge of constant rate of change*.

x	y
–2.7	
3	
5	–8.8
8.2	
23.1	

x	y
–19	
–7	3
	–4.8
3.2	
9.1	

Algebra II

Module 1: Homework
Linear Functions

54. Use the given graph to answer the questions that follow given that $\Delta y = -2 \cdot \Delta x$.

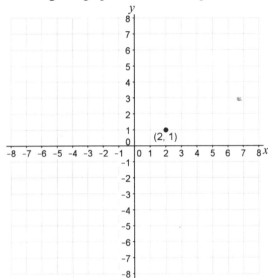

  a. What is the value of y when $x = -1$? Represent your reasoning on the graph.

  b. What is the value of y when $x = 6$? Represent your reasoning on the graph.

55. Use the given graph to answer the questions that follow given that $\Delta y = 3 \cdot \Delta x$.

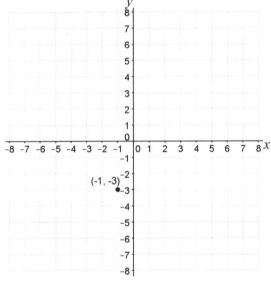

  a. What is the value of y when $x = 2.5$? Represent your reasoning on the graph.

  b. What is the value of y when $x = -3$? Represent your reasoning on the graph.

56. The amount of gas in a gas tank decreases at a constant rate of 1 gallon per 26 miles.

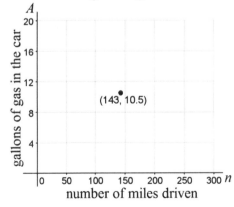

  a. What information does the point $(143, 10.5)$ tell us about this situation?

  b. If A represents the amount of gas in the tank in gallons and n represents the number of miles driven, write a statement showing the relationship between the changes in these variables.

  _____ = _____

  (*continues...*)

Algebra II

Module 1: Homework
Linear Functions

   c. What is the amount of gas in the tank after driving 221 miles? Find the answer and demonstrate your reasoning on the graph.
   d. What is the amount of gas in the tank after driving 40 miles? Find the answer and demonstrate your reasoning on the graph.
   e. What was the original amount of gas in the tank?

INVESTIGATION 6: DEVELOPING A FORMULA FOR LINEAR FUNCTIONS

57. Suppose we are given that $\Delta y = -2 \cdot \Delta x$ and that when $x = 3$, $y = 2$. We want to know the new value of y when $x = 8$. Answer the questions that follow.

    $y = -2(x-3) + 2$
    $y = -2(8-3) + 2$ a. What does $8-3$ represent?
    $y = -2(5) + 2$ b. What does $-2(5)$ represent?
    $y = -10 + 2$ c. What does $-10 + 2$ represent?
    $y = -8$ d. What does -8 represent?

58. Suppose we are given that $\Delta y = 4.5 \cdot \Delta x$ and that when $x = 1$, $y = 4$. We want to know the new value of y when $x = -4$. Answer the questions that follow.

    $y = 4.5(x-1) + 4$
    $y = 4.5(-4-1) + 4$ a. What does $-4-1$ represent?
    $y = -4.5(-5) + 4$ b. What does $4.5(-5)$ represent?
    $y = 22.5 + 4$ c. What does $22.5 + 4$ represent?
    $y = 26.5$ d. What does 26.5 represent?

59. Suppose we are given that $\Delta y = 1.5 \cdot \Delta x$ and that when $x = 5$, $y = 2$. We want to know the new value of y for <u>any</u> value of x. Answer the questions that follow.
    $y = 1.5(x-5) + 2$ a. What does $x-5$ represent?
                        b. What does $1.5(x-5)$ represent?
                        c. What does $1.5(x-5) + 2$ represent?

60. Suppose we are given that $\Delta y = -4 \cdot \Delta x$ and that when $x = 12$, $y = 1$. We want to know the new value of y for <u>any</u> value of x. Answer the questions that follow.
    $y = -4(x-12) + 1$ a. What does $x-12$ represent?
                         b. What does $-4(x-12)$ represent?
                         c. What does $-4(x-12) + 1$ represent?

61. The constant rate of change of y with respect to x is 4, and (5, 4) is a point on the graph.
    a. Write the formula for the linear function.
    b. Find the value of y when $x = 2$.

Algebra II   **Module 1: Homework**
                                                                                       Linear Functions

62. The constant rate of change of *y* with respect to *x* is –3.2, and (–3, –2) is a point on the graph.
    a. Write the formula for the linear function.
    b. Find the value of *y* when *x* = 5.

63. Write the formula for each of the linear functions described below.
    a. *y* changes at a constant rate of 4.8 with respect to *x*, and (7, 9.3) is a point on the graph.
    b. *y* changes at a constant rate of –1.9 with respect to *x*, and (4, 6) is a point on the graph.

64. Write the formula that defines the linear relationship represented in each of the following graphs.
    a.   b.

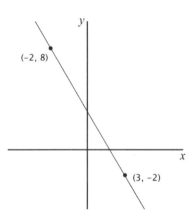 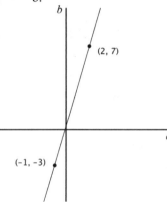

65. Write the formula that defines the linear relationship represented in each of the following graphs.
    a.   b.

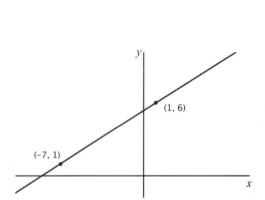 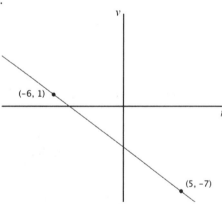

Algebra II   **Module 1: Homework**
                                                                    Linear Functions

66. Use the graph provided to answer the following questions.

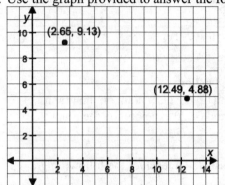

   a. Determine the change in *x* and the change in *y* between the two points. Use this to determine the constant rate of change between the quantities.
   b. Using the point (2.65, 9.13) as a reference point, illustrate on the axis the change in *y* that is necessary for *x* to change from $x = 2.65$ to $x = 6.2$. How much is this change?
   c. Write the expression that represents "the change in *x* from $x = 2.65$ to $x = 6.2$."
   d. What is the change in *y* that corresponds with the change in *x* from part (b)?
   e. Write the expression that represents "the change in *y* as *x* changes from $x = 2.65$ to $x = 6.2$."
   f. What is the value of *y* when $x = 6.2$?

67. Write the formula that defines the linear relationship given in each of the following tables.

   a.

x	y
−6	16
−1	1
2	−8
8	−26

   b.

w	d
−9	0
−4	10
1	20
14	46

68. Write the formula that defines the linear relationship given in each of the following tables.

   a.

x	y
−5	−15.5
−1	−1.5
2	9
18	65

   b.

w	d
−12	17.5
−8	11.5
3	−5
17	−26

69. The length of a burning candle decreases by 2.5 inches every hour. The candle has been burning for 1.5 hours and is currently 11.5 inches long.
   a. Write the formula that will calculate the length of candle remaining (in inches) given the number of hours the candle has been burning.
   b. Describe what the different parts of the formula calculate.
   c. Use your formula to find the length of candle remaining for each of the following number of hours burned.
      i. 3 hours   ii. 5 hours   iii. 0 hours (so prior to burning)

70. Given the formula $y = -3(x - 4) - 6$, answer the following questions.
   a. What is the constant rate of change of the linear relationship?
   b. What point do we know for sure must be on the graph of the function?
   c. For some ordered pair solution (x, y), explain what each of the following represents.
      i. $x - 4$ ii. $-3(x - 4)$ iii. $-3(x - 4) - 6$

Algebra II

Module 1: Homework
Linear Functions

71. Given the formula $y = 5(x+2) + 11$, answer the following questions.
    a. What is the constant rate of change of the linear relationship?
    b. What point do we know for sure must be on the graph of the function?
    c. For some ordered pair solution (x, y), explain what each of the following represents.
        i. $x + 2$
        ii. $5(x+2)$
        iii. $5(x+2) + 11$

INVESTIGATION 7: EXTRA PRACTICE WITH LINEAR FUNCTIONS

Investigation 7: *Extra Practice with Linear Functions* and associated homework is available online at www.rationalreasoning.net.

INVESTIGATION 8: SLOPE-INTERCEPT FORM OF A LINEAR FUNCTION

83. Consider the formula $y = 5x + 3$. Suppose we want to find the value of y when $x = 4.5$. Explain how the formula determines the value of y using the meaning of constant rate of change. (You may sketch a graph or diagram if it helps you explain.)

84. Consider the formula $y = -1.2x - 4$. Suppose we want to find the value of y when $x = -3$. Explain how the formula determines the value of y using the meaning of constant rate of change. (You may sketch a graph or diagram if it helps you explain.)

85. A certain brand of cereal is packaged so that the weight of the cereal (when dry) is 504 grams. Five servings of the cereal is equivalent to 140 grams of cereal.
    a. Assuming the weight of cereal is changing at a constant rate, what is the change in the weight of cereal remaining after 3 servings have been consumed? After 1 serving? How much cereal remains in each case?
    b. Write the formula that calculates the weight of cereal remaining in the box (in grams) in terms of the number of servings that have been consumed since opening the box. Be sure to define any necessary variables.
    c. State the vertical intercept of the graph and explain what it represents in this situation.
    d. The *horizontal intercept* in this context is the number of servings where the graph intersects the horizontal axis. Find the horizontal intercept and explain what it represents in this situation.

86. Write the formula for each linear relationship described or represented in the following exercises.
    a. the constant rate of change of y with respect to x is 3 and the graph passes through the point (0, 6.4)
    b. Write the formula for y in terms of x.

x	y
−3	11.5
0	7
2	4
8	−5

Algebra II

Module 1: Homework
Linear Functions

87. Write the formula for each linear relationship described or represented in the following exercises.
    a. the constant rate of change of *y* with respect to *x* is –2.6 and the graph passes through the point (0, –2.9)
    b. Write the formula for *y* in terms of *x*.

x	y
–5	–26.5
–2	–13.6
0	–5
6	–20.8

88. Write a formula for each of the following linear relationships in
    i) the form $y = m(x - x_1) + y_1$.
    ii) the form $y = mx + b$.

    a. *y* changes at a constant rate of 5.2 with respect to *x*, and (4, –2) is a point on the line.

    b.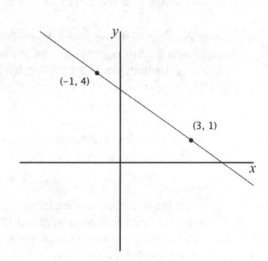

89. Write a formula for each of the following linear relationships in
    i) the form $y = m(x - x_1) + y_1$.
    ii) the form $y = mx + b$.

    a. *y* changes at a constant rate of –1.8 with respect to *x*, and (–3, 4) is a point on the line.

    b.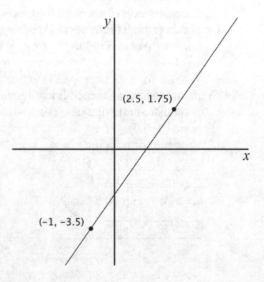

Algebra II   **Module 1: Homework**
Linear Functions

90. Write a formula for each of the following linear relationships in
    i) the form $y = m(x - x_1) + y_1$.
    ii) the form $y = mx + b$.
    a. y changes at a constant rate of 3.4 with respect to x and (0, 2.9) is a point on the line.
    b. y changes at a constant rate of -1.7 and $(3, -5)$ is a point on the line.
    c. A line that passes through the points $(-4, 7.2)$ and $(5, 3)$.
    d. A line that passes through the points $(0, 3.3)$ and $(4.8, 0)$.

91. Sketch the graph for each of the following functions.
    a. $y = -3x + 4$ b. $y = 2.5x - 2$

92. Sketch the graph for each of the following functions.
    a. $y = -2.5x + 6$ b. $y = 3x - 1$

INVESTIGATION 9: PROPORTIONALITY

93. A photo printing shop has the ability to print photos in any size that a customer requests. Suppose you take a photo with a digital camera with a 4:3 ratio. (This means that the height of the photo is $\frac{3}{4}$ times as large as the width of the photo). You would like to get your photo printed but do not want the image to be distorted or cropped.
    a. Suppose you would like the width of your photo print to be 12 inches. What should the height be so that the picture is not distorted? Explain how you are thinking about his problem to come up with the answer.
    b. Construct a table that relates the heights h of the photo print in inches to the widths w of 4 inches, 7 inches, 10 inches, and 14 inches. Give a convincing reason as to why the dimensions you found will not distort the photo print.
    c. Complete the following statement: As the width of the photo changes, the height of the photo changes so that …
    d. Write a formula that determines the height of the photo given its width such that the photo will not be distorted from the original.

94. Use the information from Exercise #93 to answer the questions.
    a. Draw a graph of the height vs. the width for undistorted photos with a width up to 14 inches. Be sure to label your axes with the appropriate quantities and units. Represent the photo widths on the horizontal axis.
    b. Explain what a point on your graph represents.
    c. What is the constant rate of change for the relationship you graphed? What does this tell us?
    d. Where does your graph cross the vertical axis? What does this point represent?
    e. Write a formula that determines the width of the photo given its height such that the photo will not be distorted from the original.

95. Use the information from Exercises #93-94 to answer the questions.
    a. What is the ratio $\frac{h}{w}$ equal to? What does this tell us?
    b. Plot the points (0, 0) and (4, 3) on the graph in Exercise #94, then calculate Δw and Δh from the first point to the second point and represent these changes on the graph.
    c. Talk about the link between $\frac{h}{w}$ and $\frac{\Delta h}{\Delta w}$ when we change away from (0, 0). Why did this happen?
    d. Is the result you found in part (c) true for all linear functions? If so, why does this happen. If not, what makes this context different? *You may want to look back at exercises in previous investigations.*

Algebra II

Module 1: Homework
Linear Functions

96. A sports trainer is preparing large containers of sports drink for a football game. The drink is prepared by mixing powder with water. The brand of sports drink being used calls for 13 cups of powder for every 5 gallons of water.
    a. How many cups of powder should be used for 2 gallons of water? Explain your reasoning.
    b. How many cups of powder should be used for 6.3 gallons of water? For 1.8 gallons of water?
    c. How many gallons of water should be used if you have 6 cups of powder? Explain your reasoning.
    d. How many gallons of water should be used if you have 15 cups of powder? 20 cups of powder?
    e. How many cups of powder should be used for 1 gallons of water? How many gallons of water should be used for one cup of powder?
    f. Suppose the trainer mixes the sports drink using 8.5 gallons of water and 24 cups of powder. Will this mixture taste right according to the brand's recommendations? If yes, explain why. If no, explain whether the mixture will taste stronger or weaker than the brand's recommended drink.
    g. Write two formulas that expresses the relationship between the number of cups of powder added to the sports drink mixture and the number of gallons of water added to the sports drink mixture. Define relevant quantities.
    h. Do the equations you created in part (g) express the same relationship between the relevant quantities? How do you know?

97. A grocery store purchases 150 Honeycrisp apples from a local farmer for $55.50.
    a. How many Honeycrisp apples can the store purchase for $250? Explain your reasoning
    b. How many apples can be purchased for $25? For $100?
    c. How much will it cost to purchase 82 apples? Explain your reasoning.
    d. How much will it cost to purchase 225 apples? 400 apples?
    e. Suppose a different farmer offers to sell the store 200 apples for $70. Is the cost charged by this farmer the same, a better deal or a worse deal for the store?

INVESTIGATION 10: EXTRA PRACTICE WITH PROPORTIONALITY

> Investigation 10: *Extra Practice with Proportionality* and associated homework is available online at www.rationalreasoning.net.

INVESTIGATION 11: INTRODUCTION TO FUNCTIONS AND FUNCTION NOTATION

112. Recall that a function has three parts:
     A set of input values referred to as the domain of the function.
     A set of output values referred to as the range of the function.
     A rule that assigns to each input value *exactly one* output value.
   Given that the function $f(t) = 3t + 22$, where $d = f(t)$ determines the number of feet you have walked from your front door t seconds since you started your walk:
    a. What is the name of the function?
    b. What is the independent or input variable and what does it represent?
    c. How is the dependent or output variable referenced in the above function?
    d. What is the rule of the function that determines the output value that is associated with each input value?
    e. Explain why we can say that f is a function.
    f. Can we also think about the quantity *time since you starting walking in seconds t* as a function of the quantity *distance from your front door in feet d*? Explain why or why not.
    g. Rewrite the formula so that it expresses a time since you starting walking in seconds t as a function of the quantity distance from your front door in feet d.

Algebra II

Module 1: Homework
Linear Functions

113. For each of the following functions:
    i. What is the function name?
    ii. What is the input?
    iii. What is the output?
    iv. What is the rule that assigns one input value to exactly one output value?
    v. Evaluate each function for the input values, 2 and 3.5.

   a. $f(x) = 3x - 4$
   b. $g(t) = 0.5(4 - t)$
   c. $j(p) = \dfrac{4(p-1)}{3}$

114. Use function notation to define a function that meets the verbal description given below:
   Example: A function *f* is defined in terms of the input variable *x*. The output of the function is 12 more than a number that is 2.5 times as large as the input. *Answer:* $f(x) = 2.5x + 12$
   a. The name of the function is *p*. The input is specified by the variable *m*. The output of this function is half the value of the input.
   b. The name of the function is *k*. The input is specified by the variable *b*. The output of this function is 4.5 less than a number that is 3 times as large as its input.

115. The following table represents the relationship between a student's name and how long it took the student to complete a quiz.

Student Name	Time to complete quiz
Mary	6 minutes
Caleb	7 minutes
William	4 minutes
Stephanie	6 minutes

   a. Is the time to complete the quiz a function of the student's name? Why or why not?
   b. Is the student's name a function of the time to complete the quiz? Why or why not?

116. An input quantity and output quantity is given for each part. Determine if the output quantity could be a function of the input quantity. If the answer is no, provide an example that shows the relationship is not a function.
   a. Input quantity: Birth date. Output quantity: Age.
   b. Input quantity: Radius of a circle. Output quantity: Area of a circle.
   c. Input quantity: Circumference of a circle. Output quantity: Area of a circle.
   d. Input quantity: Perimeter of a rectangle. Output quantity: Area of a rectangle.
   e. Input quantity: Amount of time a person walks at a constant speed. Output quantity: How far the person walks at that constant speed.
   f. Input: Distance a person walks at a constant speed. Output: Amount of time the person walks at that constant speed.

117. Let $V = g(s)$ when $g(s) = s^3$ where *V* is the volume in cubic inches of a cube and *s* is the side length in inches.
   a. What does $g(2.1)$ represent?
   b. What is the value of $g(2.1)$?
   c. What does the statement $g(4) = 64$ tell us?
   d. Create a graph of *g*, then plot and label the points $(0, g(0))$, $(1, g(1))$, and $(3, g(3))$.
   e. What does the graph of *g* represent?

Algebra II

Module 1: Homework
Linear Functions

118. Let $A = f(s)$ when $f(s) = s^2$ where A is the area of a square in square inches and s is the side length in inches.
    a. What does $f(4.1)$ represent?
    b. What is the value of $f(4.1)$?
    c. What does the statement $f(6) = 38.44$ tell us?
    d. Create a graph of f, then plot and label the points $(0, f(0))$, $(1, f(1))$, and $(3, f(3))$.
    e. What does the graph of f represent?

119. The **domain** of a function is the set of values that are reasonable to input to the function. A function's domain may be restricted because some values don't make sense for a certain quantity or in a specific function. The **range** of a function is the set of all of the output values you get when you use all of the values in your domain.
    a. What is the domain of f from Exercise #117 and what does it represent?
    b. What is the range of f from Exercise #117 and what does it represent?
    c. How do the domain and range change if we remove the context? That is, we consider the general function $g(x) = x^3$ where the input and output don't have to represent real world quantities?

120. The **domain** of a function is the set of values that are reasonable to input to the function. A function's domain may be restricted because some values don't make sense for a certain quantity or in a specific function. The **range** of a function is the set of all of the output values you get when you use all of the values in your domain.
    a. What is the domain of f from Exercise #118 and what does it represent?
    b. What is the range of f from Exercise #118 and what does it represent?
    c. How do the domain and range change if we want the make sure the square's side length is at least 2 inches long, but we also want to be able to draw the square on an 8.5" x 11" piece of paper?
    d. How do the domain and range change if we remove the context? That is, we consider the general function $f(x) = x^2$ where the input and output don't have to represent real world quantities?

121. Given that one US dollar is equal to 80.296 Japanese yen,
    a. What are the two quantities that are being related?
    b. If m is the number of US dollars and n is the number of Japanese yen, define a function that determines the number of Japanese yen when the number of US dollars is known.
    c. What does it mean to say that the number of *U.S. dollars m* is a function of the number of *Japanese yen n*?
    d. Can we also think about the *number of Japanese yen n* as a function of the number of *U.S. dollars m*? Explain why or why not. If so, what is the input (or independent) variable, and what is the output (or dependent) variable?
    e. Define a function that determines the number of US dollars when the number Japanese yen are known.

122. A candle is originally 15 inches long. When lit, it burns at a constant rate of 1.7 inches per hour.
    a. Define a function h that represents the remaining length of the candle r in terms of the number of hours t that the candle has been burning.
    b. Determine the value of $h(4.1)$ and explain what it represents in the context of the problem.
    c. If a problem says "Solve $h(t) = 2$ for t," what are you being asked to find?

Algebra II

Module 1: Homework
Linear Functions

123. Given a function g is defined by the formula, $g(x) = 3x - 8.3$,
   a. Determine the values of $g(-1)$, $g(0)$, $g(2)$, and $g(4)$.
   b. On a graph, plot the points $(-1, g(-1))$, $(0, g(0))$, $(2, g(2))$, and $(4, g(4))$.
   c. Function g is a linear function. What is its vertical intercept and constant rate of change?

124. Given a function k is defined by the formula, $k(x) = \dfrac{4x-5}{2}$,
   a. Determine the values of $k(-2)$, $k(0)$, $k(1)$, and $k(3)$.
   b. On a graph, plot the points $(-2, k(-2))$, $(0, k(0))$, $(1, k(1))$, and $(3, k(3))$.
   c. Function k is a linear function. What is its vertical intercept and constant rate of change?

125. Evaluate each function for the given input value.
   a. Given $k(t) = 0.7t - 4$, evaluate $k(0.3)$.
   b. Given $n(x) = \dfrac{-4x + 20}{7}$, evaluate $n(4)$.
   c. Given $q(s) = -0.4s - 3.4$, evaluate $q(-1)$.
   d. Given $h(x) = -3x + 5$, evaluate $h(x - 1)$. What is the meaning of $h(x - 1)$?

126. Given the function f defined by $f(x) = 3x + 7$ and the function g defined by $g(x) = \dfrac{4x+2}{5}$, evaluate:
   a. $f(-6)$ b. $g(100)$ c. $f(x+3)$ d. $f(4) + g(4)$

INVESTIGATION 12: INTRODUCTION TO PIECEWISE DEFINED FUNCTIONS

127. A rainstorm begins at 7:00 a.m. and lasts until 9:00 a.m. when it stops raining. During the storm (from 7 to 9), the rain falls steadily at a rate of 1.5 inches per hour. No rain fell prior to 6:00 a.m. on that day. Suppose we want to model the total amount of rainfall (in inches) as a function of the number of hours elapsed since 6:00 a.m. up until noon of that day. The graph of this relationship is shown here.

   a. Complete the following statements by filling in the blanks where t represents the time elapsed since 6 am in hours and $a(t)$ represents the total accumulation of rain in inches.

      i. From 6:00 a.m. to 7:00 a.m. we can model the total rainfall by $a(t) = $ _____.
         However, this formula only works for values of t such that ____ $\leq t \leq$ ____.

      ii. From 7:00 a.m. to 9:00 a.m. we can model the total rainfall by $a(t) = $ _____.
         However, this formula only works for values of t such that ____ $\leq t \leq$ ____.

      iii. From 9:00 a.m. to 12:00 p.m. we can model the total rainfall by $a(t) = $ _____.
         However, this formula only works for values of t such that ____ $\leq t \leq$ ____.

(continues...)

Algebra II

Module 1: Homework
Linear Functions

b. Let's summarize:

$$a(t) = \begin{cases} \underline{\hspace{2cm}} & \text{if } \underline{\hspace{0.5cm}} \leq t \leq \underline{\hspace{0.5cm}} \\ \underline{\hspace{2cm}} & \text{if } \underline{\hspace{0.5cm}} \leq t \leq \underline{\hspace{0.5cm}} \\ \underline{\hspace{2cm}} & \text{if } \underline{\hspace{0.5cm}} \leq t \leq \underline{\hspace{0.5cm}} \end{cases}$$

c. Evaluate each of the following.
   i. $a(0.7)$ ii. $a(1.4)$ iii. $a(2.9)$ iv. $a(4.5)$

128. A snowstorm begins at 6:00 p.m. and lasts until 8:30 p.m. when it stops snowing. The snow remains accumulated on the ground. At 9 p.m. it begins snowing again and continues until 10:30 p.m. when it stops for the day.

During the first part of the storm (from 6 to 8:30), the snow falls steadily at a rate of 3 inches per hour. During the second part of the storm (from 9 to 10:30), the snow falls steadily at a rate of 2 inches per hour. There is no snow on the ground before the storm begins.

Suppose we want to model the depth of the snow on the ground (in inches) as a function of the number of hours elapsed since 6:00 p.m. up until midnight of that day. The graph of this relationship is shown here.

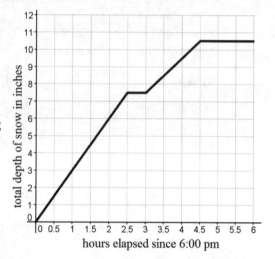

a. Complete the following statements by filling in the blanks where t represents the time elapsed since 6 pm in hours and $d(t)$ represents the total depth of the snow in inches.

   i. From 6:00 p.m. to 8:30 p.m. we can model the depth of snow on the ground by
      $d(t) = \underline{\hspace{2cm}}$. However, this formula only works for values of t such that
      $\underline{\hspace{0.5cm}} \leq t \leq \underline{\hspace{0.5cm}}$.

   ii. From 8:30 p.m. to 9:00 p.m. we can model the depth of snow on the ground by
      $d(t) = \underline{\hspace{2cm}}$. However, this formula only works for values of t such that
      $\underline{\hspace{0.5cm}} \leq t \leq \underline{\hspace{0.5cm}}$.

   iii. From 9:00 p.m. to 10:30 p.m. we can model the depth of snow on the ground by
      $d(t) = \underline{\hspace{2cm}}$. However, this formula only works for values of t such that
      $\underline{\hspace{0.5cm}} \leq t \leq \underline{\hspace{0.5cm}}$.

   iv. From 10:30 p.m. to midnight we can model the depth of snow on the ground by
      $d(t) = \underline{\hspace{2cm}}$. However, this formula only works for values of t such that
      $\underline{\hspace{0.5cm}} \leq t \leq \underline{\hspace{0.5cm}}$.

b. Let's summarize:

$$d(t) = \begin{cases} \underline{\hspace{2cm}} & \text{if } \underline{\hspace{0.5cm}} \leq t \leq \underline{\hspace{0.5cm}} \\ \underline{\hspace{2cm}} & \text{if } \underline{\hspace{0.5cm}} \leq t \leq \underline{\hspace{0.5cm}} \\ \underline{\hspace{2cm}} & \text{if } \underline{\hspace{0.5cm}} \leq t \leq \underline{\hspace{0.5cm}} \\ \underline{\hspace{2cm}} & \text{if } \underline{\hspace{0.5cm}} \leq t \leq \underline{\hspace{0.5cm}} \end{cases}$$

(*continues...*)

Algebra II
Module 1: Homework
Linear Functions

c. Evaluate each of the following.
   i. $d(1.3)$ ii. $d(2.7)$ iii. $d(3.4)$ iv. $d(5)$

129. An online retailer has different shipping costs based on the amount of merchandise purchased.

Merchandise Total	Shipping & handling
Up to $15.00	$5.95
$15.01 to $30.00	$6.95
$30.01 to $50.00	$8.95
$50.01 to $100.00	$11.95
$100.01 and up	Free shipping

   a. What is the total amount a customer pays for shipping if the merchandise total is:
      i. $45.23 ii. $0.25 iii. $101.00 iv. $15.02
   b. Define a function h that determines the shipping cost in terms of the total amount of the purchase.
   c. Evaluate:
      i. $h(5.50)$ ii. $h(58.60)$ iii. $h(220.47)$

130. Jane is on one end of a seesaw and John is on the other end. The given graph represents Jane's height from the ground (in feet) with respect to the elapsed time (in seconds) since she got onto one end of the seesaw.

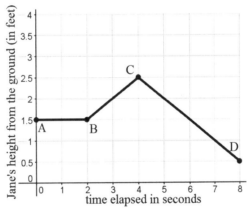

   a. Give the coordinates of the points A, B, C, and D and describe what each point represents in the context of this problem.
   b. Define a formula to represent the height of Jane above the ground (in feet)
      i. for the first 2 seconds after she got onto the seesaw.
      ii. as the number of seconds since she got onto the seesaw increases from 2 seconds to 4 seconds.
      iii. as the number of seconds since she got onto the seesaw increases from 4 seconds to 8 seconds.
   c. Define a piecewise function j to describe Jane's height from the ground in terms of the time elapsed (in seconds) for the first 8 seconds after she got onto the seesaw.
   d. If Jane and John started riding on the seesaw from the same height above the ground, but on opposite ends of the seesaw, construct a graph of John's height from the ground (in feet) in terms of the number of seconds since Jane boarded the seesaw.

Algebra II

Module 1: Homework
Linear Functions

131. Most income tax systems (both state and federal) use a graduated income tax. Generally speaking, this means that the more money you earn per year the greater percentage of your income you owe in taxes. A single person's federal income tax is calculated using the formula below, where $T(i)$ is the amount an individual owes in state income tax (in dollars) and i is the person's taxable annual income (in dollars).

$$T(i) = \begin{cases} 0.10i & \text{if } 0 \leq i \leq 8{,}700 \\ 0.15(i-8{,}700)+870 & \text{if } 8{,}700 < i \leq 35{,}350 \\ 0.25(i-35{,}350)+4{,}867.50 & \text{if } 35{,}350 < i \leq 85{,}650 \\ 0.28(i-85{,}650)+17{,}442.50 & \text{if } 85{,}650 < i \leq 178{,}650 \\ 0.33(i-178{,}650)+43{,}482.50 & \text{if } 178{,}650 < i \leq 388{,}350 \\ 0.35(i-388{,}350)+112{,}683.50 & \text{if } i > 388{,}350 \end{cases}$$

a. What percent of a person's income is owed in federal taxes if he/she makes $6,600 per year? How is this represented in the function?
b. Evaluate each of the following, then explain what the value represents.
   i. $T(22{,}300)$ ii. $T(139{,}950)$ iii. $T(72{,}600)$
c. In the second part of the formula, what does the expression $i-8{,}700$ represent?
d. In the second part of the formula, what does the expression $0.15(i-8{,}700)$ represent?
e. In the second part of the formula, what does 870 represent? Why is it needed?
f. Determine if you agree with the following statement, then explain your thinking: People who make $100,000 per year have to pay 28% of their income in federal taxes.

132. Most income tax systems (both state and federal) use a graduated income tax. Generally speaking, this means that the more money you earn per year the greater percentage of your income you owe in taxes. A married couple's federal income tax is calculated using the formula below, where $T(i)$ is the amount an individual owes in state income tax (in dollars) and i is the couple's taxable annual income (in dollars).

$$T(i) = \begin{cases} 0.10i & \text{if } 0 \leq i \leq 17{,}400 \\ 0.15(i-17{,}400)+1{,}740 & \text{if } 17{,}400 < i \leq 70{,}700 \\ 0.25(i-70{,}700)+9{,}735 & \text{if } 70{,}700 < i \leq 142{,}700 \\ 0.28(i-142{,}700)+27{,}735 & \text{if } 142{,}700 < i \leq 217{,}450 \\ 0.33(i-217{,}450)+48{,}749 & \text{if } 217{,}450 < i \leq 388{,}350 \\ 0.35(i-388{,}350)+105{,}146 & \text{if } i > 388{,}350 \end{cases}$$

a. What percent of a couple's income is owed in federal taxes if they make $14,000 per year? How is this represented in the function?
b. Evaluate each of the following, then explain what the value represents.
   i. $T(62{,}300)$ ii. $T(139{,}950)$ iii. $T(222{,}600)$
c. In the third part of the formula, what does the expression $i-70{,}700$ represent?
d. In the third part of the formula, what does the expression $0.25(i-70{,}700)$ represent?
e. In the third part of the formula, what does 9,735 represent? Why is it needed?
f. Determine if you agree with the following statement, then explain your thinking: Couples who make $100,000 per year have to pay 25% of their income in federal taxes.

Algebra II

Module 1: Homework
Linear Functions

INVESTIGATION 13: PIECEWISE DEFINED FUNCTIONS – EXTRA PRACTICE

> Investigation 13: *Piecewise Defined Functions – Extra Practice* and associated homework is available online at www.rationalreasoning.net.

INVESTIGATION 14: ABSOLUTE VALUE

141. A potato chip manufacturing plant produces bags of chips and states that the bag contains 8 ounces of chips. Ideally every bag of chips weighs exactly 8 ounces, but this precision is not possible. The company allows the weight of the bag of chips to be within 0.2 ounces of the ideal weight.
    a. Let w represent the actual weight of a bag of chips. List some possible values for w that represent allowable bags of chips, then represent the possible values on the number line.
    b. Write a mathematical statement (inequality) to represent the possible values of w in part (a).
    c. We will call the difference between the actual weight of the bag of chips and the desired weight the *error*. Write an expression to represent the error.
    d. Write a mathematical statement (inequality) that represents errors within our tolerance of 0.2 ounces.
    e. Write a mathematical statement (inequality) that represents actual weights of chip bags w that are not usable, then represent these values on a number line.

142. An exit poll was completed during a certain election. They found that 54.2% of people voted for the leading candidate. The margin of error for the poll was 5%.
    a. Let p represent the actual percentage of people who voted for the leading candidate. List some possible values for p that represent the actual percentage of people, then represent the possible values on the number line.
    b. Write a mathematical statement (inequality) to represent the possible values of p in part (a).
    c. We call the difference between the actual percentage of people who voted for the leading candidate and the measured percentage of people the *error*. Write an expression to represent the error.
    d. Write a mathematical statement (inequality) that represents errors within our tolerance of 5%.
    e. Write a mathematical statement (inequality) that represents actual percentage of people p that are not within the margin of error, then represent these values on a number line.

143. Suppose we want to choose an x value that is within 6.2 units of 8.
    a. List some values of x that meet this condition.
    b. Write a mathematical statement (inequality) to describe the possible changes in x away from 8.
    c. Write a mathematical statement (inequality) to describe the possible values of x.
    d. Represent the possible values of x on the number line.

144. Suppose we want to choose an x value that is within 3.5 units of 7.
    a. List some values of x that meet this condition.
    b. Write a mathematical statement (inequality) to describe the possible changes in x away from 7.
    c. Write a mathematical statement (inequality) to describe the possible values of x.
    d. Represent the possible values of x on the number line.

Algebra II

Module 1: Homework
Linear Functions

145. Represent each of the following on a number line.
    a. All numbers x whose distance from 7 is less than or equal to 1.5.
    b. All numbers x whose distance from 9 is less than 3.
    c. All numbers x whose distance from –2 is less than 4.

146. For each inequality describe a distance from a specific point that are being represented by the given inequalities
    a. $-1.2 < x - 8 < 1.2$
    b. $-4 < x - (-9.1) < 4$

147. Suppose we want the change in x away from 7 to be positive but no more than 4.
    a. List some values of x that make this true.
    b. Write a mathematical statement (inequality) to describe the possible changes in x away from 7.
    c. Write a mathematical statement (inequality) to describe the possible values of x.
    d. Represent the possible values of x on the number line.

148. Suppose we want to choose a value of x that is at least 3 units away from $x = -2$.
    a. List some values of x that are at least 3 units away from $x = -2$, then represent all possible such values on the number line.
    b. Write a mathematical statement (inequality) that describes values of x that are at least 3 units away from $x = -2$. Try to think of at least two different ways to write this.

149. Use absolute value to rewrite the inequalities from Exercises #141d, #143b, #145a, #145b, and #145c.

150. Use absolute value to rewrite the inequalities from Exercises #142d, #144b, #146a, #146b, and #148b.

151. Use BOTH absolute value notation and a number line to represent the values of x described below.
    a. All numbers x whose distance from 7 is less than 3.
    b. All numbers x whose distance from – 4 is less than 2
    c. All numbers x whose distance from 3.5 is less than 0.01.

152. Describe what each of the following represents using the definition of absolute value inequalities.
    a. $|x-3| < 1$
    b. $|x+4| \leq 0.5$

153. Use absolute value notation and number line notation, to:
    a. Represent all measurements of a board x that are greater than 0.5 inch away from 8.5 inches.
    b. Represent all weights x of a blackbird model airplane that are greater than 0.07 ounces away from its ideal weight of 28.56 ounces.

Algebra II

154. Represent the solution set of the following absolute value inequalities by:
    i. Describing the solutions in words
    ii. Illustrating the solutions on a number line
    iii. Writing an inequality (with no absolute values)
    a. $|x-3|>2$
    b. $|x+5|>1.5$

155. Represent the solution set of the following absolute value statements by:
    i. Describing the solutions in words
    ii. Illustrating the solutions on a number line
    iii. Writing an equality or inequality (with no absolute values)
    a. $|x|=4$
    b. $|x|\leq 4$
    c. $|x|>4$

156. Represent the solution set of the following absolute value statements by:
    i. Describing the solutions in words
    ii. Illustrating the solutions on a number line
    iii. Writing an equality or inequality (with no absolute values)
    a. $|x|=2.5$
    b. $|x|<12.5$
    c. $|x|>0.5$

157. Use the definition of absolute value to explain why:
    a. $|x|=-5$ has no solution
    b. $|x|<-3$ has no solution

158. Solve the following absolute value inequalities for x.
    a. $|x-3|=1.5$
    b. $|x-1|<3$
    c. $|x-2|=7$

159. Solve the following absolute value inequalities for x.
    a. $|x+3|<2$
    b. $-|x-9|=-1$
    c. $|x-9|=1$

Algebra II

Module 2: Investigation 0
Supporting Materials for Module 2

This investigation contains practice with skills and ideas that will help you be successful in Module 2. Your teacher will direct you to these exercises during the module as necessary.

Part 1: The Meaning of Exponents
(Recommended use: Prior to Investigation 4)

1. What does an exponent tell us? For example, what does the expression 3^4 represent?

2. Rewrite each of the following using exponents.
   a. $6 \cdot 6 \cdot 6 \cdot 6 \cdot 6 \cdot 6 \cdot 6$
   b. $x \cdot x \cdot x \cdot x$
   c. $5 \cdot 5 \cdot 5 \cdot 8 \cdot 8 \cdot 8 \cdot 8 \cdot 8$

3. Rewrite each of the following as a product of factors (so without using exponents). *Do not simplify your answer.*
   a. $10^2 \cdot 9^3$
   b. $4p^5 t^4$
   c. $\left(5^2\right)^4$
   d. $\left(7x^3 y^2\right)\left(2x^2 y^4\right)$

4. Simplify your results in Exercises #3c and #3d as much as possible.

Part 2: Radicals
(Recommended use: Prior to Investigation 4)

5. What does the square root of a number represent? For example, what is the value of $\sqrt{64}$ and what does it represent?

6. a. Why are there two solutions to the equation $x^2 = 25$?
   b. What are the solutions to the equation $x^2 = 81$? What about $w^2 = 144$?

7. a. Are there also two solutions to the equation $x^3 = 8$? Explain.
   b. What are the solutions to the equation $r^3 = 27$? What about $p^3 = -125$?

8. The *cube root* of a number $\left(\sqrt[3]{\#}\right)$ represents the number that, when raised to an exponent of 3, returns the original number. For example, $\sqrt[3]{8} = 2$ because $2^3 = 8$. Find the value of each of the following.
   a. $\sqrt[3]{64}$
   b. $\sqrt[3]{729}$
   c. $\sqrt[3]{-343}$

9. For some number x, what does each of the following represent?
   a. $\sqrt[5]{x}$
   b. $\sqrt[4]{x}$
   c. $\sqrt[10]{x}$

10. Solve each of the following equations.
   a. $2b^2 = 200$
   b. $r^4 = 1,296$
   c. $x^7 = -128$

   d. $5x^3 = 320$
   e. $64z^8 = 0.25$
   f. $12x^5 + 4 = 2,920$

Part 3: Percent Change and Factors
(Recommended use: Prior to Investigation 5)

11. Suppose we are buying an item on sale for 20% off of its original price of $65.
   a. What number can we multiply $65 by to find out the discount in dollars? What is the discount in dollars?

   b. Use the result of part (a) to determine the price we are paying.

   c. What percent of the original price are we paying?

   d. What number can we multiply $65 by to determine the sale price of the item?

12. For each sale described, do the following.
   i) State the number we can multiply the original price by to find the sale price.
   ii) Find the sale price in dollars.
   a. original price: $150, sale: 40% off
   b. original price: $19, sale: 10% off
   c. original price: $915.99, sale: 15% off
   d. original price: $22.99, sale: 12.5% off

13. Suppose a store is raising the price of a $42 item by 15%.
   a. By how many dollars is the price increasing? What number can we multiply $42 by to determine this?

   b. What is the new price of the item?

   c. What number can we multiply $42 by to find the new price of the item?

14. For each price increase described, do the following.
   i) State the number we can multiply the original price by to find the new price.
   ii) Find the new price in dollars.
   a. original price: $52, increase: 10%
   b. original price: $13, increase: 50%
   c. original price: $14.99, increase: 2.3%
   d. original price: $1,499.99, increase: 100%

15. A city with a population of 480,560 people at the end of the year 2000 grew by 6.2% over 10 years. What was its population at the end of 2010?

16. Student Council reported that attendance at Prom this year was 4.8% less than attendance last year. If 818 people attended Prom last year, how many people attended Prom this year?

Algebra II

Module 2: Investigation 1
Introduction to Sequences and Limits of Sequences

A **sequence** in mathematics is an ordered set of objects (usually numbers), such as 2, 4, 6, 8, 10 or 7, 5, 6, 4, 5, 3. The objects in the sequence are called **terms**.

Finite vs. Infinite Sequences

A **finite sequence** is a sequence with a set number of terms. Finite sequences are written like 2, 4, 6, 8, 10 or 2, 4, 6, ..., 32.

An **infinite sequence** is a sequence with infinitely many terms where the pattern generating the sequence is repeated without end. Infinite sequences are written like 5, 10, 15, ...

---

1. Consider the sequence 1, 2, 4, 8, 16, ...
   a. Describe the pattern.

   b. Find the next three terms.

   c. Is it difficult to find the value of the 1000th term? If so, explain why. If not, find its value.

2. Consider the sequence 3, 6, 11, 18, 27, ...
   a. Describe the pattern.

   b. Find the next three terms.

   c. Is it difficult to find the value of the $2,500^{th}$ term? If so, explain why. If not, find its value.

3. Consider the sequence $\frac{2}{1}, \frac{5}{4}, \frac{10}{9}, \frac{17}{16}, \frac{26}{25}, ...$
   a. Describe the pattern.

   b. Will any of the following be terms in this sequence? If so, which one(s)?

   $\frac{101}{100}$ $\frac{116}{115}$ $\frac{161}{160}$ $\frac{226}{225}$

   c. Is it difficult to find the value of the 50th term? If so, explain why. If not, find its value.

Algebra II

Module 2: Investigation 1
Introduction to Sequences and Limits of Sequences

4. Sequences can be thought of as functions in which the input quantity is the **term position** (consisting of positive integers) and the output quantity is the **term value** (consisting of the terms of the sequence). Let's take another look at the sequence in Exercise #3.

input: term position	output: term value
1	2/1
2	5/4
3	10/9
4	17/16
5	26/25

   Like other functions, we can create the graph of a sequence. By convention we track the term position on the horizontal axis and the term value on the vertical axis.

   a. Before graphing, consider the following question. When you plot the sequence, should the points be connected? Explain.

   b. Graph the sequence. Use at least the first six terms.

   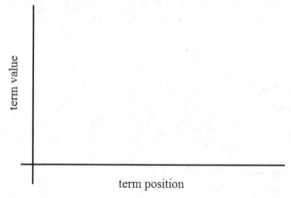

   c. What does the graph suggest about the term values as the term position increases?

Limit of a Sequence

The behavior you noted in Exercise #4 means that this sequence has a *limit*, or a constant value that the term values approach as the term position increases.[1]

Note: Only infinite sequences (sequences with infinitely many terms) are said to have limits. Finite sequences (sequences with a definitive end) cannot possess a limit.[2]

In Exercise #4, the *limit is 1*.

---

[1] The formal definition of limit (paraphrased) requires that the term values be able to get as close as we want to the limiting value as the term position increases for the limit to exist. In other words, if we say that a sequence *has a limit of 4*, it means that eventually the term values must be within (for example) 0.1 of 4, and within 0.003 of 4, and within 0.00000007 of 4, etc.

[2] Even if the term values of a finite sequence are getting closer to a constant value, eventually the sequence stops and the last term's value is the closest we can get to this constant. Therefore, the term values can't get arbitrarily close to the constant value. For example, consider the sequence 7.9, 7.99, 7.999, 7.9999. The term values are certainly getting closer to 8, but the final term is 0.0001 away from 8. We could never get within 0.000001 of 8, for example, or within 0.00000003 of 8. Therefore "8" doesn't fit the requirement to be a limit of the sequence.

Algebra II

Module 2: Investigation 1
Introduction to Sequences and Limits of Sequences

5. A new sequence is generated by taking the difference of 3 and the term values from the sequence in Exercise #4. For example, the first term of the new sequence is $3 - \frac{2}{1} = 1$. The second term of the sequence is $3 - \frac{5}{4} = \frac{7}{4}$.

    a. Graph this new sequence (use at least 6 points).

    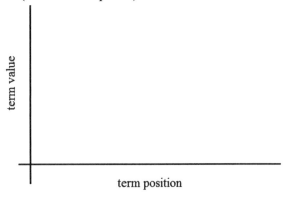

    b. Does the sequence appear to have a limit? If so, what is the limit?

    c. What is the relationship between the limit of this sequence and the limit of the sequence in Exercise #4? (*That is, is there a mathematical reason why the sequence has term values approaching this limit while the term values of the other sequence approach 1?*)

For Exercises #6-9, do the following.
   a) Examine the pattern and then write the next three terms.
   b) Graph the sequence (use at least six points).
   c) Does the sequence appear to have a limit? If it does, state the limit.

6. 7, 9, 11, 13, ...

7. $1, \frac{1}{2}, \frac{1}{3}, \frac{1}{4}, \ldots$

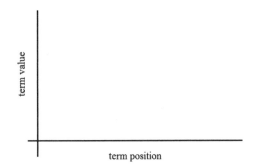

8. 5.1, 4.9, 5.01, 4.99, 5.001, 4.999, ...

9. 1, −1, 2, −2, 4, −4, 8, −8, ...

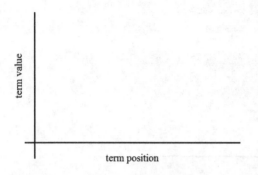
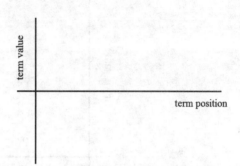

10. A sequence is formed by choosing any real number *x* to be the first term, then generating the sequence by taking each term value and dividing by 10 to get the next term value.
    a. Choose a few possible values of *x* and explore the kinds of sequences produced by following this pattern.

    b. Do all sequences formed in this manner have a limit? If so, what is the limit? Why does this happen?

    c. What if the pattern was instead formed by multiplying the term value by 10 to get the subsequent term value. Would sequences formed in this manner have a limit? Explain.

11. Explain, in your own words, what it means for a sequence to have a limit.

Algebra II

Module 2: Investigation 2
Formulas for Sequences

In Exercises #1-2, find the indicated term value for the given sequence.
1. $15, 17, 19, 21, 23, 25, 27, \ldots, 73$; find a_6
2. $44, 22, 11, \ldots$; find a_5

3. If a_n represents the value of the nth term in some sequence, how can we represent each of the following?
   a. the value of the term before the nth term

   b. the value of the term after the nth term

   c. the value of the term two terms before the nth term

We've seen that a sequence can be thought of as a function where each term position maps to a single term value. Therefore, it's not surprising that we can also write formulas that relate these quantities.

Recursive Formula for a Sequence's Term Values

4. Consider the sequence $17, 23, 29, 35, \ldots, 323$. One of the most common ways to describe this sequence is by saying something similar to "The sequence begins with 17, and the value of every term is 6 more than the value of the previous term."
   a. Using sequence notation, how can you communicate to someone that the value of the first term in the sequence is 17?

   b. Using sequence notation, how can you communicate to someone that the value of any term (such as the nth term) is always 6 more than the value of the previous term?

Recursive Formula for a Sequence's Term Values

A *recursive formula* defines the value of a term a_n based on the value of the previous term or terms.

In Exercises #5-7, use the given formula to write the first four terms of each sequence.

5. $a_1 = 3$
   $a_n = a_{n-1} - 5$

6. $a_1 = 40$
   $a_n = \dfrac{a_{n-1}}{4}$

7. $a_1 = 1$
   $a_2 = 2$
   $a_n = (a_{n-1})(a_{n-2})$

Algebra II

Module 2: Investigation 2
Formulas for Sequences

In Exercises #8-9, write a recursive formula to define each sequence, then find a_9.

8. 11, 8, 5, 2, ...

9. −7, 21, −63, 189, ...

10. John wasn't feeling well, so he went to the doctor yesterday. The doctor gave him a prescription for antibiotics and told him to take one 450 mg dose every 8 hours. John's body metabolizes the drug such that 33% of the medicine remains in his body by the time he takes the next dose.
    a. Why might doctors create a schedule where you take a new dose before the previous dose is completely removed from the body?

    b. Write a recursive formula that will tell you how much medicine is in John's body after taking n doses of the medicine.

    c. Suppose John never stops taking the medicine.
        i. What will happen to the amount of medicine in his body after each dose? Perform some calculations if necessary to explore this question.

        ii. Is there a limit to the maximum amount of antibiotics in John's system? If so, what is the limit?

11. What's the biggest drawback of using a recursive formula to describe a sequence?

Algebra II

Module 2: Investigation 2
Formulas for Sequences

12. Consider the sequence 4, 8, 12, 16, …. Each term value is exactly 4 times as large as the corresponding term position, so we could think of the sequence as 4(1), 4(2), 4(3), 4(4), … and say that $a_n = 4n$. Find the value of the 25th term.

13. Consider the sequence 1, 8, 27, 64, …. Each term value is the third power of the corresponding term position, so we could think of the sequence as $1^3, 2^3, 3^3, 4^3, …$ and say that $a_n = n^3$. Find the value of the 25th term.

14. How are the formulas $a_n = 4n$ and $a_n = n^3$ different from the recursive formulas we wrote earlier in this investigation?

Explicit Formula for a Sequence's Term Values

An ***explicit formula*** defines the value of a term a_n based on its position n.

Explicit Formula for a Sequence's Term Values

In Exercises #15-17, find the value of the 12th term in each sequence.

15. $a_n = \dfrac{2}{n}$

16. $b_n = 0.5(n-1)$

17. $c_n = \dfrac{3n}{n+1}$

Algebra II

Module 2: Investigation 2
Formulas for Sequences

In Exercises #18-19, write the explicit formula defining the sequence.
18. 12, 24, 36, ... , 192
19. 0, 3, 8, 15, 24, ... , 288

20. How many terms are in the sequences in Exercises #18 and #19?

21. When a certain ball is dropped, it bounces back up to ¾ of the distance it fell. Suppose the ball is initially dropped from a height of 12 meters.
    a. Write the first three terms of the sequence describing the height the ball will return to after n bounces.

    b. Write an explicit formula for the sequence that will tell you the height the ball will bounce up to after n bounces.

    c. Is there a limit to this sequence?

Algebra II **Module 2: Investigation 3**
Arithmetic Sequences

1. Write the recursive formula for each sequence.
   a. −6, 2, 10, 18, ...
   b. 13, 11, 9, 7, ...
   c. 4, 12, 36, ...
   d. 3.74, 3.79, 3.84, ...

   e. $-\frac{1}{2}, \frac{3}{2}, -\frac{9}{2}, ...$
   f. 4.1, 3.3, 2.5, ...
   g. $6, 2, \frac{2}{3}, ...$
   h. $\frac{10}{3}, -\frac{20}{9}, \frac{40}{27}, ...$

2. The sequences in Exercise #1 can be organized into two broad categories based on how the patterns are formed.
   a. What are the categories and which sequences belong to each category?

   b. Provide another sequence for each of the two categories you described and then provide a sequence that doesn't fit into either of the two categories.

Categorizing Common Sequences

Common Difference:

Arithmetic Sequence:

Common Ratio:

Geometric Sequence:

3. An arithmetic sequence has a common difference of –3 and the value of its initial term is 8. Write the first several terms of the sequence and the recursive formula that defines the sequence.

4. What is the recursive formula that defines the value of a_n for an arithmetic sequence with $a_1 = k$ and a common difference d?

In the previous exercises we thought about generating a term value in an arithmetic sequence by adding the common difference to the value of the previous term. Another way of thinking about how to find the term value in an arithmetic sequence is to think about what must be added to the initial term to generate other terms in the sequence.

5. Suppose an arithmetic sequence begins with –7 and has a common difference of 3. Complete the following statements.

   $a_1 = -7$

   $a_2 = -7 +$ _____ = _____

   $a_3 = -7 +$ _____ = _____

   $a_4 = -7 +$ _____ = _____

   $a_5 = -7 +$ _____ = _____

   $a_6 = -7 +$ _____ = _____

   $a_n = -7 +$ _____

6. We can create a general formula that serves as a model for the explicit formulas for *all* arithmetic sequences. What is the explicit formula that defines the value of a_n for an arithmetic sequence with an initial term value a_1 and a common difference d?

Arithmetic Sequence Formulas

Recursive Formula: **Explicit Formula:**

Algebra II

Module 2: Investigation 3
Arithmetic Sequences

7. Your friends learned about sequences in another class and don't understand how or why the explicit formula for arithmetic sequences $\left(a_n = a_1 + d(n-1)\text{ or }a_n = d(n-1)+a_1\right)$ works to find any term value (they just memorized the formula for a test). Explain to your friends what each of the following represents in the formula. *Hint: Think about linear functions.*
   a. n
   b. a_1

   c. a_n
   d. $n-1$

   e. $d(n-1)$
   f. $a_1 + d(n-1)$ or $d(n-1)+a_1$

8. a. Each of the following are portions of arithmetic sequences. Fill in the blanks and then write an explicit formula for each sequence.
      i. 15, ____, 20, ...
      ii. 49, –7, ____, ...

   b. Find the 6th term in each sequence.

9. a. Each of the following is a portion of an arithmetic sequence. Fill in the blanks and then write an explicit formula for each sequence.
      i. ____, 32, 40, ...
      ii. 3, ____, ____, ____, –6, ...

   b. Find the 12th term in each sequence.

Algebra II

Module 2: Investigation 3
Arithmetic Sequences

10. Auditoriums are often designed so that there are fewer seats per row for rows closer to the stage. Suppose you are sitting in Row 22 at an auditorium and notice that there are 68 seats in your row. It appears that the row in front of you has 66 seats and the row behind you has 70 seats. Assume this pattern continues throughout the auditorium.

    a. How many seats are in Row 1?

    b. The last row has 100 seats. How many rows are in the auditorium?

    c. Write a recursive formula that defines the value of a_n, the number of seats in Row n.

    d. Write an explicit formula that defines the value of a_n, the number of seats in Row n.

11. An arithmetic sequence has terms $a_6 = 85$ and $a_{31} = 10$. Write the explicit formula defining the value of a_n.

12. In this module we've seen that some sequences have a limit. However, most infinite arithmetic sequences don't have a limit. Explain why an infinite arithmetic sequence with $d \neq 0$ will never have a limit.

Algebra II

Module 2: Investigation 4
Geometric Sequences

[*Investigation 0 contains review/practice with exponents, the meaning of radical expressions, and how to solve basic equations involving exponents. You can review these concepts as needed.*]

1. A geometric sequence has a common ratio of –3 and the value of its initial term is 4. Write the first several terms of the sequence and the recursive formula that defines the sequence.

2. We can create a general formula that serves as a model for the recursive formulas for *all* geometric sequences. What is the recursive formula that defines the value of a_n for a geometric sequence with an initial term value *k* and a common ratio *r*?

3. Suppose $450 is invested in an account earning 2.8% interest compounded annually. Furthermore, suppose that no additional deposits or withdrawals are made.
   a. To find the account value after the first year we multiply $450 by 1.028. What is the value of the account after 1 year?

   b. If we make a list of the account balance at the end of each year since the initial deposit was made, why will this list form a geometric sequence?

   c. What is the recursive formula that calculates a_n, the value of the account *n* years since the initial deposit was made?

In the previous exercises we thought about generating a term value in a geometric sequence by multiplying the common ratio by the value of the previous term. Another way of thinking about how to find the value of a term in a geometric sequence is to think about what must by multiplied by the initial term value to generate other terms in the sequence.

4. Suppose a geometric sequence begins with 5 and has a common ratio of 4. Complete the following statements.

   $a_1 = 5$

   $a_2 = 5 \cdot$ _____ = _____

   $a_3 = 5 \cdot$ _____ = _____

   $a_4 = 5 \cdot$ _____ = _____

   $a_5 = 5 \cdot$ _____ = _____

   $a_6 = 5 \cdot$ _____ = _____

   $a_n = 5 \cdot$ _____

Algebra II

Module 2: Investigation 4
Geometric Sequences

5. We can create a general formula that serves as a model for the explicit formulas for *all* geometric sequences.
   a. What is the explicit formula that defines the value of a_n for a geometric sequence with an initial term value a_1 and a common ratio r?

   b. Explain what each of the following represents in the formula (be clear and specific).
      i. n
      ii. a_n

      iii. $n-1$
      iv. r^{n-1}

      v. $a_1 \cdot r^{n-1}$

Geometric Sequence Formulas
<u>Recursive Formula:</u>   <u>Explicit Formula:</u>

6. a. Each of the following is a portion of a geometric sequence. Fill in the blanks and then write an explicit formula for each sequence.
      i. 2, 8, ____, ...
      ii. 81, ___, 9, ...

   b. Find the value of the 6th term of each sequence.

Algebra II

Module 2: Investigation 4
Geometric Sequences

7. a. Each of the following is a portion of a geometric sequence. Fill in the blanks and then write an explicit formula for each sequence.
   i. ____, 60, 15, ...
   ii. 2, ____, ____, 250, ...

   b. Find the value of the 12th term of each sequence.

8. A pattern is formed according to the following instructions. Beginning with an equilateral triangle the midpoints of the sides are connected forming four smaller equilateral triangles. The middle triangle is then colored black to create the diagram in Stage 2. This process is repeated with all of the white triangles at each stage to form the diagram in the next stage.

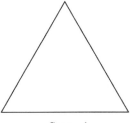
Stage 1

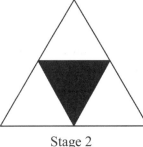
Stage 2

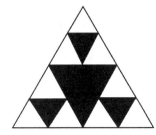
Stage 3

Stage 4   ...

a. Create a sequence that shows the total number of white triangles a_n at stage n and then write the explicit formula for a_n.

b. Does the sequence from part (a) have a limit? What does this represent in the context?

c. Assume that the original white triangle has an area of 8 cm^2. Write the first several terms b_n of the sequence representing the area of one white triangle at stage n and then write the explicit formula for b_n.

Algebra II

Module 2: Investigation 4
Geometric Sequences

d. Still assuming that the original white triangle has an area of 8 cm², write the first several terms c_n of the sequence representing the total area of all of the white triangles at stage n and then write the explicit formula for c_n.

e. Does the sequence from part (d) have a limit? What represent in the context?

f. The term values of a new sequence are defined by $d_n = 8 - c_n$.
   i. What does this new sequence represent?

   ii. Does this sequence have a limit? What does this represent in the context?

9. A geometric sequence has terms $a_9 = 45,927$ and $a_{15} = 33,480,783$. Write the explicit formula defining the value of a_n.

Algebra II

Module 2: Investigation 5
Introduction to Series and Partial Sums

1. After graduating from college, Andy received job offers from two firms. Firm A offered Andy an initial salary of $52,000 for the first year with a 6% pay raise guaranteed for the first five years. Firm B offered Andy an initial salary of $55,000 with a 4% pay raise guaranteed for the first five years. Andy plans to work for five years and then quit and go to graduate school. He really likes both firms, and the hours and job responsibilities are similar, so he will make his decision based on salary.

   a. *Make a prediction:* Which offer do you think he should accept?

   b. What number do we multiply $52,000 by to find out Andy's salary during his second year if he works for Firm A? What number do we multiply $55,000 by to find out Andy's salary during his second year for Firm B?

   c. Write out the sequences that represent his annual salaries with each firm for the first five years.

Year n	1	2	3	4	5
Salary with Firm A in year n					
Salary with Firm B in year n					

   d. Compare Andy's salary in the fifth year working for each firm. Does this tell him which firm he should choose? Why or why not?

   e. Fill in the following table that keeps track of Andy's total salary after working for 1, 2, 3, 4, and 5 years with each firm.

year	total salary earned while working for Firm A by the end of the year	total salary earned while working for Firm B by the end of the year
1		
2		
3		
4		
5		

   f. Pick one row from the table and explain what it represents.

   g. According to Andy's criteria, which firm should he choose?

Algebra II

Module 2: Investigation 5
Introduction to Series and Partial Sums

Series

Series: _____

Ex: Given the sequence 3, 6, 9, 12, 15, 18, the corresponding series is

_____

---

2. Write the corresponding series for each given sequence and give the total sum.
   a. 1, 6, 3, 14, 10, 29
   b. –7, –2, 2, 5, 7, 8, 8

When working with series we are often curious about the "running total" of the sums, or the sum of the terms up to a certain point. We call these the *partial sums* and denote them in the form S_n, where S_n is the sum of the first n terms of the sequence.

Partial Sum of a Series

The nth Partial Sum: [represented by S_n] The sum of the terms of a sequence from the first term to the nth term. The partial sums are like a "running total" for calculating the sum of a series.

Ex: Given the sequence 5, 10, 20, 40, 80, 160,
   a) the third partial sum is given by $S_3 = 5 + 10 + 20 = 35$.
   b) the fifth partial sum is given by $S_5 = 5 + 10 + 20 + 40 + 80 = 155$.
   c) the first partial sum is just the first term, or $S_1 = 5$

In Exercises #3-6, find the indicated partial sum for the given sequence.

3. 10, 8, 6, 4, …, find S_5

4. $a_n = (n+1)^2 - 3$, find S_3

5. $\frac{1}{3}, \frac{2}{3}, 1, \frac{4}{3}, \ldots$, find S_7

6. $a_1 = 6$
   $a_n = 0.5 a_{n-1}$, find S_4

Algebra II

Module 2: Investigation 5
Introduction to Series and Partial Sums

Sequence of Partial Sums

Sequence of Partial Sums: A sequence of partial sums is a sequence where the nth is the partial sum S_n.

Ex: For the sequence 5, 10, 20, 40, 80, 160, the sequence of partial sums is S_1, S_2, S_3, S_4, S_5, S_6, or 5, 15, 35, 75, 155, 315.

---

7. Given the sequence 1, 4, 9, 16, …, write the first five terms for the sequence of partial sums

8. If $a_n = -7 + \frac{4}{5}(n-1)$, write the first five terms for the sequence of partial sums

9. A basketball tournament is held that includes 64 teams. Each round the teams are paired off and play with the loser being eliminated from the tournament.
   a. The sequence representing the number of games played in each round of the tournament begins 32, 16, …. Complete the sequence and explain why it's a finite sequence.

   b. Turn the sequence into a series and find the sum of the series. Explain what this number represents and why it might be useful to the people running the tournament.

   c. Write the sequence of partial sums and explain what it represents in this context.

Module 2: Investigation 5

© 2014 Carlson, O'Bryan & Joyner

Algebra II

Module 2: Investigation 5
Introduction to Series and Partial Sums

10. Suppose a pattern is formed using blocks as follows. Let *n* represent the step number, let a_n represent the number of blocks added at Step *n*, and let S_n represent the total number of blocks at Step *n*.

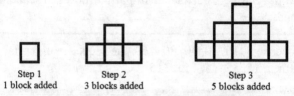

Step 1
1 block added

Step 2
3 blocks added

Step 3
5 blocks added

Then $a_n = 2n - 1$ represents the number of blocks added at Step *n*.

a. Write the first five terms for a_n and S_n, then explain what the terms of these sequences represent.

b. Write a formula that determines S_n given a step number *n*.

c. How many blocks will be added at step 9? How many total blocks are in Step 9?

d. Which step number has 225 total blocks?

Algebra II

Module 2: Investigation 6
Finite Arithmetic Series

Finite Arithmetic Series

A *finite arithmetic series* is the sum of the terms of a finite arithmetic sequence.

Ex: –4, –1, 2, 5, 8 is a finite arithmetic sequence because it has a common difference ($d = 3$) and only finitely many terms. Then (–4) + (–1) + 2 + 5 + 8 is the corresponding finite arithmetic series with a sum of 10.

Throughout history, mathematicians in many cultures discovered clever methods for quickly finding the sum of different types of series. One famous story involves Carl Friederich Gauss while he was an elementary school student. One day Gauss was misbehaving, so his teacher gave him the task of adding up the integers from 1 to 100 to keep him busy. Almost instantly Gauss produced the correct answer (much to the shock of his teacher!). Let's see if we can figure out what made the problem so easy for him.

1. The first 100 natural numbers form a sequence defined by $a_n = n$. Explain why this is an arithmetic sequence.

2. Evaluate each of the following sums of two terms from the sequence.
   a. $a_1 + a_{100}$
   b. $a_2 + a_{99}$
   c. $a_3 + a_{98}$
   d. $a_4 + a_{97}$
   e. $a_5 + a_{96}$

3. Write two other pairs of terms whose sums match the sums from Exercise #2.

4. Explain why this pattern occurs. In other words, why must the pairs $a_1 + a_{100}$ and $a_2 + a_{99}$, and any other pairs of the form $a_{1+k} + a_{100-k}$ for $0 \leq k \leq 49$, all have the same sum? (*Hint: Think about the common difference of the sequence.*)

5. How many pairs of numbers from 1 to 100 have this sum?

6. Using your previous results, find the sum of the first 100 natural numbers.

Algebra II

Module 2: Investigation 6
Finite Arithmetic Series

7. Use this approach to answer the following questions.
   a. Find the sum of the integers from 1 to 250.

   b. One method of exercising with weights is to complete 10 repetitions, rest for 30 seconds, and then perform the same exercise completing 9 repetitions. After a 30 second rest, 8 repetitions are completed. This process continues until the person only lifts the weight once. How many total repetitions are completed during such a routine?

8. The approach you've been practicing seems to work for finite arithmetic series that have an even number of terms and a common difference of 1. Write a convincing argument why this approach will also work for *any* finite arithmetic sequence with an even number of terms and a common difference of d.

9. Recall the following context from Investigation 3. *Auditoriums are often designed so that there are fewer seats per row for rows closer to the stage. Suppose you are sitting in Row 22 at an auditorium and notice that there are 68 seats in your row. It appears that the row in front of you has 66 seats and the row behind you has 70 seats. Assume this pattern continues throughout the auditorium.*
   a. Go back to Investigation 3 and locate the following information: the number of rows, the number of seats in the first row, and the number of seats in the last row.

   b. How many seats are in the auditorium?

Algebra II

Module 2: Investigation 6
Finite Arithmetic Series

10. Using mathematical symbols and notation, write the formula showing how to find the sum S_n of the first n terms of a finite arithmetic series with an even number of terms. Explain what the parts of your formula represent.

11. Find the sum of each of the following finite arithmetic series.
    a. $4+7+10+13+16+19+22+25$
    b. $15+13+11+9+7+5+3+1-1-3-5-7$

    c. the sum of the even numbers from 2 to 300
    d. $6+13+20+...+181$

12. Shilah recently decided to start jogging each morning and wants to work up to 3.5 miles per day. Tomorrow she will run 0.5 miles, and she will increase the distance she runs by 0.2 miles each day until she reaches 3.5 miles per day.
    a. How many days will she have been jogging when she reaches 3.5 miles per day?

    b. How many total miles will Shilah have run during this time?

Algebra II

Module 2: Investigation 6
Finite Arithmetic Series

13. Is there a problem applying the reasoning we've developed to a finite arithmetic series with an odd number of terms? Explain your thinking.

14. Find the sum of each arithmetic series.
    a. $105 + 141 + 177 + 213 + 249 + 285 + 321$
    b. $-126 - 117 - 108 - 99 - 90 - 81 - 72 - 63 - 54$

    c. $14 + 10 + 6 + \ldots - 138$
    d. the sum of the positive four-digit multiples of 20

15. Write a finite arithmetic series with a sum of 80 that has 10 terms and a common difference $d \neq 0$. *(Instead of using "guess and check" try to think about a strategy for solving this problem using the ideas you've learned in this investigation.)*

Algebra II

Module 2: Investigation 7
Finite Geometric Series

Finite Geometric Series

A *finite geometric series* is the sum of the terms of a finite geometric sequence.

Ex: 1, –2, 4, –8 is a finite geometric sequence because it has a common ratio ($r = -2$) and only finitely many terms. Then $1 + (-2) + 4 + (-8)$ or $1 - 2 + 4 - 8$ is the corresponding finite geometric series with a sum of –5.

For Exercises #1-3, consider the finite geometric series $5 + 15 + 45 + 135 + 405 + 1{,}215$ with a sum $S_6 = 1{,}820$.

1. a. Suppose each of the terms of the series is multiplied by 2. Write out the terms of the new series, then find its sum and compare this to the sum of the original series.

   b. Suppose each of the terms of the series is multiplied by 4. Write out the terms of the new series, then find its sum and compare this to the sum of the original series.

2. If we multiply each term of the original series by 3, we get the following new series.
$$3(5) + 3(15) + 3(45) + 3(135) + 3(405) + 3(1{,}215)$$
$$15 + 45 + 135 + 405 + 1{,}215 + 3{,}645$$

   a. How does the sum of this new series compare to the sum of the original series?

   b. Compare the terms of the new series and the terms of the original series. What do you notice?

   c. Why do so many of the terms from the original series reappear in the new series when the term values are multiplied by 3, but not by 2 or 4?

Algebra II

Module 2: Investigation 7
Finite Geometric Series

3. Let S_6 represent the sum of the original series and $3 \cdot S_6$ represent the sum of the new series.
   a. What is the value of $S_6 - (3 \cdot S_6)$?

   b. One of your classmates claims that he can calculate $S_6 - (3 \cdot S_6)$ without knowing the sums of the series. He says that $a_1 - 3a_6$ will have the same value as $S_6 - (3 \cdot S_6)$ and shows the following work. Is he correct?

   $$S_6 = 5 + \cancel{15} + \cancel{45} + \cancel{135} + \cancel{405} + \cancel{1,215}$$
   $$-\left(3 \cdot S_6 = \cancel{15} + \cancel{45} + \cancel{135} + \cancel{405} + \cancel{1,215} + 3,645\right)$$
   $$S_6 - 3 \cdot S_6 = 5 - 3,645$$
   $$S_6 - 3 \cdot S_6 = -3,640$$

   c. Use the equation $S_6 - 3 \cdot S_6 = -3,640$ to solve for S_6.

4. Let's practice the technique from Exercise #3 with the finite geometric series $2 + 10 + 50 + 250 + 1,250 + 6,250 + 31,250$ with a sum S_7 and a common ratio $r = 5$. If we multiply the terms of the series by 5 we get the following series with a sum of $5 \cdot S_7$.

   $$2(5) + 10(5) + 50(5) + 250(5) + 1,250(5) + 6,250(5) + 31,250(5)$$
   $$10 + 50 + 250 + 1,250 + 6,250 + 31,250 + 156,250$$

   Complete the steps to find S_7 without actually adding up the terms of the series.

   $$S_7 = 2 + 10 + 50 + 250 + 1,250 + 6,250 + 31,250$$
   $$-\left(5 \cdot S_7 = 10 + 50 + 250 + 1,250 + 6,250 + 31,250 + 156,250\right)$$
   $$S_7 - 5 \cdot S_7 = \underline{}$$

   $$\underline{} S_7 = \underline{}$$

   $$S_7 = \underline{}$$

Algebra II Module 2: Investigation 7
Finite Geometric Series

Sum of a Finite Geometric Series

If S_n is the sum of a finite geometric series with a first term a_1, a last term a_n, and a common ratio r, then

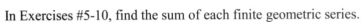

We then solve the equation for S_n to find the sum.

This technique allows us to find the sum
- without adding up all of the term values and
- even if we don't know all of the term values or the number of terms.

In Exercises #5-10, find the sum of each finite geometric series.

5. $2 + 4 + 8 + \ldots + 2{,}048$

6. $1 + 6 + 36 + \ldots + 10{,}077{,}696$

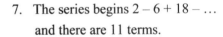

7. The series begins $2 - 6 + 18 - \ldots$ and there are 11 terms.

8. $\frac{1}{16} - \frac{1}{4} + 1 - \ldots - 1024$

Algebra II

9. $4 + \frac{4}{3} + \frac{4}{9} + \frac{4}{27} + \ldots + \frac{4}{2{,}187}$

10. Rising healthcare costs are a constant issue facing families and businesses. Suppose the cost of insuring an individual is $8,900 this year and is expected to increase by 7% per year for the next 9 years. How much do we expect it to cost to cover this person over the entire 10-year period?

Algebra II
Module 2: Investigation 8
Sigma Notation

So far we have written series in a form like $4+7+10+13+16$. However, this form has shortcomings. Series with many terms are very common in math, and in these cases we'd need to either write out the entire long series or shorten it (like writing $12+17+22+...+122$) which can hide important information. Mathematicians needed a way to write the sum of long series in a condensed way that also provides maximum information. Their solution was **sigma notation**.

The best way to introduce sigma notation is through an example. Consider the series $4+7+10+13+16$. This series is made up of the terms of the sequence 4, 7, 10, 13, 16 described with the explicit formula $a_n = 4+3(n-1)$. The first term is $a_1 = 4$. The last term is $a_5 = 16$. We use the capital Greek letter sigma (Σ) to mean *sum of*.

We want to represent the statement "the sum of the terms of the sequence a_n from a_1 to a_5" in some way other than $4+7+10+13+16$. We can write this series in sigma notation as $\sum_{n=1}^{5}[4+3(n-1)]$.

Sigma Notation

We can write a series with terms a_n in sigma notation as follows.

$$\sum_{n=1}^{5} a_n$$

- "the sum of"
- "to a_5" (above sigma)
- "from a_1" (below sigma)
- "the term values a_n of the sequence"

If we know the **explicit** formula defining the terms of the series (such as $a_n = 4+3(n-1)$), then we can replace a_n with its explicit formula.

$$\sum_{n=1}^{5}[4+3(n-1)]$$

Here are two important points about sigma notation.

I) $\sum_{n=1}^{5} a_n = a_1 + a_2 + a_3 + a_4 + a_5$. We can translate sigma notation into the expanded form by substituting $n=1$, then continuing to substitute values of n until we reach the index value written above sigma (in this case $n=5$).

II) The number written above sigma is **not** intended to represent *the number of terms*. For example, consider $\sum_{n=5}^{8} a_n$ (4 terms) and $\sum_{n=2}^{3} a_n$ (2 terms).

$$\sum_{n=5}^{8} a_n = a_5 + a_6 + a_7 + a_8 \qquad \sum_{n=2}^{3} a_n = a_2 + a_3$$

In Exercises #1-4, write out the terms of the series and find the sum.

1. $\sum_{n=1}^{4}[22-2(n-1)]$

2. $\sum_{n=1}^{5} 4(3)^{n-1}$

3. $\sum_{n=1}^{3} 2^{n+2}$

4. $\sum_{n=3}^{7} \frac{n^2}{2}$

In Exercises #5-8, write each series in sigma notation. (*Hint: You might first need to determine the number of terms in the series.*)

5. $3+15+75+375+1{,}875+9{,}375$

6. $2+6+10+14+\ldots+90$

7. $1+\frac{4}{3}+\frac{5}{3}+2+\ldots+33$

8. $\frac{1}{2^3}+\frac{1}{2^4}+\frac{1}{2^5}+\ldots+\frac{1}{2^{10}}$

9. Recall the following context from Investigations 3 and 6. *Auditoriums are often designed so that there are fewer seats per row for rows closer to the stage. Suppose you are sitting in Row 22 at an auditorium and notice that there are 68 seats in your row. It appears that the row in front of you has 66 seats and the row behind you has 70 seats. Assume this pattern continues throughout the auditorium.*

    Represent the total number of seats in the auditorium using sigma notation. *(Feel free to refer back to Investigations 3 and 6 to find any additional information you may need.)*

10. Recall the following context from Investigation 5. *A basketball tournament is held that includes 64 teams. Each round the teams are paired off and play with the loser being eliminated from the tournament. The sequence representing the number of games played in each round of the tournament begins 32, 16, ….*

    Represent the total number of games played in the tournament using sigma notation. *(Feel free to refer back to Investigation 5 to find any additional information you may need.)*

Algebra II
Module 2: Investigation 8
Sigma Notation

In Exercises #11-14, do the following.
   a) State whether the series is arithmetic, geometric, or neither.
   b) If the series is arithmetic or geometric, give the common difference or common ratio.
   c) Write down the first and last terms of the series.

11. $\sum_{n=1}^{14}[3+2n]$

12. $\sum_{n=1}^{6} 7^n$

13. $\sum_{n=1}^{11} \frac{1}{n^2}$

14. $\sum_{n=5}^{9}[4-6(n-1)]$

For Exercises #15-18, find the sum of the series.

15. $\sum_{i=1}^{10} 4(3)^{i-1}$

16. $\sum_{n=1}^{105}[6-4(n-1)]$

17. $\sum_{n=1}^{5}\left[n^2-6\right]$

18. $\sum_{n=1}^{40} 3(n-3)$

19. When writing a series using sigma notation we use the explicit formula and not the recursive formula. For example, $12+17+22+\ldots+122$ is written as $\sum_{n=1}^{28}[12+5(n-1)]$ and not as $\sum_{i=1}^{28}[a_{n-1}+5]$. What are the advantages of using the explicit formula for the term values instead of the recursive formula in sigma notation?

Algebra II

Module 2: Investigation 9
Infinite Series

1. Explain what it means for a sequence to have a limit.

2. Explain what the sequence of partial sums represents for a series.

3. Explain what each of the following represents.

    a. $\sum_{n=6}^{10}[3n+2]$

    b. $\sum_{n=1}^{\infty}[40+n^2]$

4. Using a ruler, draw a square on a piece of paper. Make sure that the square is reasonably large – at least 4 inches per side. We'll call the area of this square "one square unit." Now consider the series
$$\tfrac{1}{2}+\tfrac{1}{4}+\tfrac{1}{8}+\tfrac{1}{16}+\ldots$$
Let's imagine these terms each represent areas (in square units).
    a. Write the first six terms of the sequence of partial sums.

    b. Cut your square in half parallel to one of the sides. You are now holding two rectangles, each with an area of ½ square unit. Set one of these to the side. This represents the value of the first term in your sequence of partial sums.
    Take the second rectangle and cut it in half. You are now holding two squares, each with an area of ¼ square unit. Set one of these aside next to the rectangle with an area of ½ square unit. Repeat this process several more times. How does this activity represent the sequence of partial sums?

    c. Does the sequence of partial sums have a limit? If so, what is the limit?

Algebra II

Module 2: Investigation 9
Infinite Series

5. Consider the series $6 + 2 + \frac{2}{3} + \frac{2}{9} + \ldots$. This time, let's imagine the terms represent lengths.
   a. Write the first six terms in the sequence of partial sums.

   b. On the line below, a bolded line segment is drawn measuring 6 units. This represents the first term in the sequence of partial sums. Extend this line segment to generate a segment whose length is the second term in the sequence of partial sums.

   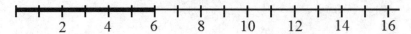

   c. Repeat part (b) to represent the next several terms in the sequence of partial sums.

   d. Does the sequence of partial sums have a limit? If so, what is the limit?

For Exercises #6-11, do the following.
   a) State whether the series is arithmetic or geometric and identify the common difference or common ratio.
   b) Graph the first six terms of the sequence that makes up the series. Does there appear to be a limit? If so, what is it?
   c) Generate and graph the first six terms of the sequence of partial sums for the series.
   d) Does the sequence of partial sums appear to have a limit? If so, what is it?

6. $\sum_{n=1}^{\infty} 2^{n+3}$

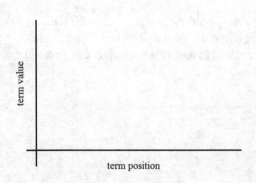
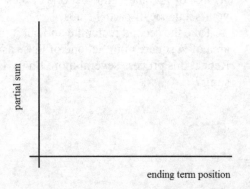

7. $-1+2+5+...$

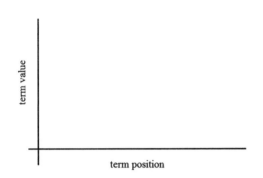 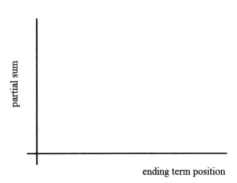

8. $\sum_{n=1}^{\infty} 3\left(-\frac{1}{3}\right)^{n-1}$

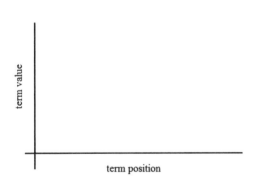 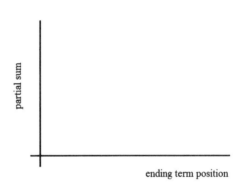

9. $\sum_{n=1}^{\infty} -2n+6$

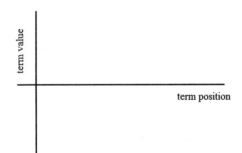 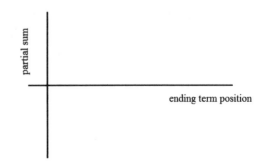

10. $3 + \frac{3}{2} + \frac{3}{4} + \frac{3}{8} + \ldots$

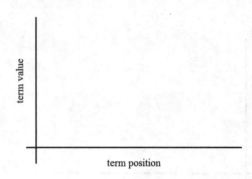 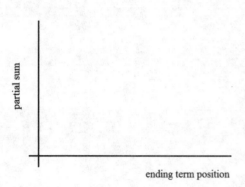

11. $4 - 8 + 16 - 32 + \ldots$

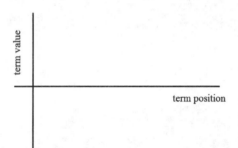 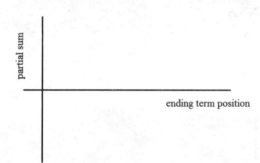

12. What does it mean when the sequence of partial sums for an infinite series has a limit? (In other words, why is part (d) important in Exercises #4-9?)

13. TRUE or FALSE: For an infinite geometric series with a common ratio $r \neq 0$ and $a_1 \neq 0$, the sequence of partial sums always has a limit. Justify your answer.

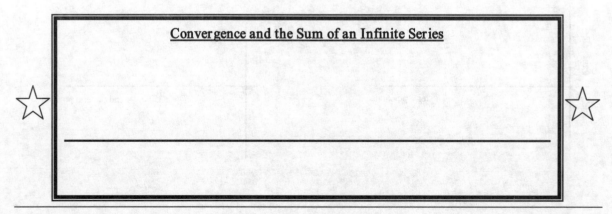

Algebra II

Module 2: Investigation 10
Infinite Geometric Series

Sum of an Infinite Geometric Series

If S is the sum of an infinite geometric series with a first term a_1 and a common ratio r, then

[ ]

We solve the equation for S to find the sum.

Remember:
- The sum of an infinite series is the limit of the sequence of partial sums.
- An infinite geometric series will only have a sum if $|r|<1$.

In Exercises #1-3, find the sum if possible.

1. $\sum_{n=1}^{\infty} 24\left(\frac{2}{3}\right)^{n-1}$
2. $\sum_{n=1}^{\infty} (1.2)^n$
3. $\sum_{n=1}^{\infty} 7(-0.4)^{n-1}$

4. A pendulum is made up of a string or solid arm with a weight attached to the end. Because of outside factors such as air resistance and gravity, each swing of a pendulum is a little shorter than the previous one. Suppose the length of the pendulum's swing follows a geometric sequence with the first swing being 100 cm long and the second swing length being 99% of the previous swing length.

   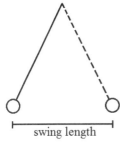

   a. What is the common ratio for the sequence of pendulum swing lengths? Write out the first several terms of the sequence of swing lengths.

   b. Suppose an infinite series is formed from this sequence. Find the sum of the series (if possible).

   c. Does the sum of the series have a real-world meaning? If so, describe the meaning. If not, explain your thinking.

Algebra II   **Module 2: Investigation 10**
                                                  Infinite Geometric Series

For Exercises #5-8, do the following.
  a) Write the series in sigma notation.
  b) Find the sum of the series (if possible).

5. $120 + 30 + \frac{30}{4} + \frac{30}{16} + \ldots$

6. $-10 + 5 - 2.5 + 1.25 - \ldots$

7. $1.25 - 2.5 + 5 - 10 + \ldots$

8. $1 + x + x^2 + x^3 + \ldots$ where $-1 < x < 1$

In Exercises #9-11, create a series that meets the stated requirements. Write the series using sigma notation. (*Note: Do not use any of the series we have provided in this investigation.*)

9. an infinite geometric series with a negative common ratio that has a sum

10. an infinite geometric series that does not have a sum

11. an infinite geometric series with a sum between 0 and 1

Algebra II

Module 2: Investigation 10
Infinite Geometric Series

12. A doctor prescribes a pain reliever to a patient with severe back pain. The doctor tells the patient to take a 500 mg does every 8 hours and that she will reevaluate the patient in the future. After some research about the medication, the patient learns that his body likely eliminates 90% of the drug in his system over the course of 8 hours.

   a. How much of the drug is in the patient's system right after he takes the second dose? Show the expression that calculates this value.

   b. How much of the drug is in the patient's system right after he takes the third dose? Show the expression that calculates this value.

   c. How much of the drug is in the patient's system right after he takes the fourth dose? Show the expression that calculates this value.

   d. Explain how the amount of drug in the patient's system after the nth dose forms a geometric series.

   e. Suppose the patient stays on the medication indefinitely. What eventually happens to the amount of drug in his system when he takes a new dose?

13. Suppose the doctor prescribes an ibuprofen regimen to another patient to alleviate problems with swollen joints. The doctor tells the patient to take 600 mg of ibuprofen every 12 hours, and during each 12-hour period the patient's body eliminates 85% of the drug in the patient's system.

   a. Find the amount of drug in the patient's system right after taking the 7th dose.

   b. Suppose the patient stays on the medication indefinitely. What eventually happens to the amount of drug in his system?

Algebra II

Module 2: Homework
Sequences and Series

INVESTIGATION 1: SEQUENCES AND THE LIMITS OF SEQUENCES

For Exercises #1-6, do the following.
a) Describe the pattern.
b) Complete the table showing the relationship between the term positions and the term values.
c) Graph the sequence (use at least six points).
d) Does the sequence appear to have a limit? If so, what is it?

1. 40, 52, 64, ...

input: term position	output: term value
1	
2	
3	
4	
5	
6	

2. 15, 5, $\frac{5}{3}$, ...

input: term position	output: term value
1	
2	
3	
4	
5	
6	

3. 1, $\frac{2}{3}$, $\frac{4}{9}$, ...

input: term position	output: term value
1	
2	
3	
4	
5	
6	

4. −6, −7.8, −9.6, ...

input: term position	output: term value
1	
2	
3	
4	
5	
6	

5. 0, 1, 4, 9, ...

input: term position	output: term value
1	
2	
3	
4	
5	
6	

6. 6, 9, 10.5, 11.25, ...

input: term position	output: term value
1	
2	
3	
4	
5	
6	

7. A sequence is formed by choosing any real number x to be the first term, and then generating the sequence by taking each term value and dividing by 2 to get the next term value.
   a. Choose a few possible values of x and explore the kinds of sequences produced by following this pattern.
   b. Do all sequences formed in this manner have a limit? If so, what is the limit? Why does this happen?

8. A sequence is formed by choosing any real number x to be the first term, then generating the sequence by taking each term value and multiplying by 3.5 to get the subsequent term value.
   a. Choose a few possible values of x and explore the kinds of sequences produced by following this pattern.
   b. Do all sequences formed in this manner have a limit? If so, what is the limit? Why does this happen?

Algebra II

Module 2: Homework
Sequences and Series

9. A pyramid is formed from blocks according to the given pattern. Write the next three terms in the sequence showing the total number of blocks in the pyramid at each stage.

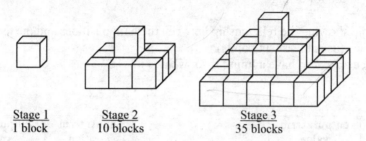

Stage 1: 1 block
Stage 2: 10 blocks
Stage 3: 35 blocks

10. A sequence is formed by determining the number of diagonals that can be drawn inside of regular polygons as shown below. (A diagonal is a line segment drawn inside of the figure from one vertex to another.) Determine the pattern formed by this sequence and write the next three terms of the sequence.

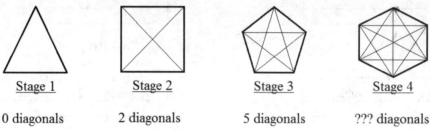

Stage 1: 0 diagonals
Stage 2: 2 diagonals
Stage 3: 5 diagonals
Stage 4: ??? diagonals

11. An infinite sequence begins 40, 60, ...
    a. Determine a possible pattern for this sequence such that the sequence *will not have a limit*. Write the next three terms using this pattern.
    b. Determine another possible pattern for this sequence such that sequence *will not have a limit*. Write the next three terms using this pattern.
    c. Determine another possible pattern for this sequence such that sequence *will have a limit*. Write the next three terms using this pattern.

12. An infinite sequence begins −8, −3 , ...
    a. Determine a possible pattern for this sequence so that the sequence *will not have a limit*. Write the next three terms using this pattern.
    b. Determine another possible pattern for this sequence so that sequence *will not have a limit*. Write the next three terms using this pattern.
    c. Determine another possible pattern for this sequence so that sequence *will have a limit*. Write the next three terms using this pattern.

INVESTIGATION 2: FORMULAS FOR SEQUENCES

In Exercises #13-16, find the indicated term value for the given sequence.

13. $1, -3, 9, ...$; find a_5

14. $104, 107, 110, ...$; find a_6

15. $\frac{1}{2}, \frac{1}{3}, \frac{1}{4}, ...$; find a_{10}

16. $1, -5, 25, -125, 625, ...$; find a_5

Algebra II

Module 2: Homework
Sequences and Series

17. If a_n represents the value of the nth term in some sequence, how could you represent each of the following?
    a. the value of the term three term positions before the nth term
    b. the value of the term two term positions after the nth term
    c. the term position one term prior to the nth term

18. If a_n represents the value of the nth term in some sequence, how could you represent each of the following?
    a. the term position four terms after the nth term
    b. the value of the term five term positions before the nth term
    c. the value of the term three term positions after the nth term

In Exercises #19-24, use the given formula to write the first four terms of each sequence.

19. $a_1 = -2$, $a_n = a_{n-1} + 3$

20. $a_1 = 360$, $a_n = \dfrac{a_{n-1}}{3}$

21. $a_1 = 10$, $a_n = -2a_{n-1} + 4$

22. $a_1 = 4$, $a_n = \dfrac{128}{a_{n-1}}$

23. $a_1 = -1$, $a_2 = 4$
    $a_n = a_{n-1} + a_{n-2}$

24. $a_1 = 1$, $a_2 = 2$, $a_3 = 4$
    $a_n = 2a_{n-3} + 3a_{n-2}$

In Exercises #25-28, write a recursive formula to define each sequence.

25. 25, 30, 35, ...

26. 1, −7, −15, ...

27. 3, 18, 108, ...

28. $1, -\frac{1}{8}, \frac{1}{64}, ...$

In Exercises #29-30, write the first four terms of the sequence.

29. $a_n = \dfrac{5(n-1)}{4n}$

30. $a_n = 7(-3)^n$

In Exercises #31-32, find the value of the 15th term in the sequence.

31. $a_n = -4 - 15(n-1)$

32. $a_n = \dfrac{14}{(n+3)^2}$

In Exercises #33-34, write the explicit formula defining the sequence and then find the number of terms in the given sequence.

33. 1, 4, 9, 16, 25, ..., 225

34. 10, 13, 16, 19, ..., 178

35. After a car is purchased its value generally decreases over time. Suppose a car's original purchase price is $24,679 and that is the value of the car in any given year is always 0.85 times as large as the value of the car the year before.
    a. Write a sequence showing the car's value at the end of each of the first five years since its original purchase.
    b. Write a recursive formula for the sequence in part (a).
    c. Write an explicit formula for the sequence in part (b).

36. Suppose we have 57 micrograms of a particular bacteria population in a petri dish. We place this petri dish in an incubator and the bacteria grows in such a way that its mass triples each day.
   a. Write a sequence showing the bacteria's weight at the end of each of the first five days since it was placed in the incubator.
   b. Write a recursive formula for the sequence in part (a).
   c. Write an explicit formula for the sequence in part (b).

INVESTIGATION 3: ARITHMETIC SEQUENCES

For Exercises #37-44, do the following.
   a) Write the recursive formula for the arithmetic sequence.
   b) Write the explicit formula for the arithmetic sequence.
   c) Determine the number of terms in the arithmetic sequence.

37. 13, 15, 17, ..., 251

38. 14, 13.7, 13.4, ..., −10.6

39. 62, 63.1, 64.2, ..., 124.7

40. −30, −26.5, −23, ..., 488

41. −33, −41, −49, ..., −825

42. 1077, 1077.1, 1077.2, ..., 1085.4

43. All of the 3-digit multiples of 7

44. All of the 5-digit multiples of 25

45. Interior angles of convex polygon are formed by two adjacent sides of the polygon opening into the interior of the figure. The sum of the measures of all of the interior angles of a polygon follow the pattern shown below.

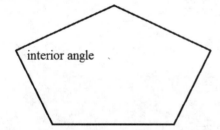

number of sides n	sum of the measures of the interior angles a_n
3	180°
4	360°
5	540°
6	720°

   a. We can think of the sequence beginning with a_3 instead of a_1. Write the recursive formula for the sequence.
   b. Write the explicit formula for the sequence.
   c. Regular polygons are polygons where all of the interior angles are congruent and all of the sides are the same length. What is the measure of *each* interior angle in a regular polygon with 18 sides?

46. Auditoriums are often designed so that there are fewer seats per row for rows closer to the stage. Suppose you are sitting in Row 18 at an auditorium and notice that there are 73 seats in your row. It appears that the row in front of you has 70 seats and the row behind you has 76 seats. Suppose this pattern continues throughout the auditorium.
   a. How many seats are in the first row?
   b. How many seats are in the 30th row?
   c. Write an explicit formula for the sequence where a_n is the number of seats in the nth row.

Algebra II

Module 2: Homework
Sequences and Series

47. An arithmetic sequence has the following two terms: $a_{13} = 58$ and $a_{28} = 140.5$. Write the explicit formula defining the value of the nth term a_n.

48. An arithmetic sequence has the following two terms: $a_6 = 13.5$ and $a_{24} = 51.3$. Write the explicit formula defining the value of the nth term a_n.

49. A finite arithmetic sequence has the following two terms: $a_9 = 52.9$ and $a_{27} = 121.3$. The value of the final term is 934.5. How many terms are in the sequence?

50. A finite arithmetic sequence has the following two terms: $a_5 = 71.6$ and $a_{23} = 33.8$. The value of the final term is -167.8. How many terms are in the sequence?

INVESTIGATION 4: GEOMETRIC SEQUENCES

For Exercises #51-56, do the following.
   a) Write the recursive formula for the geometric sequence.
   b) Write the explicit formula for the geometric sequence.
   c) Determine if the sequence has a limit. If so, state the limit.

51. 3, 6, 12, 24, ...

52. 38, 19, 9.5, 4.75, ...

For Exercises #53-56, see directions on previous page

53. −15, 60, −240, 960, ...

54. 88, −11, $\frac{11}{8}$, ...

55. 90, 30, 10, $\frac{10}{3}$, ...

56. 12, 18, 27, 40.5, ...

57. Inflation describes the tendency of prices for goods to rise over time. Historically, prices of goods have risen about 3% per year. Suppose a loaf of bread costs $2.99 in January 2012 and that its cost will rise by 3% per year over the next several years.
   a. Write a recursive formula for the cost of a loaf of bread a_n in dollars n years since January 2012.
   b. Write an explicit formula for the cost of a loaf of bread a_n in dollars n years since January 2012.
   c. Assuming this pattern continues, what do you expect to be the cost of a loaf of bread in 2025?

58. In 2010, a university's budgeting committee decided that they would set the annual tuition rate at $14,000 per year and that they would raise this rate by 2.5% each year following the 2010-2011 school year.
   a. Write a recursive formula for the tuition rate a_n in dollars n years since the 2010-2011 school year.
   b. Write an explicit formula for the tuition rate a_n in dollars n years since the 2010-2011 school year.
   c. Assuming this pattern continues, what do you expect to be the tuition rate for the 2018-2019 school year?

Algebra II

Module 2: Homework
Sequences and Series

For Exercises #59-62, do the following.
    a) Fill in the blanks to make the sequence arithmetic.
    b) Fill in the blanks to make the sequence geometric.

59. 12, 36, _____

60. _____, 1, −4, ...

61. 14, _____, 224, ...

62. 20, _____, 45, ...

63. A geometric sequence has the following two terms: $a_3 = 72$ and $a_7 = 93{,}312$. Write the explicit formula defining the value of the nth term a_n.

64. A geometric sequence has the following two terms: $a_5 = 2048$ and $a_8 = 131{,}072$. Write the explicit formula defining the value of the nth term a_n.

65. A finite geometric sequence has the following two terms: $a_7 = 288$ and $a_{12} = 9{,}216$. The last term in the sequence is $37{,}748{,}736$. How many terms are in the sequence?

66. A finite geometric sequence has the following two terms: $a_4 = 250$ and $a_8 = 156{,}250$. The last term in the sequence is $2{,}441{,}406{,}250$. How many terms are in the sequence?

INVESTIGATION 5: INTRODUCTION TO SERIES AND PARTIAL SUMS

Suppose a series is formed using the terms from the sequences defined in Exercises #67-70 below. For each, find the given partial sum.

67. 103, 206, 309, ..., find S_4

68. 9, 16, 25, 36, ..., find S_6

69. $a_n = \dfrac{(8-n)^2}{n^3}$, find S_1

70. $a_1 = 64$, $a_n = \tfrac{1}{4} a_{n-1}$, find S_5

In Exercises #71-74, suppose a series is formed using the terms from the given sequence. Write the first five terms for the sequence of partial sums.

71. $a_n = 20 - 6n$

72. $3, \tfrac{9}{2}, \tfrac{27}{4}, \ldots$

73. $a_1 = 2.5$
$a_n = 6 \cdot a_{n-1} - 2$

74. $a_1 = 1,\ a_2 = 1$
$a_n = a_{n-2} + a_{n-1}$

75. The table below shows the terms of a sequence where a_n is the power bill (in dollars) for a single family home during the nth month of 2011.

n	a_n	S_n
1	53.15	
2	52.90	
3	57.44	
4	73.80	
5	98.17	
6	152.82	

n	a_n	S_n
7	212.63	
8	175.10	
9	118.02	
10	68.98	
11	56.65	
12	53.70	

    a. Fill in the column showing the partial sums S_n.
    b. Explain what the partial sums are keeping track of in this context.

Algebra II

Module 2: Homework
Sequences and Series

76. The table below shows the terms of a sequence where a_n is the number of DWI (Driving While Intoxicated) tickets given out in Travis County for the nth month of 2012.

n	a_n	S_n
1	62	
2	48	
3	55	
4	41	
5	54	
6	62	

n	a_n	S_n
7	67	
8	70	
9	63	
10	59	
11	56	
12	48	

   a. Fill in the column showing the partial sums S_n.
   b. Explain what the partial sums are keeping track of in this context.

In Exercises #77-79 below you are given information about the terms of arithmetic series and the sequence of partial sums for each series. Complete the tables.

77.

n	a_n	S_n
1		
2		
3	11	
4	15	
5	19	
6		
7		

78.

n	a_n	S_n
1		
2		
3		
4		
5		5
6		0
7		−7

79.

n	a_n	S_n
1		
4		
5		
10	23.5	
13	31	
16		
18		

In Exercises #80-82 below you are given information about the terms of geometric series and the sequence of partial sums for each series. Complete the tables.

80.

n	a_n	S_n
1		
2		
3	54	
4	162	
5		

81.

n	a_n	S_n
1		−20
2		10
3		
4		
5		

82.

n	a_n	S_n
1		
4		
5		
10	2,048	
13	16,384	

INVESTIGATION 6: FINITE ARITHMETIC SERIES

For Exercises #83-84, find the sum of the series.

83. $121+128+135+...+233$ (17 terms)

84. $83+81+79+...-19$ (52 terms)

For Exercises #85-88, do the following for each finite arithmetic series.
   a. Find the number of terms in the series.
   b. Find the sum of the series.

85. $17+21+25+...+89$

86. $223+214+205+...+(-56)$

87. $6+6.3+6.6+...+25.2$

88. $-10+(-8.2)+(-6.4)+...+42.2$

Algebra II

Module 2: Homework
Sequences and Series

For Exercises #89-90, do the following for each finite arithmetic series.
  a. Give the first three terms of the series and the last term of the series.
  b. Find the number of terms in the series.
  c. Find the sum of the series.

89. the sum of all 4-digit numbers

90. the sum of the 3-digit multiples of 2

91. Auditoriums are often designed so that there are fewer seats per row for rows closer to the stage. Suppose you are sitting in Row 23 at an auditorium and notice that there are 65 seats in your row. It appears that the row in front of you has 63 seats and the row behind you has 67 seats. Suppose this pattern continues throughout the auditorium. You count a total of 42 rows in the auditorium. How many seats does the auditorium contain?

92. A child builds a pyramid out of multiple decks of playing cards. As he starts from the bottom-up, the layers are numbered starting at the base. Suppose he creates the sixth layer by using 12 cards. His friend notes that the fifth layer is made out of 16 cards, and the boy is planning to use 8 cards to create the seventh layer. Suppose this pattern continues throughout the entire pyramid. The boy who is building says that his pyramid will have a total of eight layers. How many cards does it take to build the pyramid?

93. A finite arithmetic series has 73 terms with $a_7 = 21.4$ and $a_{13} = 32.8$. What's the sum of the series?

94. A finite arithmetic series has 90 terms with $a_5 = 182$ and $a_{86} = 291.6$. What's the sum of the series?

95. A finite arithmetic series has a sum of 81,674 with a first term of 912 and a final term of -70. How many terms are in the series?

96. A finite arithmetic series has a sum of 104,030 with a first term of -12 and a final term of 1042. How many terms are in the series?

97. Write a finite arithmetic series with a sum of 200 that has 8 terms with a common difference $d \neq 0$. (*Instead of using "guess and check" try to think about a strategy for solving this problem using the ideas you've learned in this worksheet.*)

98. Write a finite arithmetic series with a sum of 542 that has 12 terms with a common difference $d \neq 0$. (*Instead of using "guess and check" try to think about a strategy for solving this problem using the ideas you've learned in this worksheet.*)

INVESTIGATION 7: FINITE GEOMETRIC SERIES

99. The finite geometric series $10 + 40 + 160 + ... + 163,840$ has 8 terms and a sum S_8. Show the process of multiplying the sum by the common ratio and finding the difference of S_8 and $r \cdot S_8$ to find the sum. (In other words, justify why $S_n - r \cdot S_n = a_1 - r \cdot a_n$ is true for finite geometric series.)

100. The finite geometric series $3 + 9 + 27 + ... + 2,187$ has 7 terms and a sum S_7. Show the process of multiplying the sum by the common ratio and finding the difference of S_7 and $r \cdot S_7$ to find the sum. (In other words, justify why $S_n - r \cdot S_n = a_1 - r \cdot a_n$ is true for finite geometric series.)

Algebra II

Module 2: Homework
Sequences and Series

In Exercises #101-106, find the sum of each finite geometric series.

101. $5 + 25 + 125 + \ldots + 390{,}625$

102. $-3 + 6 - 12 + \ldots - 786{,}432$

103. $4 + 12 + 36 + \ldots + 78{,}732$

104. $-128 + 64 - 32 + \ldots + 1$

105. $1 + \frac{1}{10} + \frac{1}{100} + \ldots + \frac{1}{100{,}000{,}000}$

106. $36 + 12 + 4 + \ldots + \frac{4}{2{,}187}$

107. The local power company plans to raise rates to cover the increased costs of producing power. In a recent newsletter, they notified their customers to expect power costs to increase by 4.2% per year for the next 10 years (including this year). (The power costs one year are 1.042 times as large as the costs in the previous year.)
    a. If you paid a total of $1031.60 for power last year, what do you expect to pay for power over the next 10 years (assuming you continue to live in your current residence and that your power usage does not change)?
    b. How does this compare to the total price if power costs remained the same over the next 10 years?

108. A bouncing ball reaches heights of 16 cm, 12.8 cm, and 10.24 cm on three consecutive bounces.
    a. If the ball was dropped from a height of 25 cm, how many times has it bounced when it reaches a height of 16 cm? 10.24 cm?
    b. Write the first five terms of the sequence representing the bounce height after *n* bounces.
    c. How much total distance *in the downward direction only* has the ball traveled after five bounces? *(Be careful!)*
    d. How much total distance *in the upward direction only* has the ball traveled after five bounces (and reaching the top of its fifth bounce)?
    e. What is the *total distance* traveled by the ball when it reaches the top of its fifth bounce?

109. Suppose you had the option of receiving $100 today, $200 tomorrow, $300 the next day, and so on for a month, or choosing to get $0.01 today, $0.02 tomorrow, $0.04 the next day, $0.08 the day after that, and so. Which would you choose? How much more money do you get with your chosen option? (*Assume "one month" means "30 days"*).

INVESTIGATION 8: SIGMA NOTATION FOR SERIES

In Exercises #110-113, write out the terms of each series.

110. $\sum_{n=1}^{7} 5(2)^{n+1}$

111. $\sum_{n=1}^{5} \frac{n^2 - 3}{10}$

112. $\sum_{n=4}^{8} \frac{4^n + n}{3n}$

113. $\sum_{n=2}^{4} \left(2(5)^n + 3(4)^{n-2} \right)$

In Exercises #114-117, write each series using sigma notation. (*Hint: You might first need to determine the number of terms in the series.*)

114. $10 + 16 + 22 + 28 + 34 + 40 + 46 + 52$

115. $3 - 12 + 48 - 192 + 768 - 3072$

116. $-6 + (-4) + (-2) + \ldots + 142$

117. $5 + 20 + 80 + \ldots + 5{,}242{,}880$

Algebra II

Module 2: Homework
Sequences and Series

INVESTIGATION 9: INFINITE SERIES

For Exercises #118-121 do the following.
a) State whether the series is arithmetic or geometric and identify the common difference or common ratio.
b) Graph the first six terms of the sequence that makes up the series. Does there appear to be a limit? If so, what is it?
c) Generate and graph the first six terms of the sequence of partial sums for the series.
d) Does the sequence of partial sums appear to have a limit? If so, what is it?

118. $\sum_{n=1}^{\infty} 7\left(\frac{4}{3}\right)^{n-1}$

119. $\sum_{n=1}^{\infty} -0.3n + 7$

120. $-1 + 0.2 + (-0.04) + 0.008 + (-0.0016) + 0.00032$

121. $\frac{16}{4}, 5, \frac{25}{4}, \frac{125}{16}, \frac{625}{64}, \frac{3125}{256}$

INVESTIGATION 10: INFINITE GEOMETRIC SERIES

In Exercises #122-124, find the sum if possible.

122. $\sum_{n=1}^{\infty} 28\left(-\frac{5}{4}\right)^{n-1}$

123. $\sum_{n=1}^{\infty} \left(\frac{5}{8}\right)^{n}$

124. $\sum_{n=1}^{\infty} -5(0.95)^{n-1}$

For Exercises #125-128 do the following.
a) Write the series in sigma notation.
b) Find the sum of the series (if possible).

125. $-12 + (-6) + (-3) + (-1.5) + ...$

126. $2 - 2.5 + 3.125 - \frac{125}{32} + ...$

127. $2.1 - 2.1x + 2.1x^2 - 2.1x^3 + ...$; where $1 < x < 1.2$

128. $4.5 + 4.5a + 4.5a^2 + ...$; where $|a| < 1$

In Exercises #129-134, create a series that meets the stated requirements. Write the series using sigma notation. (*Note: Do not use any of the series we have provided in the investigations.*)

129. an infinite geometric series with a positive common ratio that does not have a sum

130. an infinite geometric series that has a negative common ratio and does not have a sum

131. an infinite geometric series with a positive common ratio that does have a sum

132. an infinite geometric series with a negative common ratio that does have a sum

133. an infinite geometric series with a sum between 0 and 1

134. an infinite geometric series with a sum between 0 and −1

Algebra II **Module 3: Investigation 0**
Supporting Materials for Module 3

This investigation contains practice with skills and ideas that will help you be successful in Module 3.

Part 1: Review of Function Notation

1. Let *p* represent the population of a certain city (in thousands) and *n* represent the number of years since the beginning of the year 2000. Suppose we define the function $p = h(n)$.
   a. What's the input to function *h*? What's the output?

   b. What does $h(11)$ represent?

   c. What does it mean to say that $h(13) = 34$?

   d. Suppose you are asked to find *n* such that $h(n) = 29.4$. What would the answer tell us?

   e. What does the expression $h(8) - h(1)$ represent?

   f. What's the difference between saying "*h*" and "$h(n)$"?

2. Given the function $f(x) = 7(x-2) + 1$ where $y = f(x)$, answer each of the following questions.
   a. Evaluate $f(8.4)$ and explain what your answer represents.

   b. Evaluate $f(-1.5)$ and explain what your answer represents.

   c. Given that $f(x) = -55$, solve for *x* and explain what your answer represents.

3. Given the function $g(w) = \dfrac{w+3}{4}$ where $v = g(w)$, answer each of the following questions.

   a. Evaluate $g(-7)$ and explain what your answer represents.

   b. Evaluate $g\left(\tfrac{1}{2}\right)$ and explain what your answer represents.

   c. Given that $g(w) = 2.05$, solve for w and explain what your answer represents.

Part 2: Percent Change and Factors

4. Suppose we are buying an item on sale for 30% off of its original price of $80.
   a. What percent of the original price is the new sale price?

   b. What number can we multiply $80 by to determine the sale price of the item?

   c. What is the percent change in the price of the item?

5. Due to rising costs, a car company must increase the price of its cars this year. Suppose one specific car model was priced at $15,700 last year. This year the price will be 4% higher.
   a. What percent of last year's price is the new price of the car?

   b. What number can we multiply $15,700 by to determine the new price of the car?

   c. What is the percent change in the price of the car?

6. The population of a city increased from 45,641 to 48,316 people.
   a. What number do we multiply the original population (45,641) by to get the new population (48,316)?

   b. What is the percent change in the population?

Algebra II

7. A company's profits last year were $15.9 million and its profits this year were $14.2 million.
    a. What number do we multiply last year's profits ($15.9 million) by to get this year's profits ($14.2 million)?

    b. What is the percent change in the company's profits?

Part 3: Finding the Common Ratio and Missing Terms in a Geometric Sequence

In Exercises #8-11, you are given portions of a geometric sequence. Find the common ratio, then fill in the missing terms of the sequence. *Note: Assume all common ratios are positive for these exercises.*

8. 7, _____, 63, ...

9. 9, _____, _____, 72, ...

10. -1, _____, _____, _____, -3.8416, ...

11. 3, _____, _____, _____, _____, 50,421, ...

Algebra II

Module 3: Investigation 0
Supporting Materials for Module 3

Part 4: Negative Exponents

We know that whole number exponents represent repeated multiplication by some factor. For example, 4^3 means we have a product with 3 factors of 4. That is, $4^3 = 4 \cdot 4 \cdot 4$.

Negative integer exponents, however, mean something different: they represent repeated *division* by some number. For example, $5 \cdot (4)^{-3}$ means to begin with the number 5 and divide by 4 a total of 3 times. This is the same as dividing by 4^3, so $5 \cdot (4)^{-3} = \dfrac{5}{4^3}$.

Some more examples follow.

$$(6)^{-4} = 1 \cdot (6)^{-4}$$
$$= \dfrac{1}{6^4}$$

$$20 \cdot (3)^{-5} \cdot (3)^4 = 20 \cdot (3)^{-5+4}$$
$$= 20 \cdot (3)^{-1}$$
$$= \dfrac{20}{3}$$

12. Simplify each of the following expressions.

   a. $1000 \cdot (5)^{-3}$
   b. 2^{-4}
   c. $72 \cdot \left(4^{-1}\right)^2$

13. Rewrite each expression using only positive exponents.

   a. $x^{-2} y^3$
   b. $x^3 \cdot \left(x^{-2}\right)^4$
   c. $\dfrac{x^{-4}}{x^2}$

Algebra II

Module 3: Investigation 1
Introduction to Exponential Functions

Suppose that CNN posted a breaking news story on their website this morning. The website administrator noticed that at noon about 60,000 people had read the story, and that the number seemed to be increasing by about 15% each hour. Suppose that this pattern continues for the next 10 hours.

hours since noon t	expected number of people who have read the story P
0	60,000
1	
2	
5	
10	

1. Complete the table giving the expected number of people P that have read the story t hours since noon.

2. Write the formula for a function f that we can use to find the expected number of people who have read the story P for a given value of t.

3. Three of your classmates were asked to draw graphs of the relationship between the expected number of people who have read the story and the number of hours since noon. These are the results.

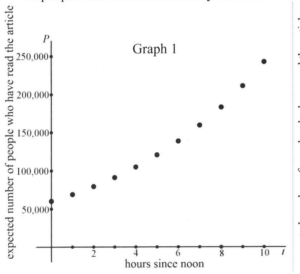

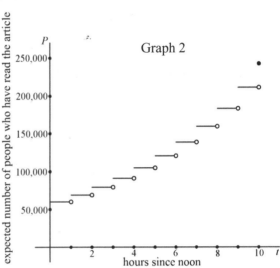

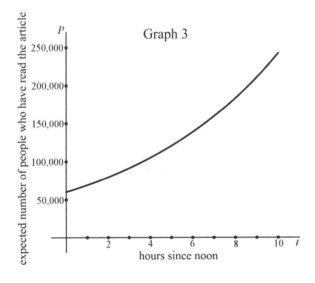

a. How is each student thinking about the relationship in order to create his or her graph?

Algebra II

Module 3: Investigation 1
Introduction to Exponential Functions

b. Which graph do you think is the most accurate representation of the relationship between *P* and *t*? Explain your choice.

In Module 2 we studied geometric sequences. A geometric sequence is an example of an exponential function where the inputs had to be whole numbers. In general, exponential functions are continuous functions, meaning that we don't have to restrict inputs to only whole numbers.

4. Use the results of previous exercises to answer the following questions.
   a. How many people do we estimate will have read the story by 12:30? What value of t do we use to find this?

   b. How many people do we estimate will have read the story by 3:15? What value of t do we use to find this?

   c. What is the change in the expected number of people who have read the article from noon to 5:00 pm?

   d. What is the change in the expected number of people who have read the article from 5:00 pm to 10:00 pm?

5. a. Will the change in the number of people who have read the article per hour be the same from noon to 10:00 pm? Explain.

   b. Explain how your answer in part (a) shows up in your answers to Exercises #4c and #4d.

Algebra II

Module 3: Investigation 1
Introduction to Exponential Functions

6. Suppose that our prediction was inaccurate, and that instead of increasing by 15% per hour, the number of people who have read the article increased by 24% per hour from noon to 10:00 pm.
   a. By what number do we multiply the number of people who have read the article by 4:00 pm to find the number of people who have read the article by 5:00 pm?

   b. How does this change the function formula from Exercise #2?

Exponential Function

A function is **exponential** if, for equal changes in the input value, the ratio of the corresponding output values is constant. *In other words, every time the input changes by equal increments the output value is multiplied by a constant.*

The general formula for an exponential function is

[ ]

where

   a is

   b is

---

7. What do we mean by "hourly growth factor" in the CNN news story context?

8. Suppose that our prediction was inaccurate, and that the following table shows the number of people who had read the story for different values of t.

hours since noon t	expected number of people who have read the story P
0	60,000
1	64,800
2	69,984
3	75,583

   a. How can we determine the hourly growth factor given this new information?

   b. What is the hourly growth factor based on the information in the table? What is the percent change in the number of people who have read the article each hour?

Algebra II

Module 3: Investigation 1
Introduction to Exponential Functions

9. Given an exponential function that grows by some percentage every year, and the output values for year *n* and year *n* + 1, how can we find the annual (yearly) growth factor?

10. Given an exponential function of the form $f(x) = a \cdot b^x$ and given that $f(7) = 240$ and $f(8) = 144$, answer the following questions.
    a. What is the value of *b*?

    b. What are the values of $f(6)$ and $f(9)$?

Algebra II

Module 3: Investigation 2
Practice with Exponential Functions

In Exercises #1-8, determine the requested information for each exponential function.

1. $k(x) = 12(1.28)^x$

    1-unit Growth Factor:

    1-unit Percent Change:

    Initial Value:

2. $t(n) = 4.35(0.93)^n$

    1-unit Decay Factor:

    1-unit Percent Change:

    Initial Value:

3.
x	0	1	2	3
f(x)	108	36	12	4

    1-unit Decay Factor:

    1-unit Percent Change:

    Initial Value:

    Formula:

4.
x	1	2	3
g(x)	5	6.25	7.8125

    1-unit Growth Factor:

    1-unit Percent Change:

    Initial Value:

    Formula:

5.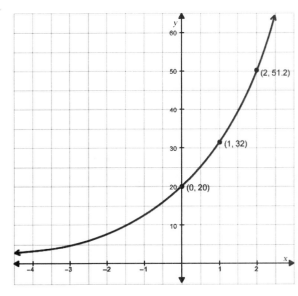

    1-unit Growth or Decay Factor:

    1-unit Percent Change:

    Initial Value:

    Formula:

Module 3: Investigation 2

6.

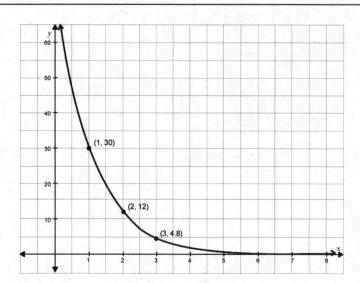

1-unit Growth or Decay Factor:

1-unit Percent Change:

Initial Value:

Formula:

7. An investment of $500 increases by 2.3% each month.

   1-month Growth Factor: 1-month Percent Change:

   Initial Value: Formula:

8. After starting with 78 micrograms, the mass of bacteria decreases by 35% every hour.

   1-hour Decay Factor: 1-hour Percent Change:

   Initial Value: Formula:

9. Suppose that John agrees to take a 5% pay cut this year to help his company make it through the year without needing to lay off any workers. If the company gives him a 5% raise next year, how will his new salary compare to his original salary before the pay cut? Defend your choice.
   I) The salaries are the same.
   II) His new salary is more than his original salary.
   III) His new salary is less than his original salary.

Algebra II

Module 3: Investigation 3
Partial Growth Factors of Exponential Functions

[Investigation 0 in Module 2 and Investigation 0 in Module 3 contain practice/review in working with radicals and roots. Refer to these investigations as necessary.]

1. A biologist first measures the mass of bacteria in an experiment to be 50 micrograms (μg). After 7 hours the mass grows to 100 μg. This pattern of doubling mass occurs every 7 hours following a continuous exponential model.

    a. Which quantities are changing in this situation? Which quantities are not changing?

Elapsed Time (hours)	Number of times as large as the original 50 μg	Mass of Bacteria (μg)
0		
7		
14		
21		
28		
35		
42		

    b. Complete the given table.

    c. Define a function f that expresses the bacteria mass M (in μg) after n 7-hour time intervals since the biologist first measured the mass.

2. Given that the bacteria mass (in μg) doubles every 7 hours (i.e., the 7-hour growth factor is 2) following a continuous exponential model, which of the following two statements best describes what happens every hour? Explain your reasoning.

    (i) Every hour the bacteria mass increases by a constant amount. That is, a constant amount is added to the bacteria mass at the end of the previous hour to find the new mass.

    (ii) Every hour the bacteria mass is multiplied by a constant factor of $\sqrt[7]{2} \approx 1.104$. That is, 1.104 is multiplied by the bacteria mass at the end of the previous hour to find the new mass.

3. Suppose we want to know how the bacteria mass changes for different time intervals.
    a. What is the 1-hour growth factor? What does this number tell us?

b. What is the 2-hour growth factor? What does this number tell us?

c. What is the 5-hour growth factor? What does this number tell us?

4. Given that the bacteria mass (in μg) in a Petri dish doubles every 7 hours (starting with 50 μg), do the following.

Elapsed Time (hours)	Number of times as large as the original 50 μg	Bacteria Mass (μg)
0		
1		
2		
3		
4		
5		
6		
7		

a. Complete the given table (round to the nearest hundredth).

b. Define a function g that expresses the bacteria mass M (in μg) present t hours since the biologist first measured the mass.

c. Define a function h that expresses the bacteria mass M (in μg) present x minutes since the biologist first measured the mass.

5. Suppose that once the bacteria mass reaches 210 micrograms, laboratory researchers apply a substance they call *Chemical X* to study its effects on the bacteria. They observe that the chemical kills $\frac{1}{3}$ of the remaining living bacteria every 4 minutes.

   a. Define a function f that expresses the mass of living bacteria L present after n 4-minute time intervals since adding *Chemical X*.

Algebra II

Module 3: Investigation 3
Partial Growth Factors of Exponential Functions

    b. Define a function g that expresses the mass of living bacteria *L* present *t* minutes after adding *Chemical X.*

    c. What is the mass of the living bacteria 3 minutes after adding *Chemical X?*

    d. What is the mass of the living bacteria 28 minutes after adding *Chemical X?*

    e. Does a greater mass of bacteria die from 8 to 9 minutes after adding *Chemical X* or from 15 to 16 minutes after adding *Chemical X?* Explain.

6. Wilson's investment account increases in value by 4% every year according to a continuous exponential model.
    a. What is the annual (1-year) growth factor, and what does it mean in this context?

    b. What is the monthly (1-month) growth factor, and what does it mean in this context?

7. Due to economic challenges, a city finds that it's losing some of its population. Over the course of 5 years the city's population decreased from 87,130 people to 79,490 people. Assume that the city's population follows a continuous exponential model.
    a. What is the 5-year decay factor, and what does it mean in this context?

    b. What is the 2-year decay factor, and what does it mean in this context?

Algebra II
Module 3: Investigation 4
Developing and Applying the Properties of Rational Exponents

Part I

In Investigation 3 we learned how to determine partial growth factors. For example, if we know the annual growth factor for an exponential function is 1.13, then the monthly growth factor is $\sqrt[12]{1.13} \approx 1.0102$. If the 4-minute decay factor is 0.72, then the 1-minute decay factor is $\sqrt[4]{0.72} \approx 0.9212$. In this investigation we will learn how to use exponents instead of radicals to represent these calculations.

Note: For this investigation, assume that all variables represent positive numbers.

1. Simplify each expression using the rules of exponents.

   a. $\left(2^3\right)^4$ b. $x^2 \cdot x^5$ c. $(3x)^4$ d. $\dfrac{10x^7}{2x^3}$ e. $\left(2x^2y\right)^3\left(5xy^4\right)$

2. Simplify each expression using the rules of exponents.

   a. $5^{1/2} \cdot 5^{1/2}$ b. $2^{1/4} \cdot 2^{1/4} \cdot 2^{1/4} \cdot 2^{1/4}$ c. $7^{1/3} \cdot 7^{1/3} \cdot 7^{1/3}$ d. $\left(5^{1/6}\right)^6$

3. In Exercise #2 we saw that $7^{1/3} \cdot 7^{1/3} \cdot 7^{1/3} = 7$ according to the rules of exponents. This statement is generally true: that is, $x^{1/3} \cdot x^{1/3} \cdot x^{1/3} = x$. Write a convincing argument for why it must be true that $7^{1/3} = \sqrt[3]{7}$ (and more generally $x^{1/3} = \sqrt[3]{x}$).

 Rational Exponents of the form 1/*n*
If $x \geq 0$ and $n > 0$, then $x^{1/n} = \boxed{}$.

4. Explain what $10^{1/8}$ represents. (*You don't need to calculate its value.*)

5. Rewrite each of the following expressions using radicals.

   a. $5^{1/2}$ b. $11^{1/6}$ c. $y^{1/9}$ d. $x^{1/n}$

6. Rewrite each of the following expressions using rational exponents.

   a. $\sqrt[5]{7}$ b. $\sqrt[7]{150}$ c. $\sqrt[w]{4}$

Algebra II
Module 3: Investigation 4
Developing and Applying the Properties of Rational Exponents

7. Recall the following context from Investigation 3. *A biologist first measures the mass of bacteria in an experiment to be 50 micrograms (μg). After 7 hours the mass grows to 100 μg. This pattern of doubling mass occurs every 7 hours following a continuous exponential model.*
   a. Represent the one-hour growth factor of the bacteria mass using a rational exponent.

   b. Use your calculator to find the approximate decimal value of the one-hour growth factor.

8. The given table shows the population of Rockton city during the first 50 years of the 20th century.

Year	Approximate Population of Rockton City
1900	59,640
1910	74,850
1920	93,930
1930	117,890
1940	147,950
1950	185,680

   a. Is an exponential function a good fit to model this data? Explain.

   b. By approximately what percent did the population of Rockton City change each year from 1900 to 1950?

Part II

9. Rewrite each of the following expressions using the rules of exponents.
   a. $\left(2^{1/3}\right)^5$ b. $\left(2^5\right)^{1/3}$ c. $\left(8^{1/5}\right)^2$ d. $\left(8^2\right)^{1/5}$

In Exercise #9 we used the rules of exponents to conclude that $\left(2^{1/3}\right)^5 = 2^{5/3}$ and $\left(2^5\right)^{1/3} = 2^{5/3}$. Therefore, $\left(2^{1/3}\right)^5 = \left(2^5\right)^{1/3}$ and both expressions are identical to $2^{5/3}$. We can generalize these results.

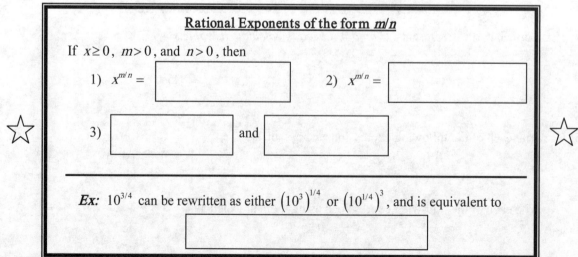

Rational Exponents of the form *m/n*

If $x \geq 0$, $m > 0$, and $n > 0$, then

1) $x^{m/n} =$ ☐   2) $x^{m/n} =$ ☐

3) ☐ and ☐

Ex: $10^{3/4}$ can be rewritten as either $\left(10^3\right)^{1/4}$ or $\left(10^{1/4}\right)^3$, and is equivalent to ☐

Algebra II

Module 3: Investigation 4
Developing and Applying the Properties of Rational Exponents

10. Consider $9^{5/2}$, which can be rewritten as $\left(9^{1/2}\right)^5$ or $\left(9^5\right)^{1/2}$.

    a. Rewrite $\left(9^{1/2}\right)^5$ using radicals.

    b. Rewrite $\left(9^5\right)^{1/2}$ using radicals.

    c. Which of the final expressions in parts (a) and (b) is easier to use to simplify $9^{5/2}$ as a whole number? Why?

11. Consider $32^{4/5}$, which can be written as $\left(32^{1/5}\right)^4$ or $\left(32^4\right)^{1/5}$.

    a. Rewrite $\left(32^{1/5}\right)^4$ using radicals.

    b. Rewrite $\left(32^4\right)^{1/5}$ using radicals.

    c. Which of the final expressions in parts (a) and (b) is easier to use to simplify $32^{4/5}$ as a whole number? Why?

In Exercises #10-11 we saw that, although it's possible to rewrite $x^{m/n}$ as $\left(x^{1/n}\right)^m$ or $\left(x^m\right)^{1/n}$, one of the forms might be easier to work with depending on the context.

12. Simplify the given expression as a whole or rational number if possible. If it's not possible, rewrite the expression without using exponents (if possible) and simplify the radical as much as you can.

    a. $25^{3/2}$

    b. $8^{5/3}$

    c. $2^{7/4}$

    d. $2^{3/5}$

    e. $(x^{5/3})^{3/5}$

    f. $125^{7/3}$

    g. $7^{31/14}$

    h. $\left(\frac{81}{4}\right)^{3/2}$

    i. $y^{9/2}$

Algebra II

Module 3: Investigation 4
Developing and Applying the Properties of Rational Exponents

13. Rewrite each radical expression using rational exponents.

    a. $\sqrt[7]{3}$

    b. $\sqrt[6]{5^4}$

    c. $\left(\sqrt[3]{x}\right)^2$

    d. $\sqrt[4]{xy^3}$

    e. $10^3 \cdot \sqrt[5]{10^7}$

    f. $\left(\sqrt[4]{x}\right)^3 \left(\sqrt[4]{x^2}\right)^2$

14. Use the rules of exponents to simplify the following expressions as whole or rational numbers if possible, or in simplified radical form.

    a. $9^{3/2} \cdot 9^{11/2}$

    b. $64^{1/2} \cdot 64^{5/2}$

    c. $\dfrac{11^{1/4}}{11^{1/6}}$

    d. $2^{1/4} \cdot 4^{3/4}$

15. $f(t) = 13.7(1.017)^t$ models the population of a certain country (in millions of people) t years after the end of the year 2000.

    a. What do we know about this country's population around the year 2000?

    b. If we want to find the population of the country 26 months after the end of the year 2000, what value should we use for t?

    c. By what factor does the country's population grow every 26 months? Represent your answer using rational exponents and find the approximate decimal value.

    d. According to the model, by what percent is the country's population growing every month?

16. In a laboratory experiment, a biologist placed 200 micrograms of bacteria in a culture. The mass of the bacteria then began to increase exponentially. After 3.5 days the culture contained 1000 micrograms of bacteria. Define a function f to model the bacteria mass (in micrograms) in the culture t days since the experiment began.

Algebra II

Module 3: Investigation 5
n-unit Growth and Decay Factors (n≠1)

It's common to be given a specific growth or decay factor in a real-world situation. For example, banks usually report interest or investment returns in terms of an annual growth rate. But you might be interested in knowing how your money will grow every month, or maybe over a 2 or 5 year period. In this investigation we will focus on finding growth and decay factors for a variety of changes in the input value of a function relationship.

1. a. Determine the specified growth factors for the exponential function represented in the table.

number of days, x	0	1	2	3	4	5	6	7
bacteria mass (in micrograms), M	4	12	36	108	324	972	2916	8748

   i) 1-day growth factor: ii) 2-day growth factor:

   iii) 3-day growth factor: iv) ½-day (12-hour) growth factor:

   v) ¼-day (6-hour) growth factor:

   b. Determine the initial bacteria mass.

   c. Define a function to determine the bacteria mass in terms of the number of days x since we started measuring the bacteria mass.

   d. Define a function to determine the bacteria mass in terms of the number of weeks n since we started measuring the bacteria mass.

2. a. Determine the specified growth factors for the exponential function represented in the table. *(For this exercise, assume that 1 month = 30 days.)*

months since first measuring the mass, n	1	3	5	7	9
amount of radioactive substance (in µg), R	70.3125	28.125	11.25	4.5	1.8

   i) 2-month decay factor: ii) 6-month decay factor:

   iii) 1-month decay factor: iv) 1-day decay factor:

   b. Determine the amount of radioactive substance when it was first measured.

   c. Define a function f to determine the amount of radioactive substance R in terms of the number of months n since the mass of the radioactive substance was first measured.

Algebra II   **Module 3: Investigation 5**
                                              n-unit Growth and Decay Factors (n≠1)

d. Define a function g to determine the amount of radioactive substance R in terms of the number of 6-month periods x since the radioactive substance was first measured.

3. Determine the specified growth factors for the exponential function represented in the graph, and then determine the function formula that models the data

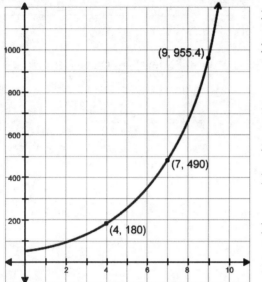

3-unit Growth Factor: ≈ 2.722

2-unit Growth Factor: ≈ 1.950

1-unit Growth Factor: ≈ 1.396

½-unit Growth Factor: ≈ 1.182

5-unit Growth Factor: ≈ 5.308

Initial Value: ≈ 47.4

Formula: $f(x) \approx 47.4(1.396)^x$

4. Determine the specified decay factors for the exponential function represented in the graph, and then determine the function formula that models the data.

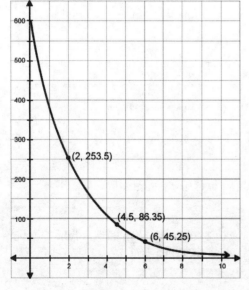

4-unit Decay Factor: ≈ 0.1785

1.5-unit Decay Factor: ≈ 0.524

1-unit Decay Factor: 0.65

1/3-unit Decay Factor: ≈ 0.866

7-unit Decay Factor: ≈ 0.0490

Initial Value: 600

Formula: $f(x) = 600(0.65)^x$

Algebra II
Module 3: Investigation 5
n-unit Growth and Decay Factors (n≠1)

5. The number of fruit flies in an experiment t days after its start is given by $f(t) = 237(3)^{t/2}$.
   a. What are some other ways to write the formula?

   b. Find the 2-day growth factor.

   c. Find the 1-day growth factor.

   d. Circle the statement(s) below that describe the behavior of f:
      i) An initial number of 237 fruit flies triples every ½-day.
      ii) An initial number of 237 fruit flies triples every 2 days.
      iii) An initial number of 237 fruit flies increases by about 73.2% every 1 day.

In Exercises #6-7 write your answers in exponential and decimal form (round to the nearest thousandth).

6. The population of Springerville is currently 34,725 and is growing by 3.9% every 3 years.
   a. Find the 6-year growth factor.

   b. Find the percent change over a 6-year period.

   c. Find the 1-year growth factor.

   d. Find the percent change over a 1-year period.

7. The amount of caffeine in your body decreases by 32% every 2 hours.
   a. Find the 24-hour decay factor.

   b. Find the percent change over a 24-hour period.

   c. Find the 1-hour decay factor.

   d. Find the percent change over a 1-hour period.

Algebra II

Module 3: Investigation 5
n-unit Growth and Decay Factors (n≠1)

8. Suppose one country has a population of about 64 million people and that the population has been growing by 2.3% per year and that another country has a population of 58 million people and that the population has been growing by 3.1% per year. If the populations continue to grow at their present rates how long will it take for the two countries to have equal populations?

9. Repeat Exercise #8 if, instead of growing by 2.3% per year, the first country's population is decreasing by 2.3% per year.

10. Find the solution to the system involving $f(x) = 12(2.5)^x$ and $g(x) = 40(0.95)^x$.

11. Find the solution to the system involving $f(x) = 14.6(0.88)^x$ and $g(x) = 0.9(2)^x$.

Algebra II

Module 3: Investigation 6
Introducing Logarithms

For Exercises #1-7, let $f(x) = 4^x$.

1. Complete the table of values and graph f.

x	f(x)
−3	
−2	
0	
2	
2.5	
3	

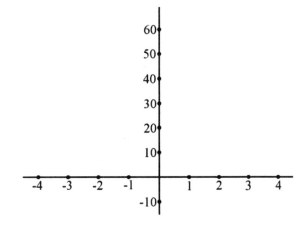

2. Evaluate $f(1.5)$ and explain what the answer represents in terms of exponents and bases.

3. Solve for x given that $f(x) = \frac{1}{4}$ and explain what the answer represents in terms of exponents and bases.

4. Suppose we want to construct a function g that gives the same information as f but that switches the input and output values. Create a table of values and graph for g.

x	g(x)

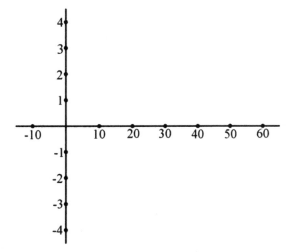

5. Function f inputs an exponent and outputs the result of raising 4 to that exponent. Explain what function g does.

Algebra II

Module 3: Investigation 6
Introducing Logarithms

6. Evaluate $g(64)$ and explain what the result represents in terms of exponents and bases.

7. How can we use the table and graph of f (see Exercise #1) to evaluate $g(64)$?

8. The population of Indianapolis, IN was 781,870 at the beginning of the year 2000 and increased by 5.6% each year.
   a. Define a function f that determines the population P of Indianapolis in terms of the number of years since 2000.

   b. In what year does your model estimate the population of Indianapolis to be 1,000,000 people?

So far we don't know how to solve equations with an unknown in the exponent without using graphs. For example, to solve the equation $5^x = 125$ for x means we need to write the equation in the form "$x =$". Algebraically solving such equations requires a new function called a ***logarithm***.

A logarithm is a function that inputs the result of raising a number to some exponent and outputs the exponent needed to get that result. Function g in Exercise #4 is a logarithmic function with a base of 4.

Logarithm

A ***logarithm of base b*** is a function that inputs the result of raising b to some exponent and outputs the exponent needed to get this result. We write this as $y = \log_b(x)$, and it's the equivalent of saying $b^y = x$. Note that $b > 0$, $b \neq 1$, and $x > 0$.

For example, we can write $\log_4(64) = 3$, which says that the exponent on 4 necessary to get 64 is 3. This is equivalent to saying $4^3 = 64$.

$$\underbrace{\log_4}_{\text{the base is 4}} (\overbrace{64}^{\text{the result of raising 4 to some exponent is 64}}) = \underbrace{3}_{\substack{\text{the exponent} \\ \text{on 4 needed} \\ \text{to get 64 is 3}}}$$

Therefore, a logarithm is an exponent!

Algebra II
Module 3: Investigation 6
Introducing Logarithms

A good way to think about a logarithm is that it answers a question. In general, $y = \log_b(x)$ (read as "y equals the log base b of x") asks "What exponent on b is necessary to get x?" and the answer is the value of y.

For example, $\log_2(32) = y$ asks "What exponent on 2 is necessary to get 32?" The answer is "$y = 5$" because $2^5 = 32$. Therefore we say $\log_2(32) = 5$.

In Exercises #9-17, find the value of the unknown that makes each statement true. *If the answer isn't an integer, then state the two integers that the answer falls between (such as "between 2 and 3").*

9. $\log_2(8) = y$

10. $\log_6\left(\frac{1}{36}\right) = x$

11. $\log_4(4) = a$

12. $\log_7(12) = m$

13. $\log_2(24) = x$

14. $\log_4(5) = n$

15. $\log_{10}(10,000) = w$

16. $\log_8(500) = u$

17. $\log_2(-4) = k$

18. Complete the table of values and graph for function g given that $g(x) = \log_3(x)$.

x	$g(x)$
$\frac{1}{9}$	
$\frac{1}{3}$	
1	
3	
9	
27	

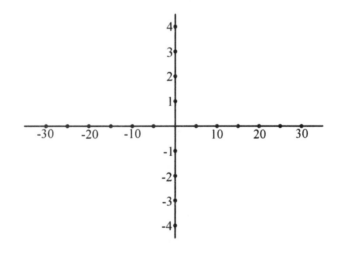

19. Pick a point on your graph and explain what it represents in terms of exponents and bases.

Module 3: Investigation 6

Algebra II

Module 3: Investigation 7
Properties of Logarithms

Remember that when we evaluate a logarithm we are finding an exponent. For example, the expression $\log_2(32)$ represents the exponent on 2 that produces a result of 32, and its value is 5 since $2^5 = 32$. So the statement $\log_2(32) = 5$ using logarithmic notation is equivalent to the statement $2^5 = 32$ using exponential notation.

Even though we can have logarithms with any positive base, many calculators only work with the two bases used most often in math and science: base 10 and base e. A base 10 logarithm, such as $\log_{10}(50)$, is called a ***common log***, and we usually leave off "10" and just write $\log(50)$. The **LOG** button on your calculator is used to compute common logs.

A base e logarithm, such as $\log_e(14)$, is called a ***natural log*** and is written as $\ln(14)$. The number e is a constant and $e \approx 2.72$. The **LN** button on your calculator is used to compute natural logs. You will learn more about e and its uses later.

Common and Natural Logs

Common Log: A base 10 logarithm, such as $\log_{10}(50)$. We write this as $\log(50)$ by convention.

Natural Log: A base e logarithm, such as $\log_e(14)$. We write this as $\ln(14)$ by convention. *Note that e is a constant and $e \approx 2.72$.*

---

1. Convert each equation to exponential form and state the value of the unknown that makes the equation true. Check your answer using your calculator.

   a. $\log(10,000) = x$ b. $\log\left(\frac{1}{10}\right) = x$ c. $\log(1100) = y$ d. $\ln(7) = z$

Recall the following properties of exponents:

I) $b^x \cdot b^y = b^{x+y}$ II) $\dfrac{b^x}{b^y} = b^{x-y}$ III) $\left(a^x\right)^y = a^{x \cdot y}$

Since a logarithm is just an exponent, these same rules apply to logarithms. Because of the notation, however, the rules will look a little different.

The first property of logarithms states

$$\log_b(m) + \log_b(n) = \log_b(m \cdot n).$$

That is, if two logs *of the same base b* are being added together, we can rewrite them as a single logarithm to simplify an expression or equation. For example, $\log_2(4) + \log_2(8)$ asks to find the exponent on 2 to get a result of 4 and the exponent on 2 to get a result of 8 and then to add them together. The answer is $2 + 3 = 5$. We could have instead looked at $\log_2(4 \cdot 8) = \log_2(32)$ and determined that 5 is the exponent on 2 needed to get a result of 32.

Algebra II

Module 3: Investigation 7
Properties of Logarithms

2. Combine the logarithms and determine (or estimate) the value of each expression.
   a. $\log_4(2) + \log_4(32)$
   b. $\log(25) + \log(4)$
   c. $\log_4(5) + \log_4(2) + \log_4(25.6)$

   d. $\log_7(12) + \log_7(2.4)$
   e. $\log_2(y) + \log_2(y)$
   f. $\ln(x) + \ln(x) + \ln(x)$

Properties of Logarithms

For any $b > 0$, $b \neq 1$ and $m, n > 0$,

Log Property #1: $\log_b(m \cdot n) = \log_b(m) + \log_b(n)$

Log Property #2: $\log_b\left(\dfrac{m}{n}\right) = \log_b(m) - \log_b(n)$

Log Property #3: $\log_b(n^c) = c \cdot \log_b(m)$

---

3. Use the three properties of logarithms to rewrite each of the following as a single logarithm. Then evaluate (or estimate) the value of the resulting expression.
   a. $\log_3(36) - \log_3(4)$
   b. $3\log_2(4) - \log_2(5)$

   c. $\log(6) + \tfrac{1}{2}\log(121)$
   d. $2\log(10) - \tfrac{1}{2}\log(25) + \log(9)$

4. Rewrite each of the following as sums and differences of logarithms.
   a. $\log_2(4x)$
   b. $\log_3\left(\dfrac{xy}{27}\right)$
   c. $\log_5(4 \cdot 10^x)$

5. Rewrite the following in logarithmic form.
   a. $y = 4^t$
   b. $y = 2.3^t$
   c. $y = 2(6.5)^t$
   d. $y = 19.43^{5t}$

6. Explain the relationship between the functions $f(x) = 2^x$ and $g(w) = \log_2(w)$.

Algebra II

Module 3: Investigation 8
Solving Exponential Equations using Logarithms

Suppose we want to solve the equation $5^x = 190$ for x. We could use our calculator to graph $y = 5^x$ and $y = 190$ and find the x-value of the intersection point. If you have an advanced calculator that can compute logs of any base, then you can instead write $5^x = 190$ as $x = \log_5(190)$ using logarithms, then use your calculator to evaluate $\log_5(190)$.

If we want to solve the equation without graphing, however, and if you don't have an advanced calculator, then we need to use a little different approach. In investigation #7 we learned some properties of logarithms including Property #3: $\log_b(m^c) = c \cdot \log_b(m)$. Let's see how we can use this property to solve the equation $5^x = 190$.

$5^x = 190$

$\log(5^x) = \log(190)$ • convert each side into an exponent on a base of 10

$x \cdot \log(5) = \log(190)$ • apply the property $\log_b(m^c) = c \cdot \log_b(m)$

$x = \dfrac{\log(190)}{\log(5)}$ • divide by $\log(5)$

$x \approx 3.26$ • evaluate with a calculator

Our calculator tells us that $x \approx 3.26$. We can easily check our solution by verifying that $5^{3.26} \approx 190$.

Solve each equation in Exercises #1-6 and then check your answer to show that it makes sense.

1. $6^x = 45$
2. $2^y = 314$
3. $0.05 = 21^x$

4. $5(3)^p = 60$
5. $37 = 10(8)^x$
6. $196.8 = 35.4(1.65)^t$

7. Susan started a college savings fund for her granddaughter with a $4,500 deposit. Suppose the account value increases 5.5% each year.
   a. Define a function to model this context. Be sure to define any necessary variables.

Algebra II
Module 3: Investigation 8
Solving Exponential Equations using Logarithms

b. how many years will it take for the account to have a value of $15,000?

8. An initial amount of 80 mg of caffeine is metabolized in the body and decreases by 17% every hour.
   a. Define a function to model this context. Be sure to define any necessary variables.

   b. How many hours will it take until only 13 mg of caffeine is left in the body?

9. You read a news story about a bird whose habitat is threatened by human activity and pollution. Four years ago about 2,420 of the birds lived in the area, but today only about 1,870 live in the area.
   a. Suppose the number of these birds living in the area follows an exponential model. How many years will it take before only 800 of the birds remain?

   b. What are some of the challenges in making long-term predictions of a threatened species as we did in this exercise?

Algebra II

Module 3: Investigation 8
Solving Exponential Equations using Logarithms

In Investigation 5 you solved the systems described in Exercises #10-13 graphically or with a table. Solve each system algebraically and compare the solutions with your answers in Investigation 5.

10. Suppose one country has a population of about 64 million people and that the population has been growing by 2.3% per year and that another country has a population of 58 million people and that the population has been growing by 3.1% per year. If the populations continue to grow at their present rates how long will it take for the two countries to have equal populations?

11. Repeat Exercise #10 if, instead of growing by 2.3% per year, the first country's population is decreasing by 2.3% per year.

12. Find the solution to the system involving $f(x) = 12(2.5)^x$ and $g(x) = 40(0.95)^x$.

Algebra II
Module 3: Investigation 8
Solving Exponential Equations using Logarithms

13. Find the solution to the system involving $f(x) = 14.6(0.88)^x$ and $g(x) = 0.9(2)^x$.

14. Solve the following systems involving logarithmic functions.
    a. $f(x) = \log_4(2x)$ and $g(x) = \log_4(x+8)$
    b. $f(x) = \log_5(x-3)$ and $g(x) = \log_5\left(\frac{x}{4}\right)$

Algebra II

Module 3: Investigation 9
Function Composition: Chaining Together Two Function Processes

1. Noelle often purchases rice as a side dish when having people over for dinner. Noelle finds that she can get 3 servings for every 1 pound of rice she purchases.

Number of servings of rice desired	Weight of rice needed (in pounds)
3	
5	
9	
11	
n	

    a. Complete the table of values showing the number of pounds of rice Noelle needs based on the number of servings desired.

    b. Define a function f that determines the weight of rice (in pounds) needed based on n, the number of servings of rice desired.

    c. The current cost of her favorite rice brand is $1.79 per pound.
        i. How much will it cost Noelle to purchase 4 pounds of rice? 6.1 pounds of rice?

        ii. Define a function g that determines the cost of rice (in dollars) based on the weight of rice needed (in pounds).

    d. Explain how you can determine the cost of the rice if Noelle wants to buy enough for 7 servings.

    e. Complete the table of values showing the total cost of rice (in dollars) based on the number of servings she needs.

Number of servings of rice	Total cost of rice (in dollars)
3	
5	
9	
11	

    f. Write a formula to determine the cost of n servings of rice.

Algebra II

Module 3: Investigation 9
Function Composition: Chaining Together Two Function Processes

Use the following context for Exercises #2-6.
A pebble is thrown into a lake and the radius of the ripple travels outward at 0.5 meters per second. Your goal will be to determine the area inside the ripple based on the number of seconds elapsed since the pebble hit the water.

2. Draw a picture of the situation and label the quantities. Imagine how the quantities are changing together.

3. a. As the time since the pebble hit the water increases, how does the radius of the circle change?

   b. As the radius of the circle increases, how does the area enclosed by the circular ripple change?

   c. As the time elapsed since the pebble hit the water increases, how does the area enclosed by the ripple change?

4. a. Describe in words how you can determine *the area inside the ripple* assuming you are able to measure *the time since the pebble hit the water*.

   b. Define a function *f* that outputs the radius length r (measured in meters) when we input the time elapsed since the pebble hit the water t (measured in seconds).

   c. Define a function *g* that outputs the area A (measured in square meters) when we input the radius length of the circular ripple r (in meters).

d. Graph f and g.

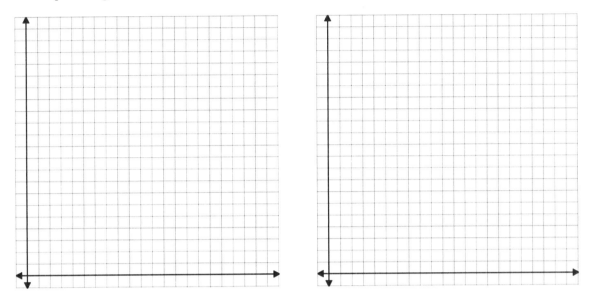

e. Use the graphs in part (d) to estimate the area 3 seconds after the pebble hits the water. Explain your process.

For the circular ripple context, we want to *represent* the two-step process of:
I. starting with the amount of time that has elapsed (in seconds) to get a radius length for the circular ripple (in meters), then
II. using that radius length to determine the area enclosed by ripple (in square meters).

The conventional way of representing this two-step process follows. When an input to a function is itself a function, we simply write the function output in place of the variable it defines. So, to represent the area of the ripple as a function of the amount of time elapsed, we would name a new function h and write $h(t) = g(f(t))$ where $h(t)$ is the area of the ripple in square meters.

Note that the output $f(t)$, which represents the radius of the circular ripple after t seconds, is the input to function g. The diagram below illustrates this convention.

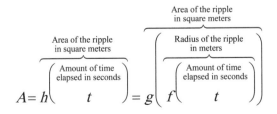 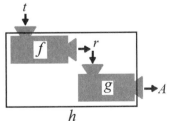

Algebra II

Module 3: Investigation 9
Function Composition: Chaining Together Two Function Processes

5. a. Complete the table to *represent* the values for each quantity in the above context. *Do not calculate the value of the expressions*, just show the calculations you would perform to find each value.

Amount of time elapsed in seconds t	Radius of the ripple in meters $f(t)$	Area of the ripple in square meters $g(f(t))$
2		
3		
6.7		
a		

b. Circle the part of your entries in the third column that correspond to $f(t)$.

c. The final step is to define one function h to determine the area of the ripple as a function of (or in terms of) the amount of time elapsed in seconds. The combining together of two function processes where the output quantity of the first function is the input quantity of the second function is called **function composition**. The new function (in this case h) is said to be the composed function and is written as

$$h(t) = g(f(t)) \quad \text{where} \quad A = h(t)$$

What rule defines function h? That is, $h(t) = g(f(t)) = $ _____.

d. Evaluate $h(3)$. How does this answer relate to the answer you obtained in Exercise #4e? Explain your reasoning.

Algebra II

Module 3: Investigation 9
Function Composition: Chaining Together Two Function Processes

6. Stefan was asked to conduct a study for a factory to determine the impact of employing fewer workers each day while having them work longer shifts. The following tables show data from his research in the form of two functions.

Function *f*

input	output
Average length of workers' shifts (in hours)	Number of expected industrial accidents per year
7	5
8	6
9	7
10	8
11	10
12	13
13	18

Function *g*

input	output
Number of workers employed each day	Average length of workers' shifts (in hours)
60	13
65	12
70	11.5
80	10
90	9
100	8
120	6.75

a. Does the expression $f(g(80))$ have a real-world meaning in this context? If so, determine its value then explain what the value represents. If not, explain your reasoning.

b. Does the expression $g(f(9))$ have a real-world meaning in this context? If so, determine its value then explain what the value represents. If not, explain your reasoning.

c. Consider the equation $f(g(x)) = 13$. Explain how to use the tables to solve this equation for *x*, then explain what the solution represents.

7. Use the following functions to answer the given questions: $f(x) = 2^x$, $g(x) = x - 3$, and $h(x) = \sqrt{x}$
   a. Evaluate $f(5.1)$.
   b. Evaluate $f(g(2))$.
   c. Evaluate $h(f(13))$.
   d. Solve the equation $g(f(x)) = 13$ for *x*.

Algebra II

Module 3: Investigation 9
Function Composition: Chaining Together Two Function Processes

e. Function m is defined as $m(x) = f(g(x))$. Write the rule for m.

f. Function p is defined as $p(x) = g(h(x))$. Write the rule for p.

In this investigation we've expressed function composition using standard function notation. In some cases you might see the alternative notation $(h \circ f)(13)$ instead of $h(f(13))$ to represent the process of evaluating $f(13)$, then using this value as an input to h. This notation comes from how we *name* the composed function: $h \circ f$.

8. Use the following functions to answer the questions below: $f(x) = 2x - 5$ and $g(x) = \frac{3x-6}{9}$

   a. Evaluate $(f \circ g)(20)$

   b. Evaluate $(g \circ f)(4)$

   c. Evaluate $(f \circ f)(x)$

   d. Evaluate $(g \circ f)(x)$

9. The area A of a square with side length s is defined by the function $f(s) = s^2$, where $A = f(s)$. The side of a square s that has a perimeter p is defined by the function $g(p) = \frac{p}{4}$, where $s = g(p)$.

   a. Evaluate $(f \circ g)(5)$ if possible and describe what your answer represents.

   b. Evaluate $f(g(5))$ if possible and describe what your answer represents.

   c. Evaluate $(g \circ f)(5)$ if possible and describe what your answer represents.

   d. Find the formula that defines values of $(f \circ g)(x)$ and describe what the formula calculates.

   e. What other notation can be used to express $(f \circ g)(x)$?

Algebra II

Module 3: Investigation 10
Introduction to Function Inverse

1. Consider the relationship defined by $y = 2^x$, and the functions $f(x) = 2^x$ and $g(y) = \log_2(y)$.
   a. Explain the relationship between the inputs and outputs of functions *f* and *g*.

   b. Complete the following tables and explain what you observe.

x	f(x)
−3	
0	
1	
4	
7	

y	g(y)
$\frac{1}{8}$	
1	
2	
16	
128	

$f(x) = 2^x$ and $g(y) = \log_2(y)$ are examples of **inverse functions**.

Inverse Functions

Two functions are **inverses** if they express the same relationship between two quantities but with the input and output quantities reversed. In other words, an inverse function *reverses the process* represented by an original function.

This means that if (*a*, *b*) is a solution of one function, then (*b*, *a*) is a solution of its inverse function and vice versa. (Note: Not all functions have inverses that are also functions.)

---

2. The function *g*, defined by $g(x) = x + 5$, describes the process of taking an input value and adding 5 to generated the corresponding output.
   a. What action will undo the process of adding of 5 to *x*?

   b. Define a function *h* that undoes the process of *g*.

   c. Explain why *g* and *h* are inverses.

Algebra II

Module 3: Investigation 10
Introduction to Function Inverse

3. Given the function k defined by $k(x) = 2x - 7$,

   a. Describe the process represented by k. (In other words, what does k do to transform an input value into an output value?)

   b. Describe the actions that will undo the process of k.

   c. Define a function m that undoes the process of k.

   d. Evaluate the following:
      i. $k(4)$
      ii. $m(1)$

4. Suppose that 1.3 U.S. dollars is equivalent to 1 Euro (the European currency), let x = the number of Euros and y = the number of U.S. dollars. Function h inputs a number of Euros and outputs the equivalent number of U.S. dollars, so $h(x) = 1.3x$ where $y = h(x)$.

   a. Determine the value of $h(42)$ and explain what it represents.

   b. Solve the equation $h(x) = 54.6$ for x, then describe the meaning of your answer.

   c. Define a function g that determines the numbers of Euros as a function of the number of U.S. dollars.

   d. Evaluate $h(28)$ and $g(36.40)$.

   e. Without performing any calculations, evaluate $h(g(36.40))$ and $g(h(28))$. What do you notice, and why does this happen?

Algebra II
Module 3: Investigation 10
Introduction to Function Inverse

Composing Inverse Functions

Suppose functions *f* and *g* are inverses. Then composing the functions (in any order) will always return the original input value because the function processes undo each other. That is,

[ ]

This provides us with a method for verifying inverses. If you think two functions are inverses you can check by composing them and showing that the result is the original input.

Inverse Function Notation

If *f* is a function, then we can use f^{-1} to name its inverse function. This helps to make their relationship clear.

NOTE: f^{-1} is just notation. The "−1" is not an exponent, and so $f^{-1} \neq \frac{1}{f}$!

5. a. In Exercise #4, how can we *represent* the fact that *g* and *h* are inverses using function notation?

   b. Create a graph of *h* (use the axes on the left) and a graph of *g* (use the axes on the right). Then do the following steps several times:
      i. Pick any input value (call it *n*) for *h*. ii. Find *h*(*n*). iii. Use h(*n*) as the input to *g*.
      iv. Determine whether *g* reverses the process of *h* for the chosen *n*.

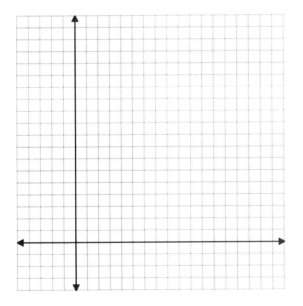
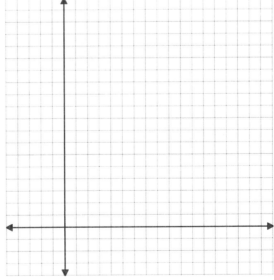

Algebra II

Module 3: Investigation 10
Introduction to Function Inverse

A common way to visualize inverse functions is shown. Function *h* inputs a number of Euros and outputs the corresponding number of U.S. dollars according to the given exchange rate, and $h(x) = 1.3x$.

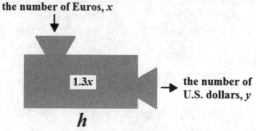

Looking at the process in reverse, we use a number of U.S. dollars as the input to determine the corresponding number of Euros according to the given exchange rate. We called this function *g* before, but using inverse notation we will now call this function h^{-1}, and so $h^{-1}(y) = \frac{y}{1.3}$.

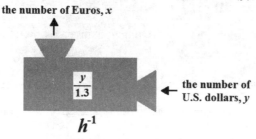

Remember that the –1 is just part of the notation – it's not an exponent. So while it's true $2^{-1} = \frac{1}{2}$ (and more generally $x^{-1} = \frac{1}{x}$ for $x \neq 0$), h^{-1} is just the name of a function and $h^{-1} \neq \frac{1}{h}$.

6. a. Define a function *g* with the following process: it adds 7 to its input, then multiplies this sum by 4.

   b. Describe the actions that will undo the process of *g*, then define this new function.

   c. Given that *y* represents the values that can be input to g^{-1}, explain why $g(g^{-1}(y)) = y$.

7. a. Given $f(x) = \log_5(x)$ where $y = f(x)$, write the formula for f^{-1}.

   b. Explain why $f^{-1}(f(x)) = x$ by discussing the processes involved.

Algebra II

Module 3: Investigation 10
Introduction to Function Inverse

8. Determine the inverse function that undoes the process of each of the given function formulas.

   a. $f(t) = 7^t$ where $d = f(t)$

   b. $g(x) = 2x + 3$ where $y = g(x)$

   c. $b(x) = \dfrac{4x+1}{3}$ where $y = b(x)$

   d. $h(s) = s^3$ where $A = h(s)$

9. Use the tables below to determine the following. Explain your thinking.

x	f(x)
−4	16
−3	13
−2	6
−1	0
0	−4
1	−7
2	−9
3	−10

x	g(x)
−8	−20
−6	−17
−4	−10
−2	−4
0	−1
2	2
4	6
6	11

   a. $f(2)$

   b. $g(f(0))$

   c. Determine x when $g(x) = -4$

   d. $g^{-1}(-4)$

   e. Explain how parts (c) and (d) are related to one another.

   f. $f^{-1}(0)$

   g. $g(f^{-1}(6))$

   h. $f(g(-2))$

   i. Solve the equation $g(f(x)) = 11$ for x.

Investigation 11: *Investment Activity: Compounding Periods & Compounding Interest Formula* and Investigation 12: *Motivating the Number e and Continuously Compounded Interest* are available online along with associated homework at www.rationalreasoning.net.

Algebra II

Module 3: Homework
Exponential & Logarithmic Functions

INVESTIGATION 1: THE MEANING OF EXPONENTS

1. The spread of diseases to people in an area often follows an exponential growth rate in its initial phase. Suppose that scientists determine that 150 people in a city have contracted a contagious disease. They expect the number of people who have the disease to increase by 10% each day for the next 10 days (after which they expect the model to change).

days since measurement t	expected number of people who have contracted the disease, P
0	150
1	
2	
5	
10	

   a. Complete the table giving the expected number of people P that have contracted the disease since the day it was initially measured t.
   b. Write the formula for a function f that we can use to find the expected number of people who have contracted the disease P for a given value of t.
   c. How many people do we estimate will have the disease 1.5 days after the measurement?
   d. How many people do we estimate will have the disease by 6.2 days after the measurement?
   e. What is the change in the expected number of people who have the disease from the day of measurement ($t = 0$) to 5 days since the measurement ($t = 5$)?
   f. Suppose that the scientists' prediction was inaccurate, and that instead of increasing by 10% per day, the number of people who contracted the disease increased by 18% per day.
      i. By what number do we multiply the number of people who have contracted the disease when $t = 4$ to find the number of people who have contracted the disease when $t = 5$?
      ii. How does this change the function formula from part (b)?

2. Suppose that city planners expect the town of Lawrence to experience a period of growth from 2010 to 2018 such that the town's population increases by 12% each year. In the year 2010, the population of the town was 18,520 (at the beginning of the year).

Years since 2010 t	Population of Lawrence, P
0	18520
1	
2	
5	
10	

   a. Complete the table giving the population of Lawrence P t years since the year 2010.
   b. Write the formula for a function f that we can use to find the population of Lawrence P for a given value of t.
   c. What do we estimate the population of Lawrence was halfway through 2010? What value of t would we use to find this?
   d. What do we estimate the population of Lawrence will be ¼ of the way through 2016? What value of t would we use to find this?
   e. What is the estimated change in population from the beginning of 2010 to the beginning of 2015?
   f. Suppose that the city planners were inaccurate, and that instead of increasing by 12% per year, the population increased by 6% per year from 2010 to 2018.
      i. By what number do we multiply the population of Lawrence in 2014 to find the population of Lawrence in 2015?
      ii. How does this change the function formula from part (b)?

Algebra II

Module 3: Homework
Exponential & Logarithmic Functions

3. Suppose that scientists' prediction was inaccurate in Exercise #1, and that the following table shows the number of people who had contracted the disease for different values of *t*.

days since measurement *t*	expected number of people who have contracted the disease, *P*
0	150
1	180
2	216
3	259

   a. How can we determine the daily growth factor given this new information?

   b. What is the daily growth factor based on the information in the table? What is the percent change in the number of people who have contracted the disease each day?

4. Suppose that city planners' prediction was inaccurate in Exercise #2, and that the following table shows the population of Lawrence for different values of *t*.

Years since 2010 *t*	Population of Lawrence, *P*
0	18520
1	21483
2	24921
3	28908

   a. How can we determine the yearly growth factor given this new information?

   b. What is the yearly growth factor based on the information in the table? What is the percent change in the population of Lawrence each year?

5. Given an exponential function that grows by some percentage every year, and the output values for year *n* and year *n* + 1, how can we find the annual (yearly) growth factor?

6. If we know that $f(3) = 67.5$ and $f(4) = 101.25$ for some exponential function of the form $f(x) = a \cdot b^x$, answer the following questions.
   a. What is the value of *b*?
   b. What are the values of $f(2)$ and $f(5)$?

7. If we know that $f(5) = 50$ and $f(6) = 40$ for some exponential function of the form $f(x) = a \cdot b^x$, answer the following questions.
   a. What is the value of *b*?
   b. What are the values of $f(4)$ and $f(7)$?

INVESTIGATION 2: PRACTICE WITH EXPONENTIAL FUNCTIONS

8. Determine the requested information for each exponential function.
   a. $f(x) = 2^x$
   1-unit Growth Factor:
   1-unit Percent Change:
   Initial Value:

   b. $f(x) = (0.98)^x$
   1-unit Decay Factor:
   1-unit Percent Change:
   Initial Value:

9. Determine the requested information for each exponential function.
   a. $f(x) = 0.56 \cdot (0.25)^x$
   1-unit Decay Factor:
   1-unit Percent Change:
   Initial Value:

   b. $f(x) = 3 \cdot (1.6)^x$
   1-unit Growth Factor:
   1-unit Percent Change:
   Initial Value:

10. Determine the requested information for each exponential function.

a.

x	0	1	2	3
f(x)	512	384	288	216

1-unit Decay Factor:
1-unit Percent Change:
Initial Value:
Function:

b.

x	1	2	3
g(x)	11.2	15.68	21.952

1-unit Growth Factor:
1-unit Percent Change:
Initial Value:
Function:

11. Determine the requested information for each exponential function.

a.

x	0	1	2	3
f(x)	24	32.64	44.39	60.3709

1-unit Growth Factor:
1-unit Percent Change:
Initial Value:
Function:

b.

x	1	2	3
g(x)	60	24	9.6

1-unit Decay Factor:
1-unit Percent Change:
Initial Value:
Function:

12. Determine the requested information for each exponential function.

a.
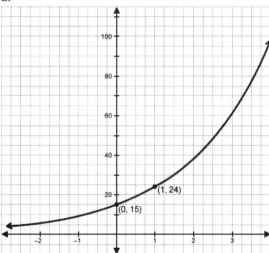

1-unit Growth Factor:
1-unit Percent Change:
Initial value:
Function:

b.
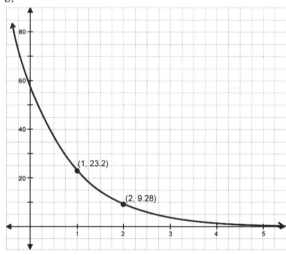

1-unit Decay Factor:
1-unit Percent Change:
Initial Value:
Function:

13. Determine the requested information for each exponential function.

a.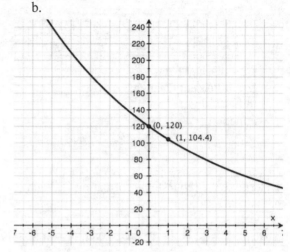

(2, 69.255)
(1, 51.3)

1-unit Growth Factor:
1-unit Percent Change:
Initial value:
Function:

b.

(0, 120)
(1, 104.4)

1-unit Decay Factor:
1-unit Percent Change:
Initial Value:
Function:

14. An investment of $9000 decreased by 2.4% each month.
    1-month Decay Factor: 1-month Percent Change:
    Initial Value: Formula:

15. After starting with a mass of 62 micrograms, the mass of bacteria increases by 12% every day.
    1-hour Growth Factor: 1-hour Percent Change:
    Initial Value: Formula:

16. Suppose that an item in a store is marked down by 15% during a sale and the new price is posted. After the sale is over, the manager tells a worker to mark up the price by 15% so that it will be what it was before the sale. Explain to the manager why this method won't work to get the item back to the original price.

INVESTIGATION 3: PARTIAL GROWTH FACTORS OF EXPONENTIAL FUNCTIONS

17. A certain strain of bacteria that is growing on your kitchen counter doubles in mass every 5 hours. Assume that you start with a mass of 7 microgram (µg).
    a. What quantities are changing in this situation? What quantities are not changing?
    b. What is the mass of the bacteria after 5 hours? 10 hours? 15 hours?
    c. Write a formula for function f that gives the mass of bacteria present after n 5-hour intervals.
    d. Represent the mass of bacteria present after 30 hours. What is the mass at that time?
    e. Determine the 1-hour growth factor. By what percent does the mass of bacteria increase every *hour*?
    f. Determine the 2-hour growth factor. By what percent does the bacteria mass increase every 2 hours?
    g. Write a formula for function g that gives the mass of bacteria in micrograms present after t hours.
    h. Write a formula for function h that gives the mass of bacteria in micrograms present after d days.

Algebra II

Module 3: Homework
Exponential & Logarithmic Functions

18. A chemical is applied to a certain strain of bacteria to study the chemical's effect. Suppose they began with a mass of 350 micrograms (μg). They find that the chemical kills 20% of the remaining living bacteria every 10 minutes.
    a. What quantities are changing in this situation? What quantities are not changing?
    b. What is the mass of the bacteria after 10 minutes? 20 minutes? 60 minutes?
    c. Write a formula for function *f* that gives the mass of bacteria present after *n* 10-minute intervals.
    d. Represent the mass of bacteria present after 2 hours. What is the mass at that time?
    e. Determine the 1-minute decay factor. By what percent does the mass of bacteria decrease every *hour*?
    f. Determine the 2-minute growth factor. By what percent does the bacteria increase every two hours?
    g. Write a formula for function *g* that gives the mass of bacteria in micrograms present after *t* minutes.
    h. Write a formula for function *k* that gives the mass of bacteria in micrograms present after *h* hours.

19. The mass of bacteria in a Petri dish was initially measured to be 14 micrograms and increased by 12% each hour over a period of 15 hours.
    a. Determine the 1-hour growth factor of the bacteria.
    b. By what total percent did the bacteria increase during the 15-hour time period?
    c. How long it will take for the bacteria mass to double?

20. A park ranger counted 27 deer living in a state park on January 1, 2000. Six months later, she counted 38 deer living in the park.
    a. Determine the 6-month growth factor and 6-month percent change for the situation.
    b. Determine the 1-month growth factor and 1-month percent change for the situation.
    c. Write a formula for function *f* that relates the number deer in the park since January 1st 2000 in terms of the number of months that have elapsed since January 1, 2000. (Assuming the number of deer continues to increase by the same percent each month.)
    d. Write a formula for function *g* that relates the number deer in the park since January 1st 2000 in terms of the number of *6-month intervals* that have elapsed since January 1, 2000. (Assuming the number of deer continues to increase by the same percent each month.)
    e. Use one of your functions to determine when the number of deer will reach approximately 125.

21. The population of Egypt was 73,312,600 in 2002 and was 78,887,000 in 2006. Assume the population of Egypt grew exponentially over this period.
    a. Determine the 4-year growth factor and percent change.
    b. Determine the 1-year growth factor and percent change.
    c. Write the formula for function *f* that gives the population of Egypt in terms of the number of years *t* that have elapsed since 2002.
    d. Re-write your function from part (c) so that it gives the population of Egypt in terms of the number of *decades d* that have elapsed since 2002.
    e. Assuming Egypt's population continued to grow according to this model, how long will it take for the population of Egypt to double?

22. An animal reserve in Arizona had 93 wild coyotes. Due to drought, only 61 coyotes remained after 3 months. Assume that the number of coyotes decreases (or decays) exponentially.
    a. Find the 3-month decay factor and percent change.
    b. Find the 1-month decay factor and percent change.
    c. Write a formula for function *f* that gives the number of coyotes in terms of the number of months *m* that have elapsed.

(continues….)

Algebra II
Module 3: Homework
Exponential & Logarithmic Functions

    d. Re-write your function formula from part (c) so that it gives the number of coyotes in terms of the number of *years y* that have elapsed.

    e. How long will it take for the number of coyotes to be one-half of the original number?

23. Sales of music CDs decreased by 6% each year over a period of 8 years.
    a. Determine the decay factor for 1 year.
    b. By what total percent did sales of the CDs decrease during the 8-year time period?
    c. How long will it take for the CD sales to be half of their original value?

24. Suppose the population of a town increased or decreased by the following percentages. For each situation, find the *annual percent change* of the population.
    a. Increases by 60% every 12 years
    b. Decreases by 35% every 7 years
    c. Doubles in size every 6 years
    d. Increases by 4.2% every 2 months
    e. Decreases by 7% every week

25. The number of asthma sufferers in the world was about 84 million in 1990 and 130 million in 2001. Let *N* represent the number of asthma sufferers (in millions) worldwide *t* years after 1990.
    a. Define a function that expresses *N* as a linear function of *t*. Describe the meaning of the slope and vertical intercept in the context of the problem.
    b. Define a function that expresses *N* as an exponential function of *t*. Describe the meaning of the growth factor and the vertical intercept in the context of the problem.
    c. The world's population grew by an annual percent change of 3.7% over those 11 years. Did the percent of the world's population who suffer from asthma increase or decrease?
    d. What is the long-term implication of choosing a linear vs. exponential model to make future predictions about the number of asthma sufferers in the world?

26. In the second half of the 20th century, the city of Phoenix, Arizona exploded in size. Between 1960 and 2000, the population of Phoenix increased about 2.76% each year. In the year 2000, the population of Phoenix was 1.32 million people.
    a. Write the formula for an exponential function *f* that gives the population *P* in Phoenix (in millions of people) where the input values *t* represent the number of years after the year 2000.
    b. Sketch a graph of this function.
    c. What input to your function will give the population of Phoenix in 1972? 1983? 1994? According to your function, approximately how many people lived in Phoenix in 1972? 1983? 1994?
    e. The change in the population of Phoenix is increasing for equal changes in time. Illustrate this on the graph of *f* for at least 3 different equal intervals of time.
    f. What is the domain and range of this function in the context of the problem?
    g. Now, define an exponential function, *g*, where *P* = *g*(*n*) gives the population of Phoenix in millions of people *n* years after 1960. Sketch a graph of this function.
    h. How does the function you created in part (g) compare to the original function created in part (a)? How are the graphs of the two functions similar and different? Explain your reasoning.
    i. What is the domain and range of the function *g*, found in part (g)?

INVESTIGATION 4: DEVELOPING AND APPLYING THE PROPERTIES OF RATIONAL EXPONENTS

27. Simplify each expression using the rules of exponents.

    a. $(3^4)^5$ b. $x^5 \cdot x^8$ c. $(2x)^5$ d. $\dfrac{12x^7}{3x^4}$ e. $(3x^3y)^2(4x^2y^3)$

Algebra II

Module 3: Homework
Exponential & Logarithmic Functions

28. Simplify each expression using the rules of exponents.
    a. $(5^4)^3$
    b. $x^2 \cdot x^6$
    c. $(3x^2)^3$
    d. $\dfrac{18x^9}{2x^2}$
    e. $(4x^2y)^3(2x^2y^4)$

29. Simplify each expression using the rules of exponents.
    a. $3^{1/2} \cdot 3^{1/2}$
    b. $5^{1/4} \cdot 5^{1/4} \cdot 5^{1/4} \cdot 5^{1/4}$
    c. $12^{1/3} \cdot 12^{1/3} \cdot 12^{1/3}$
    d. $(8^{1/7})^7$

30. In Exercise #29 we saw that $3^{1/2} \cdot 3^{1/2} = 3$ according to the rules of exponents. This statement is generally true: that is, $x^{1/2} \cdot x^{1/2} = x$. Write a convincing argument for why it must be true that $3^{1/2} = \sqrt[2]{3}$ (and more generally $x^{1/2} = \sqrt[2]{x}$).

31. Explain what $8^{1/6}$ represents. (*You don't need to calculate its value.*)

32. Rewrite each of the following expressions using radicals.
    a. $4^{1/3}$
    b. $15^{1/7}$
    c. $y^{1/12}$
    d. $x^{1/n}$

33. Rewrite each of the following expressions using radicals.
    a. $8^{1/4}$
    b. $43^{1/9}$
    c. $b^{1/3}$
    d. $p^{1/k}$

34. Rewrite each of the following expressions using rational exponents.
    a. $\sqrt[3]{9}$
    b. $\sqrt[12]{95}$
    c. $\sqrt[5]{x}$

35. Rewrite each of the following expressions using rational exponents.
    a. $\sqrt[15]{7}$
    b. $\sqrt[4]{24}$
    c. $\sqrt[b]{6}$

36. Rewrite each of the following expressions using the rules of exponents.
    a. $(4^{1/5})^6$
    b. $(3^7)^{1/9}$
    c. $(4^{1/3})^8$
    d. $(12^3)^{1/7}$

37. Rewrite each of the following expressions using the rules of exponents.
    a. $(3^{1/2})^7$
    b. $(6^2)^{1/9}$
    c. $(12^{1/3})^7$
    d. $(15^5)^{1/12}$

38. Simplify the given expression as a whole or rational number if possible. If it's not possible, rewrite the expression without using exponents (if possible) and simplify the radical as much as you can.
    a. $64^{2/3}$
    b. $16^{5/2}$
    c. $6^{5/4}$

39. Simplify the given expression as a whole or rational number if possible. If it's not possible, rewrite the expression without using exponents (if possible) and simplify the radical as much as you can.
    a. $27^{5/3}$
    b. $(y^{8/5})^{5/8}$
    c. $32^{4/5}$

40. Rewrite each radical expression using rational exponents.
    a. $\sqrt[9]{6}$
    b. $\sqrt[9]{15^5}$
    c. $(\sqrt[6]{y})^7$

41. Rewrite each radical expression using rational exponents.
    a. $\sqrt[5]{xy^2}$
    b. $10^2 \cdot \sqrt[3]{10^4}$
    c. $(\sqrt[3]{17})^7$

Algebra II
Module 3: Homework
Exponential & Logarithmic Functions

42. Use the rules of exponents to simplify each of the following expressions as a whole or rational number if possible, or in simplified radical form.

   a. $2^{5/3} \cdot 2^{7/3}$

   b. $64^{3/4} \cdot 64^{1/4}$

   c. $\dfrac{9^{3/4}}{9^{2/5}}$

   d. $3^{1/3} \cdot 9^{2/3}$

43. Use the rules of exponents to simplify each of the following expressions as a whole or rational number if possible, or in simplified radical form.

   a. $4^{1/5} \cdot 4^{9/5}$

   b. $16^{1/4} \cdot 16^{7/4}$

   c. $\dfrac{15^{1/3}}{15^{1/8}}$

   d. $8^{2/3} \cdot 4^{5/3}$

44. The population of Canada was 27,512,000 in 1990 and 30,689,000 in 2000. Assume that Canada's population increased exponentially over this time.
   a. Determine the initial value, the 10-year growth factor, and the 10-year percent change of Canada's population.
   b. By what factor does the population of Canada grow every year? Represent your answer using rational exponents and find the approximate decimal value.
   c. Define a function *f* that defines the population of Canada in terms of the number of *years* since 1990.
   d. Assuming Canada's growth rate remains consistent, predict the approximate population of Canada in 2030.

45. One year after an investment was made, the amount of money in the account was $1844.50. Two years after the investment was made, the amount of money in the account was $2001.28.
   a. Determine the 1-year growth factor and the initial value of the investment.
   b. If we want to find the value of the investment 82 months after it was made, what value should we use for *t*?
   c. By what factor does the investment grow every 82 months? Represent your answer using rational exponents and find the approximate decimal value.
   d. By what factor does the investment grow every month? Represent your answer using rational exponents and find the approximate decimal value.

INVESTIGATION 5: N-UNIT GROWTH AND DECAY FACTORS ($N \neq 1$)

Determine the specified growth factors. In addition, determine the initial value and define the exponential function that models the data.

46.

x	1	3	5	7
$f(x)$	512	384	288	216

Initial Value:
1-unit Decay Factor:
1-unit Percent Change:
3-unit Decay Factor:
1/3-unit Decay Factor:
Function Formula:

47.

x	2	5	8	11
$g(x)$	8	11.2	15.68	21.952

Initial Value:
1-unit Growth Factor:
1-unit Percent Change:
2-unit Growth Factor:
1/4-unit Growth Factor:
Function Formula:

48. Determine the specified growth factors. In addition, determine the initial value and define the exponential function that models the data.

   a. 3-unit Decay Factor:

   b. 5-unit Decay Factor:

   c. 1-unit Decay Factor:

   d. 0.6-unit Decay Factor:

   e. Initial Value:

   f. Function Formula:

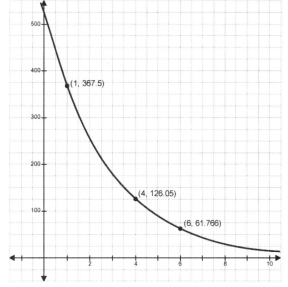

49. Determine the specified growth factors. In addition, determine the initial value and define the exponential function that models the data.

   a. 2-unit Growth Factor:

   b. ½-unit Growth Factor:

   c. 1-unit Growth Factor:

   d. 2.5-unit Growth Factor:

   e. 5-unit Growth Factor:

   f. Initial Value:

   g. Function Formula:

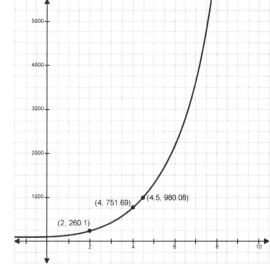

50. Determine the specified growth factors. Write your answers in exponential and decimal form (round to the nearest thousandth).

   The mass of bacteria in an experiment at time t days after its start is given by $f(t) = 95(4)^{t/5}$.

   a. 5-day growth factor: b. 1-day growth factor:

   c. Circle the statement(s) below that describes the behavior of the function above and give reasoning for why the statement(s) are true:

   i. An initial mass of 95 μg of bacteria quadruples every 1/5-day.

   ii. An initial mass of 95 μg of bacteria quadruples every 5-days.

   iii. An initial mass of 95 μg of bacteria increases by 50% every 1-day.

Algebra II

Module 3: Homework
Exponential & Logarithmic Functions

51. Determine the specified decay factors. Write your answers in exponential and decimal form (round to the nearest thousandth).
    The number of buffalo in a wildlife preserve at time t months after its initial measure is given by $f(t) = 49(0.97)^{2t}$
    a. 1-month decay factor:
    b. ½-month decay factor:
    c. Circle the statement(s) below that describe(s) the behavior of the function above:
        i. An initial number of 49 buffalo decreases by 3% every ½-day.
        ii. An initial number of 49 buffalo decreases by 3% every 2-days.
        iii. An initial number of 49 buffalo decreases by 5.91% every 1-month.

52. Determine the specified growth or decay factors. Write your answer in exponential and decimal form (round to the nearest thousandth).
    a. The population of Jackson has 57,421 people and is growing by 4.78% every 4 years.
        i) 8-year growth factor:
        ii) What is the percent change (increase) over 8-year period?
        iii) 1-year growth factor:
        iv) What is the percent change (increase) over 1-year period?
    b. The amount of drug B in your body decreases by 18% every 3 hours.
        i) 24-hour decay factor:
        ii) What is the percent change (decrease) over a 24-hour period?
        iii) 1-hour decay factor:
        iv) What is the percent change (decrease) over a 1-hour period?

53. For each function given below, find the initial value, specified factors, and percent changes.
    a. $f(x) = 2^{x/3}$
        i) Initial Value:
        ii) 1-unit Growth Factor:
        iii) 1-unit Percent Change:
        iv) 4-unit Growth Factor:
        v) 1/5-unit Growth Factor:
    b. $f(x) = 5 \cdot (0.98)^{x/4}$
        i) Initial Value:
        ii) 1-unit Decay Factor:
        iii) 1-unit Percent Change:
        iv) 3-unit Decay Factor:
        v) 1/2-unit Decay Factor:

54. For each function given below, find the initial value, specified factors, and percent changes.
    a. $f(x) = (0.25)^{2x}$
        i) Initial Value:
        ii) 1-unit Decay Factor:
        iii) 1-unit Percent Change:
        iv) 5-unit Decay Factor:
        v) 1/3-unit Decay Factor:
    b. $f(x) = 3 \cdot (1.6)^{5x}$
        i) Initial Value:
        ii) 1-unit Growth Factor:
        iii) 1-unit Percent Change:
        iv) 3-unit Growth Factor:
        v) 1/4-unit Growth Factor:

Algebra II

Module 3: Homework
Exponential & Logarithmic Functions

55. a. Determine whether the following tables represent a linear or exponential function.
    b. Fill in the blanks of the tables.
    c. Define a function for each table.

Table i

Input	Output
−2	−10
−1.5	
−1	−7.5
−0.5	
0	−5
0.5	
1	−2.5
1.5	
2	0

Table ii

Input	Output
−4	0.1111
−3	
−2	0.3333
−1	
0	1
1	
2	3
3	
4	90

Table iii

Input	Output
0	132
0.5	
1	29.04
1.5	
2	6.3888
2.5	
3	1.4055
3.5	
4	0.309

56. Suppose that the average cost for a gallon of gas is $3.59 per gallon and is increasing by about 4% per year while the average cost for a gallon of milk is $3.29 per gallon and is increasing by about 5% per year. If these trends continue into the future, how long will it take for the price of a gallon of gas and a gallon of milk to be equal?

57. The median price of a home in one area of a state is $129,400 and is decreasing by 5.6% per year. In another area of the state the median home price is $144,100 and is decreasing by 7.3% per year. If these trends continue into the future, how long will it take for the median home prices in these areas to be equal?

58. Find the solution to the system involving $f(x) = 99(0.71)^x$ and $g(x) = 54(1.06)^x$.

59. Find the solution to the system involving $f(x) = 1.9(0.95)^x$ and $g(x) = 1.5(0.96)^x$.

60. Find the solution to the system involving $f(x) = 54(0.65)^x$ and $g(x) = 62(0.68)^x$.

61. Find the solution to the system involving $f(x) = 132(1.1)^x$ and $g(x) = 160(1.18)^x$.

Algebra II

Module 3: Homework
Exponential & Logarithmic Functions

INVESTIGATION 6: INTRODUCING LOGARITHMS

62. Let $f(x) = 3^x$.
    a. Complete the table of values and graph f.

x	f(x)
–3	
–2	
0	
2	
2.5	
3	

x	g(x)

   b. Evaluate $f(4)$ and explain what the answer represents in terms of exponents and bases.
   c. Solve for x given that $f(x) = \frac{1}{9}$ and explain what the answer represents in terms of exponents and bases.
   d. Suppose we want to construct a function g that gives the same information as f but that switches the input and output values. Create a table of values and graph for g.
   e. Function f inputs an exponent and outputs the result of raising 3 to that exponent. Explain what function g does.
   f. Evaluate $g(27)$ and explain what the result represents in terms of exponents and bases.
   g. How can we use the table and graph of f to evaluate $g(27)$?

In Exercises #63-71, find the value of the unknown that makes each statement true. *If the answer isn't an integer, then state the two integers that the answer falls between (such as "between 2 and 3").*

63. $\log_8\left(\frac{1}{64}\right) = y$ 64. $\log_3(27) = x$ 65. $\log_4(60) = b$

66. $\log_6(36) = m$ 67. $\log_2\left(\frac{1}{32}\right) = a$ 68. $\log_3(143) = c$

69. $\log_3(20) = k$ 70. $\log_5(1) = r$ 71. $\log_2(30) = y$

Graph each of the following functions.
72. $f(x) = \log_4(x)$ 73. $g(x) = \log_5(x)$

74. a. Sketch the following two functions on the same set of axes: $f(x) = \log(x)$ & $g(x) = \log_5(x)$.
    b. For what values of x is $\log(x) > \log_5(x)$? Explain your reasoning.

INVESTIGATION 7: PROPERTIES OF LOGARITHMS

75. Convert each equation to exponential form and state the value of the unknown that makes the equation true. Check your answer using your calculator.
    a. $\log(1,000) = x$ b. $\log\left(\frac{1}{100}\right) = x$ c. $\log(12000) = y$

Algebra II

Module 3: Homework
Exponential & Logarithmic Functions

Properties of Logarithms

For any real numbers b, m and n with $b > 0$, $b \neq 1$ and $m, n > 0$,

Log Property #1: $\log_b(m \cdot n) = \log_b(m) + \log_b(n)$

Log Property #2: $\log_b\left(\dfrac{m}{n}\right) = \log_b(m) - \log_b(n)$

Log Property #3: $\log_b(m^c) = c \cdot \log_b(m)$

For Exercises #76-81: Use the three properties of logarithms to rewrite each of the following as a single logarithm. Then evaluate (or estimate) the value of the resulting expression.

76. $\log_4\left(\tfrac{1}{32}\right) + \log_4\left(\tfrac{1}{8}\right)$

77. $2\log_6(12) + \log_6(4)$

78. $\log_5(6.25) + \log_5(100) + \log_5(25)$

79. $\ln(6) + \ln(3) - \ln(2)$

80. $\log_5(6) - \log_5(100)$

81. $\tfrac{1}{2}\left(\log_5(8) + \log_5(3)\right)$

For Exercises #76-81: Rewrite each of the following as sums and differences of a logarithm of some number.

82. $\log_7\left(\dfrac{4}{y}\right)$

83. $\log_2\left(x^4 \cdot 12\right)$

84. $\log_5\left(\dfrac{10x^3}{y^5}\right)$

85. $\log_6\left(\dfrac{5x}{z}\right)$

86. $\log_3\left(5^x \cdot 22\right)$

87. $\log_9\left(\dfrac{ab^2}{c^3}\right)$

For Exercises #88-91: Rewrite each of the following in logarithmic form.

88. $y = 11^x$

89. $y = 5.8^x$

90. $y = 1.7(3.2)^t$

91. $y = 200(1.0027)^{12t}$

For Exercises #92-94: Solve each of the following for x.

92. $\log(0.01x) = 0$

93. $\log_5(25x^2) = 6$

94. $\ln\left(\tfrac{x}{10}\right) = 4$

95. Solve the following equation for x: $\log_2(2+x) + \log_2(7) = 3$.

INVESTIGATION 8: SOLVING EXPONENTIAL EQUATIONS USING LOGARITHMS

For Exercises #96-101, solve the given equation.

96. $4^x = 26$

97. $6^y = 209$

98. $0.97 = 12^x$

99. $4(5)^m = 60$

100. $22 = 8(4)^x$

101. $114.6 = 11.6(1.35)^t$

Algebra II

Module 3: Homework
Exponential & Logarithmic Functions

102. An initial amount of 120 mg of caffeine is metabolized in the body and decreases by 18.94% each hour.
   a. Write a formula for an exponential function that gives the amount of caffeine remaining in the body after t hours.
   b. How many hours will it take for the amount of caffeine to reach half of the initial amount?

103. The amount of medicine in a patient's bloodstream for reducing high blood pressure decreases by 27% each hour. This medicine is effective until the amount in the bloodstream drops below 1.2 mg. A doctor prescribes a dose of 85 mg.
   a. Write a formula for function A that models the amount of medicine remaining in the bloodstream after t hours.
   b. About how long until 45 mg of medicine remains in the patient's bloodstream? 1.2 mg?

104. The town of Gilbertville increased from a population of 3,562 people in 1970 to a population of 9,765 in 2000.
   a. Write a formula for an exponential function that models the town's population as a function of the number of years since 1970.
   b. What is the annual percent change?
   c. Use your function to predict the town's population in 2019.
   d. According to your function, when will the town's population reach 40,000 people?

Exercises #105-110 are repeated from Exercises #56-61. Solve each system algebraically. Compare your answers to the solutions you found previously (if applicable).

105. Suppose that the average cost for a gallon of gas is $3.59 per gallon and is increasing by about 4% per year while the average cost for a gallon of milk is $3.29 per gallon and is increasing by about 5% per year. If these trends continue into the future, how long will it take for the price of a gallon of gas and a gallon of milk to be equal?

106. The median price of a home in one area of a state is $129,400 and is decreasing by 5.6% per year. In another area of the state the median home price is $144,100 and is decreasing by 7.3% per year. If these trends continue into the future, how long will it take for the median home prices in these areas to be equal?

107. Find the solution to the system involving $f(x) = 99(0.71)^x$ and $g(x) = 54(1.06)^x$.

108. Find the solution to the system involving $f(x) = 1.9(0.95)^x$ and $g(x) = 1.5(0.96)^x$.

109. Find the solution to the system involving $f(x) = 54(0.65)^x$ and $g(x) = 62(0.68)^x$.

110. Find the solution to the system involving $f(x) = 132(1.1)^x$ and $g(x) = 160(1.18)^x$.

111. Solve the following systems involving logarithmic functions.
   a. $f(x) = \log_3(x+4)$ and $g(x) = \log_3(2x-1)$ b. $f(x) = \log(x+32)$ and $g(x) = \log(3x)$

112. Solve the following systems involving logarithmic functions.
   a. $f(x) = \log_5(4x+2)$ and $g(x) = \log_5(3x+14)$ b. $f(x) = \log\left(\frac{x}{4}+1\right)$ and $g(x) = \log(2x-6)$

Algebra II

Module 3: Homework
Exponential & Logarithmic Functions

INVESTIGATION 9: FUNCTION COMPOSITION: CHAINING TOGETHER TWO FUNCTION PROCESSES

113. John makes extra money staining wooden decks in the summer. He knows that every ounce of stain will cover 6 ft² of decking.

Number of square feet to stain	Number of ounces of stain needed
200	
385	
450	
522	
n	

  a. Complete the table of values showing the number of ounces of stain John needs based on the number of square feet to be stained.
  b. Define a function f that determines the number of ounces of stain needed based on the number of square feet stained n.
  c. The cost of his favored brand of stain is $0.29 per ounce.
   i. How much will it cost John to purchase 50 ounces of stain? 170 ounces of stain?
   ii. Define a function g that determines the cost of stain (in dollars) based on the number of ounces of stain needed s.
  d. Explain how you can determine the cost of the stain John needs to buy to cover 325 ft² of deck.

Number of square feet to stain	Total cost of stain (in dollars)
200	
385	
450	
522	

  e. Complete the table of values showing the total cost of stain (in dollars) based on the number of square feet needed to be stained.
  f. Write a formula to determine the cost of staining n square feet of deck.

114. Jackie observes that it takes her 16 minutes to travel one mile when walking at a constant rate.

Number of miles walked	Number of minutes walked
1	
2.3	
3	
4.4	
n	

  a. Complete the table of values showing the number of minutes walked based on the number of miles walked.
  b. Define a function f that determines the number of minutes walked based on the number of miles walked n.
  c. Using a heart rate monitor, Jackie determines that she burns 45 calories every 10 minutes of walking.
   i. How many calories will she burn in 25 minutes? 48 minutes?
   ii. Write a formula for function g that determines the number of calories burned based on the number of minutes walked w.
  d. Explain how you can determine the number of calories Jackie burns when walking 3.8 miles.

Number of miles walked	Number of calories burned
1	
2.3	
3	
4.4	

  e. Complete the table of values showing the number of calories burned based on the number of miles she has walked.
  f. Write a formula to determine the number of calories burned when walking n miles.

Algebra II

Module 3: Homework
Exponential & Logarithmic Functions

115. A ball is thrown into a lake, creating a circular ripple whose radius travels outward at a speed of 7 cm per second. Express the area of the circle as a function of the number of seconds that have passed since the ball hit the lake. (Recall that $A = \pi r^2$.)
    a. Draw a diagram of the situation and label the relevant quantities in the situation.
    b. Answer the following about the co-variation of the quantities.
        i. As the time since the pebble hit the water increases, how does the radius of the circle change?
        ii. As the radius of the circle increases, how does the area enclosed by the circular ripple change?
        iii. As the time elapsed since the pebble hit the water increases, how does the area enclosed by the ripple change?
    c. Write a formula for function f that outputs the radius length r (measured in centimeters) when we input the time elapsed since the pebble hit the water t (measured in seconds).
    d. Write a formula for function g that outputs the area A (measured in square centimeters) when we input the radius length of the circular ripple r (in centimeters).
    (continues…)
    e. What rule defines the function h that inputs the number of seconds since the pebble hit the water and outputs the area of the ripple in square centimeters? That is, $h(t) = g(f(t)) = $ _____?
    f. Evaluate $h(5)$ and explain its meaning in the context of the problem.

116. A spherical bubble inflates so that its radius increases by a constant rate of 5 centimeters per second. (Hint: The formula for the volume of a sphere as determined by the radius of the sphere, $V = \frac{4}{3}\pi r^3$.)
    a. Draw a diagram of the situation and label the quantities that are involved.
    b. Write a formula for function f that outputs the radius length r (measured in centimeters) when we input the time elapsed since the bubble started inflating t (measured in seconds).
    c. Write a formula for function g that outputs the volume V (measured in cubic centimeters) when we input the radius length of the bubble r (in centimeters).
    d. What rule defines the function h that inputs the number of seconds since the bubble started inflating and outputs the volume of the bubble in cubic centimeters? That is, $h(t) = g(f(t)) = $
    e. Evaluate $h(3)$ and explain its meaning in the context of the problem.

117. Suppose a 15-inch candle has a wick burning from one end which burns at a constant rate of 2.5 inches per hour.
    a. Write a formula for function f that determines the amount of candle that has burned in terms of the amount of time t that the candle has been burning.
    b. Evaluate $f(1.5)$ and describe what it represents in the context of this situation.
    c. Write a formula for function g that represents the remaining length of the 15-inch candle in terms of the total length l that has burned from the candle.
    d. Determine the value of $g(f(2))$ and explain what this represents in the context of this question.
    e. Write a formula for function h that represents the remaining length of the 15" candle after it has been burning for t hours.
    f. Evaluate: i. $h(2)$ ii. $h(2.5)$ iii. $h(3)$

Algebra II

Module 3: Homework
Exponential & Logarithmic Functions

118. Many homebuilders in the Southwest use blown-in cellulose insulation to insulate the attics of new homes. Proper insulation is important because it improves air conditioning and heating systems' efficiency and reduces the costs to heat and cool homes. One way to think about insulating an attic is to imagine the depth of insulation per bag applied to the floor of the attic. When hiring a contractor to add insulation to your attic, he provides you with the information in the following tables (showing two functions, f and g).

Function f

Number of bags of insulation applied, n	Depth of insulation (inches), $f(n)$
11	2
20	3.6
28	5
40	7.2
50	9
61	11
70	12.6
78	14

Function g

Depth of insulation (inches), d	Estimated annual heating/cooling costs (dollars), $g(d)$
15	$940
13	$1,000
11	$1,100
9	$1,250
7	$1,500
5	$1,900
4	$2,750
3	$3,600

a. Does the expression $f(g(9))$ have a real-world meaning in this context? If so, find the value and explain its meaning. If not, explain your reasoning.

b. Does the expression $g(f(28))$ have a real-world meaning in this context? If so, find the value and explain its meaning. If not, explain your reasoning.

c. Solve the equation $g(f(n))=1100$ for n, then explain what your solution represents in this context.

119. Jessie does a lot of traveling for business, so he has visited a lot of cities around the world. While traveling he likes to jog in the mornings to keep fit. Jessie has noticed that the elevation of the city he is visiting impacts how long he is able to spend running since there is less oxygen available at higher elevations. The graphs below provide information about Jessie's exercise routines recorded over many business trips.

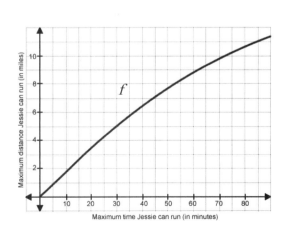
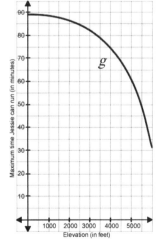

(continues...)

Algebra II

Module 3: Homework
Exponential & Logarithmic Functions

a. Does the expression $f(g(4000))$ have a real-world meaning in this context? If so, determine its value then explain what the value represents. If not, explain your reasoning.
b. Does the expression $g(f(50))$ have a real-world meaning in this context? If so, determine its value then explain what the value represents. If not, explain your reasoning.
c. Consider the equation $f(g(x)) = 5$. Explain how to use the tables to solve this equation for x, then explain what the solution represents.

120. Use the tables below to answer the following. Explain how you evaluated each expression.

x	f(x)
-3	2
-2	4
-1	0
0	-2
1	2
2	-3
3	0
4	2

x	g(x)
-3	4
-2	-1
-1	2
0	3
1	2
2	-2
3	-1
4	0

a. Evaluate the following expressions.
   i. $g(3)$ ii. $g(f(0))$ iii. $f(g(-3))$ iv. $g(g(2))$
b. Determine a value for x such that $g(x) = 3$? Explain.
c. Determine a value for x such that $g(x) = f(x)$? Explain.
d. Determine a value for x such that $f(g(x)) = -2$? Explain.

121. Use the following functions to answer (a) through (d): $f(x) = 2x + 5$, $g(x) = 2^x$, and $h(x) = \sqrt{x}$
a. Evaluate $f(3.6)$.
b. Evaluate $f(g(4))$.
c. Evaluate $h(f(11.6))$.
d. Solve the equation $h(f(x)) = 9$.

122. Use the following functions to answer (a) through (d): $f(x) = 3(x-1)$, $g(x) = \frac{x}{4}$, and $h(x) = 5\sqrt{x}$
a. Evaluate $f(2.3)$.
b. Evaluate $f(g(4))$.
c. Evaluate $h(f(11.6))$.
d. Solve the equation $g(f(x)) = 9$.

123. Given $f(x) = -2x - 5$ and $g(x) = \frac{3}{4}x + 6$, evaluate:
a. $g(f(0))$ b. $f(g(-8))$ c. $g(f(x))$ d. $f(g(x))$

Algebra II

Module 3: Homework
Exponential & Logarithmic Functions

124. Use the following functions to answer the questions below: $f(x) = 2x+3$ and $g(x) = \dfrac{3x-1}{2}$.
   a. Evaluate $(f \circ g)(4)$.
   b. Evaluate $(g \circ f)(4)$.
   c. Evaluate $(f \circ f)(x)$.
   d. Evaluate $(g \circ f)(x)$.

125. Use the following functions to answer the questions below: $f(x) = \dfrac{7}{x-1}$ and $g(x) = \dfrac{2x+5}{4}$.
   a. Evaluate $(f \circ g)(3)$
   b. Evaluate $(g \circ f)(3)$
   c. Evaluate $(f \circ f)(x)$
   d. Evaluate $(g \circ f)(x)$

126. The area A of a circle with radius r can be defined by the function $f(r) = \pi r^2$, where $A = f(r)$. The radius of a circle r that has a circumference C can be defined by the function $g(C) = \dfrac{C}{2\pi}$, where $r = g(C)$.
   a. Evaluate $(f \circ g)(7)$ if possible and describe what your answer represents.
   b. Evaluate $f(g(4))$ if possible and describe what your answer represents.
   c. Evaluate $(g \circ f)(3)$ if possible and describe what your answer represents.
   d. Evaluate $(f \circ g)(x)$ if possible and describe what your answer represents.
   e. What other notation can you use to express $(g \circ f)(x)$?

127. Consider the graphs of f and g given.

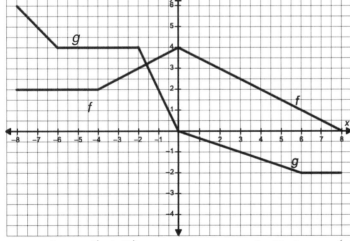

   a. Evaluate $f(g(-8))$.
   b. Evaluate $g(f(-2))$.

Algebra II

Module 3: Homework
Exponential & Logarithmic Functions

INVESTIGATION 10: INTRODUCTION TO FUNCTION INVERSE

128. $g(x) = x - 0.5$ describes a process where an output is determined by subtracting 0.5 from an input value.
   a. What process will undo the process of *g*, that is the subtracting of 0.5 from *x*?
   b. Define a function *h* that undoes the process of *g*.
   c. Explain why *g* and *h* are inverses.

129. $g(x) = 5x$ describes a process where an output is determined by multiplying an input value by 5.
   a. What process will undo the process of *g*, that is the multiplying of *x* by 5?
   b. Define a function *h* that undoes the process of *g*.
   c. Explain why *g* and *h* are inverses.

130. Let *k* be a function defined by $k(x) = \frac{x}{4} + 1$.
   a. Describe the process represented by function *k*.
   b. Describe the process that will undo the process of *k*.
   c. Define a function *m* that undoes the process of *k*.
   d. Evaluate the following: i. $k(36)$ ii. $m(10)$

131. Let *k* be a function defined by $k(x) = \frac{x+2}{3}$.
   a. Describe the process represented by function *k*.
   b. Describe the process that will undo the process of *k*.
   c. Define a function *m* that undoes the process of *k*.
   d. Evaluate the following: i. $k(10)$ ii. $m(4)$

132. Define a function that **undoes** each of the following processes.
   a. The output is found by multiplying the input by 2.5.
   b. The output is determined by subtracting 2 from the input and then multiplying by 7.
   c. We take the square root of the input to produce the output.
   d. The output is found by dividing 7 by the input.

133. Sarah can drive 53 miles for each gallon of gasoline in her Prius. Let *y* = the number of gallons of gasoline used and *x* = the number of miles Sarah can drive in her Prius. Function *h* inputs a number of miles driven by Sarah and outputs the number of gallons of gas used, so $h(x) = \frac{x}{53}$ where $y = h(x)$.
   a. Determine the value of $h(78)$ and explain what it represents.
   b. Solve the equation $h(x) = 1.47$ for *x* and then describe the meaning of your answer.
   c. Write a formula for function *g* that determines the numbers of miles driven as a function of the number of gallons of gas used.
   d. Evaluate $h(150)$ and $g(2.83)$.
   e. Without performing any calculations, evaluate $h(g(2.83))$ and $g(h(150))$. What do you notice, and why does this happen?

134. Let *V* = the volume of a cube in cubic inches and *x* = the length of the side of a cube in inches. Function *f* inputs a length of the side of a cube and outputs the cube's volume, so $f(x) = x^3$ where $V = f(x)$.
   a. Determine the value of $f(17)$ and explain what it represents.
   b. Solve the equation $f(x) = 4{,}913$ for *x* and then describe the meaning of your answer.

(*continues...*)

Algebra II

Module 3: Homework
Exponential & Logarithmic Functions

c. Write a formula for function g that determines the length of the side of a cube (in inches) given its volume (in inches).

d. Without performing any calculations, evaluate $f(g(4,913))$ and $g(f(17))$. What do you notice, and why does this happen?

135. a. Given $f(x) = \log_3(x)$ where $y = f(x)$, write the formula for f^{-1}.
   b. Explain why $f^{-1}(f(x)) = x$ by discussing the processes involved.

136. Given a function f and its inverse f^{-1}, which of the following is true? Justify your choice.
   a. f^{-1} "undoes" what f does, so $f^{-1}(f(d)) = a$ for all values of d in the domain of f.
   b. $f^{-1} = \dfrac{1}{f(d)}$

137. Write a formula for the inverse of each of the following functions.
   a. $f(t) = 3^t$ b. $g(x) = 5x - 18$ c. $b(x) = \dfrac{-4x+5}{2}$ d. $h(s) = 2x^3$

138. Write a formula for the inverse of each of the following functions.
   a. $g(x) = -4x + 5$ b. $j(x) = \dfrac{6x-5}{4}$ c. $f(t) = 5^t$ d. $h(r) = \pi r^2$

139. Use the given tables to determine the following. Explain your thinking.

x	f(x)
-4	-1
-3	3
-2	0
-1	-9
0	1
1	-6
2	-3
3	-10

x	g(x)
-9	18
-6	12
-3	6
-1	2
0	0
1	-2
3	-6
6	-12

a. $f(-2)$
b. $g(f(0))$
c. Solve $g(x) = 2$ for x.
d. Evaluate $g^{-1}(2)$ and explain how your reasoning compares to your reasoning in part (c).
e. $f^{-1}(3)$
f. $f(g^{-1}(2))$
g. $g(f^{-1}(-6))$
h. $g(f(-2))$
i. Solve the equation $g(f(x)) = 2$ for x.

Algebra II

Module 3: Homework
Exponential & Logarithmic Functions

140. Use the given tables to determine the following. Explain your thinking.

x	f(x)
−3	4
−2	−1
−1	−4
0	5
1	4
2	−2
3	−10

x	g(x)
−4	12
−2	−1
−1	2
0	0
2	−2
4	−6
5	−10

a. $f(-2)$

b. $g(f(0))$

c. Solve $g(x) = -2$ for x.

d. Evaluate $g^{-1}(-2)$ and explain how your reasoning compares to your reasoning in part (c).

e. $f^{-1}(4)$

f. $f(g^{-1}(2))$

g. $g(f^{-1}(-1))$

h. $g(f(-2))$

i. Solve the equation $g(f(x)) = -4$ for x.

INVESTIGATION 11: COMPOUNDING PERIODS & COMPOUND INTEREST FORMULA

INVESTIGATION 12: CONTINUOUSLY COMPOUNDED INTEREST

Investigation 11: *Investment Activity: Compounding Periods & Compounding Interest Formula* and Investigation 12: *Motivating the Number e and Continuously Compounded Interest* are available online along with associated homework at www.rationalreasoning.net.

Algebra II

Module 4: Investigation 1
Introduction to Quadratic Functions

1. When the side length of a square increases so does the area of the square. On the grid below we have drawn a 1 unit by 1 unit square and a 2 unit by 2 unit square. Draw squares with side lengths of 3 units, 4 units, 5 units, 6 units, and 7 units. Try to keep your diagram organized.

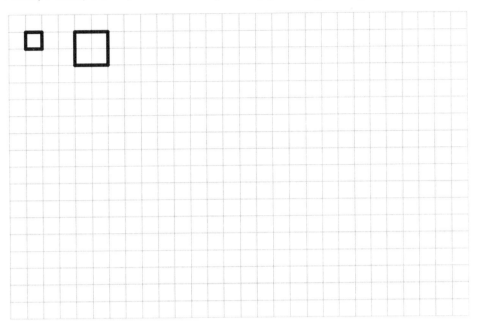

2. Imagine the larger squares being formed by the smaller squares "growing" and do the following.
   - In the 2x2 square, shade the original 1x1 square (top left corner).
   - In the 3x3 square, shade the 2x2 square.
   - Repeat this process for the remaining squares.

3. What do the unshaded portions of each square represent?

4. Is *the area of the square* a linear function of *the length of a side of the square*? Explain.

Algebra II

Module 4: Investigation 1
Introduction to Quadratic Functions

5. Complete the first two empty columns in the following table.

length of the side of the square (x units)	area of the square (A square units)	change in the area of the square (ΔA square units)	
0			
1			
2			
3			
4			
5			
6			
7			

6. Write the formula that relates the area of the square (in square units) and the length of the side of the square (in units) and draw the graph of this relationship.

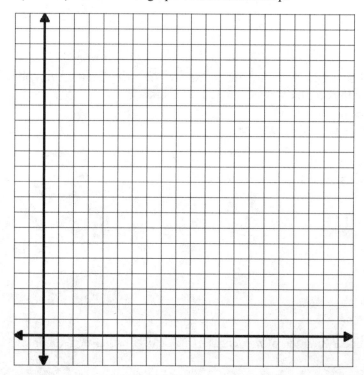

7. On your graph draw "triangles" to show changes of 1 unit in the length of the side of a square and the corresponding changes in the area of the square. Discuss how the graph supports your answer to Exercise #4.

Algebra II
Module 4: Investigation 1
Introduction to Quadratic Functions

8. a. In the table in Exercise #5 there is an unlabeled column. Label this column "change in ΔA". Explain what this column tells us about this situation and fill in the entries.

   b. Go back to your diagram in Exercise #1. How can we "see" the values of the column "change in ΔA" in our original diagram?

   c. Go back to your graph in Exercise #6. How can we "see" the values of the column "change in ΔA" in our graph?

9. Does a pattern like the one observed in Exercises #1-8 appear if we change the side lengths of the square by something other than 1 unit at a time? Use the diagram in Exercise #1 and/or the formula you wrote in Exercise #6 to complete the following tables.

length of the side of the square (x units)	area of the square (A square units)	ΔA	change in ΔA
0	0		
0.5			
1	1		
1.5			
2	4		
2.5			
3	9		
3.5			

length of the side of the square (x units)	area of the square (A square units)	ΔA	change in ΔA
0	0		
0.25			
0.5			
0.75			
1	1		
1.25			
1.5			
1.75			

Algebra II

Module 4: Investigation 1
Introduction to Quadratic Functions

For Exercises #10-11, do the following.
  a) Draw a diagram of the rectangle described.
  b) Write a formula that calculates the area of the rectangle described (you will need to define your input quantity).
  c) Complete the table.
  d) Graph the rectangle's area compared to the input quantity you defined in part (b).

10. A rectangle where the length is 4 times as large as the height.
  a.   b.

  c.   d.

0			
1			
2			
3			
4			
5			

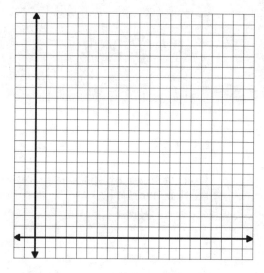

11. A rectangle where the length is 4 units more than the height.
  a.   b.

  c.   d.

0			
1			
2			
3			
4			
5			

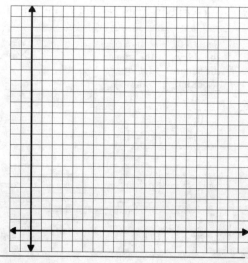

Module 4: Investigation 1
© 2014 Carlson, O'Bryan & Joyner

Algebra II

Module 4: Investigation 1
Introduction to Quadratic Functions

> **Quadratic Function**
>
> A function is **quadratic** if, for equal changes in the input value, the *change* in the output value has a constant rate of change. The function $y = x^2$ is the simplest example of a quadratic function.

We started our exploration of quadratic functions by examining the area of a square because historically the study of area was the first context where humans were forced to think about quadratic relationships. In fact, the term *quadratic* comes from the Latin word *quadratum*, meaning *square*. We also frequently refer to an exponent of 2 as "squaring" the base value because the process can be thought of as converting a length to an area by forming a square.

In Exercises #12-13, use the given formula defining a quadratic function to complete the table.

12. $y = 2(x+4)^2$

x	y	Δy	change in Δy
−11			
−8			
−5			
−2			
1			
4			

13. $y = -3x^2$

x	y	Δy	change in Δy
−4			
−2			
0			
2			
4			
6			

14. Examine each of the following tables and explain why they do *not* represent quadratic functions of y with respect to x.

x	y	Δy	change in Δy
−4	2		
−1	8		
0	18		
3	32		
4	50		
5	72		

x	y	Δy	change in Δy
0	0		
1	1		
2	8		
3	27		
4	64		
5	125		

The U-shaped graph of a quadratic function is called a **parabola**. However, not all U-shaped graphs are parabolas. In Exercises #15-16 use what you've learned in this investigation to determine if the graph can represent a quadratic function.

15.
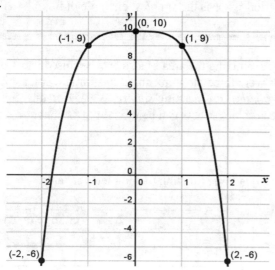

16.
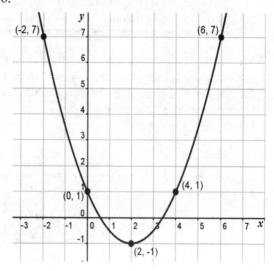

17. A classmate examined the following table that shows some ordered pairs for a quadratic function. He says that if $x=3$, then $y=13$. Do you agree? Defend your position.

x	y
0	6
2	10
4	16
6	24
8	34
10	46

One way that quadratic functions differ from linear functions is that when we examine the entire possible domain, a quadratic function has an interval where the output quantity increases and an interval where the output quantity decreases. This has to do with the pattern of changes that we've explored.

18. The following table shows ordered pairs for a quadratic function. In the table, we see that as x increases, y increases. How do we know that y must eventually decrease as x increases?

x	y	Δy	change in Δy
−2	4		
−1	15		
0	24		
1	31		
2	36		

19. Explain what you've learned about quadratic functions in this investigation.

Algebra II
Module 4: Investigation 2
Exploring Concavity

In Investigation 1 we explored the growth pattern of $f(x) = x^2$ where $A = f(x)$ for values of $x \geq 0$ by examining how the quantities *length of the side of a square* and *area of a square* change together.

length of the side of the square (x units)	area of the square (A square units)	change in the area of the square (ΔA square units)	change in ΔA
0	0		
		1	
1	1		2
		3	
2	4		2
		5	
3	9		2
		7	
4	16		2
		9	
5	25		2
		11	
6	36		2
		13	
7	49		

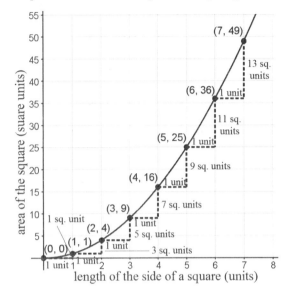

For the interval we explored in Investigation 1, A always increases as x increases. However, if we take the function out of its original context (which allows the input values to be negative) we will see that $f(x) = x^2$ is not always an increasing function.

1. Complete the table for $f(x) = x^2$ where $y = f(x)$ on the interval $-5 \leq x \leq 5$. Then graph function f.

x	y	Δy	change in Δy
−5			
−4			
−3			
−2			
−1			
0	0		
		1	
1	1		2
		3	
2	4		2
		5	
3	9		2
		7	
4	16		2
		9	
5	25		

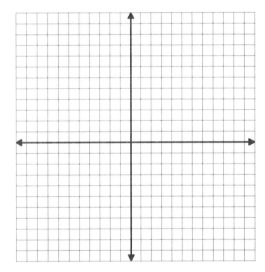

2. As x increases, does y always increase? If not, at what x value does the behavior change?

3. What do you notice about the *change* in Δy?

Algebra II

Module 4: Investigation 2
Exploring Concavity

Positive Concavity ("Concave Up")

A function f [defined by $y = f(x)$] is **concave up** (or has **positive concavity**) on an interval if, for equal increases in x, Δy is increasing. We can also say that the <u>changes</u> in Δy are positive.

Note that this must be true for all possible choices of Δx we can use to divide up the interval. In other words, it must be true whether we choose $\Delta x = 2$, $\Delta x = 0.5$, $\Delta x = 0.00001$ or any other Δx. If you can find a value of Δx that breaks the pattern, then the function doesn't have the same concavity over the entire interval.

---

4. The following is a table of values for $g(x) = -2x^2 + 8x$ where $y = g(x)$.
   a. Complete the table and graph the function by plotting the points.

x	y	Δy	change in Δy
−2	−24		
−1	−10		
0	0		
1	6		
2	8		
3	6		
4	0		

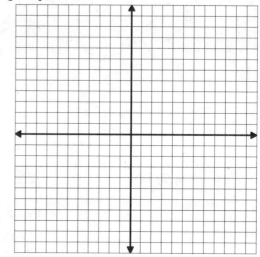

   b. List any observations you have about this function.

Negative Concavity ("Concave Down")

A function f [defined by $y = f(x)$] is **concave down** (or has **negative concavity**) on an interval if, for equal increases in x, Δy is decreasing. We can also say that the <u>changes</u> in Δy are negative.

Note that this must be true for all possible choices of Δx we can use to divide up the interval. In other words, it must be true whether we choose $\Delta x = 2$, $\Delta x = 0.5$, $\Delta x = 0.00001$ or any other Δx. If you can find a value of Δx that breaks the pattern, then the function doesn't have the same concavity over the entire interval.

© 2014 Carlson, O'Bryan & Joyner

Module 4: Investigation 2
Exploring Concavity

For each given quadratic function in Exercises #5-7, do the following.
  i) Find the missing output values, then determine the values of Δy and the *changes* in Δy.
  ii) Identify the *x*-value where the function changes from increasing to decreasing or from decreasing to increasing.
  iii) Does the function have positive concavity or negative concavity?
  iv) Sketch the function's graph.

5. $y = x^2 - 4x + 2$

x	y	Δy	
−1	7		
0			
1			
2			
3	−1		
4	2		
5	7		
6	14		

6. $y = -(x+1)(x+5)$

x	y	Δy	
−7	−12		
−5	0		
−3	4		
−1	0		
1	−12		
3			
5			
7			

7. $y = (x-4)(x+1)$

x	y	Δy	
−1			
−0.5	−2.25		
0	−4		
0.5	−5.25		
1	−6		
1.5	−6.25		
2			
2.5			

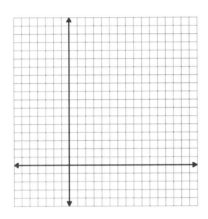
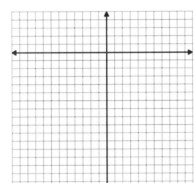
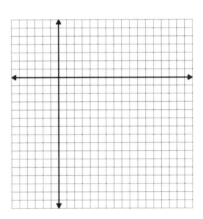

8. The table below shows a function that is *not* quadratic. Show that this function sometimes has positive concavity and sometimes has negative concavity, then use the ordered pairs to sketch a graph of the function.

x	y	Δy	change in Δy
−3	−21		
−2	0		
−1	5		
0	0		
1	−9		
2	−16		
3	−15		
4	0		

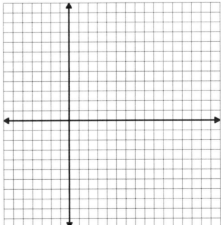

Algebra II   **Module 4: Investigation 3**
                                                  Properties of Quadratic Functions

In Exercises #1-3, complete each table. (*Note: The fourth column represents the change in* Δy.)

1. $y = x^2 - 4x + 2$

x	y	Δy	
-1	7		
0	2	-5	
1	-1	-3	
2	-2	-1	
3	-1	1	
4	2	3	
5	7	5	
6	14	7	

2. $y = -2(x+1)(x+5)$

x	y	Δy	
-5	0		
-4	6	6	
-3	8	2	
-2	6	-2	
-1	0	-6	
0	-10	-10	
1	-24	-14	
2	-42	-18	

3. $y = (x-4)(x+1)$

x	y	Δy	
-1	0		
-0.5	-2.25	-2.25	
0	-4	-1.75	
0.5	-5.25	-1.25	
1	-6	-0.75	
1.5	-6.25	-0.25	
2	-6	0.25	
2.5	-5.25	0.75	

You've probably recognized some important features of quadratic functions based on your explorations thus far. For example, every quadratic function has a *vertex*.

> ### Vertex of a Quadratic Function
>
> The **vertex** of a quadratic function is the point where the function values change from increasing to decreasing or from decreasing to increasing as the input values increase (or as we move from left to right on the graph).
>
> If the vertex contains the largest possible output value for the function, then we say the vertex is a **maximum**. If the vertex contains the smallest possible output value for the function, then we say the vertex is a **minimum**.

4. a. Examine the tables in Exercises #1-3 and determine the vertex for each function.

   b. Graph the functions on a calculator and locate the vertex to check your answers.

   c. Is the vertex a maximum or minimum for each function?

The U-shaped graph of a quadratic function is called a **parabola**. (*Caution! Not all U-shaped graphs are parabolas. Review Investigation 1 to see an example of a U-shaped graph that is not a parabola.*) All parabolas have a *line of symmetry*.

> ### Line of Symmetry
>
> The **line of symmetry** for a quadratic function is a vertical line. Values of the input equidistant from the line of symmetry produce equal output values. *Note: The line of symmetry is sometimes referred to as the axis of symmetry.*

Algebra II

Module 4: Investigation 3
Properties of Quadratic Functions

5. What is the line of symmetry for each of the functions in Exercises #1-3?

6. What is the relationship between the line of symmetry and the vertex of each parabola?

7. A partial table of values is given below for two quadratic functions. The function represented by the table on the left has a vertex at $(-1, 9)$, and the function represented by the table on the right has a vertex at $(4, 2)$.

x	y
−4	−9
−3	
−2	7
−1	
0	
1	1

x	y
−2	38
1	11
	2
	11
10	
13	83

a. What is the line of symmetry for each function?

b. Use the line of symmetry to help you complete the table of values for each quadratic function.

c. Is each function concave up or concave down? Explain how you know.

d. Is the vertex of each function a maximum or a minimum? Explain how you know.

8. If we know the concavity of a quadratic function, what does this tell us about the vertex of the parabola? Explain.

Module 4: Investigation 3
© 2014 Carlson, O'Bryan & Joyner

Algebra II

Module 4: Investigation 4
Modeling Quadratic Relationships

1. A rancher has 400 meters of fencing. He will fence off a rectangular corral next to the wall of a cliff that will form the east side of the enclosure. So, he will use all 400 meters to make just 3 sides of the rectangle (the north, south, and west sides). He wants to make a corral that will hold as many cattle as possible, so he wants to maximize the corral's area.

   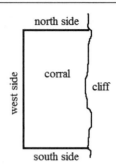

   Use the *Corral Applet* to answer the following questions.
   a. As the length of the west side of the corral increases from 0 to 20 meters, how does the length of the east side of the corral change? How does the length of the south side of the corral change? The north side?

   b. As the length of the west side of the corral increases by any amount, how does the length of the south side of the corral change?

   c. As the length of the west side of the corral increases from 0 to 10 meters, how does the area of the corral change? How does the area change as the length of the west side increases from 10 to 20 meters?

   d. What dimensions of the corral will maximize its area?

2. Write formulas to define the following functions.
   a. The length of the south side of the corral as a function of the length of the west side of the corral. Define any variables that you use.

   b. The area of the corral as a function of the length of the west side of the corral. Define any variables that you use.

Algebra II

Module 4: Investigation 4
Modeling Quadratic Relationships

3. Suppose a baseball outfielder catches a ball and throws it back towards the infield, and suppose the ball leaves his hand 6.5 feet above ground level. The baseball's height above the ground h (in feet) after t seconds since it was released can be modeled by the function $f(t) = -16t^2 + 32t + 6.5$ where $h = f(t)$.

   a. Fill in the given table and use the information to draw a graph of the function h.

seconds since the ball was released (t)	ball's height above the ground in feet (h)	change in the ball's height above the ground in feet (Δh)	change in Δh
0			
0.3			
0.6			
0.9			
1.2			
1.5			
1.8			
2.1			
2.4			

   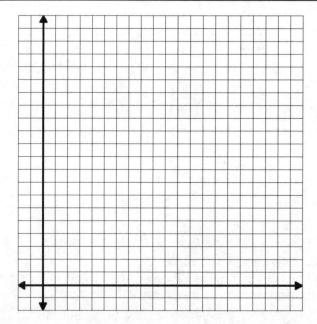

   b. Explain how the height of the baseball and the time elapsed since the baseball left the player's hand change together.

c. The function's axis of symmetry is $t = 1$.
   i) What is the baseball's maximum height above the ground?

   ii) Over what time interval(s) is the ball's height increasing as t increases? Over what time interval(s) is the ball's height decreasing as t increases?

   iii) If another player catches the ball at a height of 6.5 feet above ground level, how long was the ball in the air?

   iv) Provide two values of t for which the height will be the same. Explain your method for answering this question.

d. Using your graph, solve the equation $f(t) = 19$ for t (your answer(s) will be approximate). What does the solution represent in the context of this problem?

4. A manufacturing engineer has been given the job of designing an aluminum container having a square base and rectangular sides to hold screws and nails. It also must be open at the top. The container must be 5 inches tall, but the engineer can make the base whatever size he feels is appropriate.
   a. Sketch a diagram of this container. Let x represent the length of the side of the square base in inches.

   b. Suppose the outside of the box will be painted (including the bottom of the base). What is the surface area of the box that needs to be painted if the side of the square base is 9 inches? 4 inches? x inches?

Algebra II

Module 4: Investigation 4
Modeling Quadratic Relationships

    c. Use a calculator to graph the function that shows the surface area (in square inches) to be painted if the length of the side of the square base is x inches.
        i) What is the vertex, axis of symmetry, and horizontal intercepts for this function?

        ii) Are these features in the function's domain? Explain.

    d. What is the formula for the volume (in cubic inches) of the container as a function of the length of the side of the square base?

    e. What is the volume of the container if the square base has sides that are 6.5 inches long?

    f. What must the dimensions of the base be if the volume of the container is 423.2 cubic inches? Give two methods for finding these dimensions.

Algebra II

Module 4: Investigation 5
Zeros (Roots) of Quadratic Functions

Function f is defined by $f(x) = (2x - 5)(x + 3)$. This is written in **factored form** because the formula is expressed as a product of factors, namely the factors $(2x - 5)$ and $(x + 3)$. We can use the distributive property to rewrite $f(x) = (2x - 5)(x + 3)$ in **standard form** $f(x) = ax^2 + bx + c$.

As an example, we can write $f(x) = (2x - 5)(x + 3)$ as $f(x) = 2x^2 + x - 15$ by *expanding* $(2x - 5)(x + 3)$. We expand a product in this form by applying the distributive property. (*The distributive property says that either $a(b + c) = ab + ac$ or $(a + b)c = ac + bc$.*)

$f(x) = (2x - 5)(x + 3)$

$f(x) = \boxed{(2x - 5)}(x + 3)$

$f(x) = (2x - 5)(x) + (2x - 5)(3)$ • We distribute the number $(2x - 5)$ over the sum $x + 3$

$f(x) = (2x - 5)\boxed{(x)} + (2x - 5)\boxed{(3)}$

$f(x) = (2x)(x) + (-5)(x) + (2x)(3) + (-5)(3)$ • We perform the distributive property two more times

$f(x) = 2x^2 - 5x + 6x - 15$ • We simplify

$f(x) = 2x^2 + x - 15$ • We combine like terms

Factored Form and Standard Form of a Quadratic Function

Factored Form: The function is written as a product of factors, such as $f(x) = (2x - 5)(x + 3)$.

Standard Form: The function is written as a sum or difference of terms, such as $f(x) = 2x^2 + x - 15$.

In Exercises #1-4 you are given quadratic functions written in factored form. Determine the standard form for each function by expanding the product of factors.

1. $f(x) = (x + 1)(x + 7)$

2. $g(x) = (x - 2)(x - 5)$

3. $p(x) = -3(x + 2)(5x + 11)$

4. $f(x) = 6(x - 4)^2$

Algebra II

Module 4: Investigation 5
Zeros (Roots) of Quadratic Functions

Zeros (Roots) of a Quadratic Function

The **zeros** (also called **roots**) of a function f are the input values x such that the $f(x) = 0$.

Ex: $x = -3$ is a zero of $f(x) = (2x - 5)(x + 3)$ because $f(-3) = 0$.

Before we spend much time exploring zeros, we need to learn an important fact called *the zero-product property*.

5. a. Write four pairs of numbers such that multiplying the first number in the pair by the second number in the pair results in an answer of 0.

   b. What do these pairs of numbers have in common?

   c. Suppose we have two numbers a and b, and suppose that $a \cdot b = 0$. What does this tell us about the values of a and b?

Zero- Product Property

The *zero-product property* says that …

Let's return to function f. The zeros of f are the values of x such that $f(x) = 0$, or the values of x such that $(2x - 5)(x + 3) = 0$ since $f(x) = (2x - 5)(x + 3)$. The statement $(2x - 5)(x + 3) = 0$ says that we have two numbers, namely $(2x - 5)$ and $(x + 3)$, with a product of 0.

We therefore know that either $2x - 5 = 0$ or $x + 3 = 0$. So

$$2x - 5 = 0 \quad \text{or} \quad x + 3 = 0$$
$$2x = 5 \qquad\qquad\qquad x = -3$$
$$x = 2.5$$

The zeros of f are $x = 2.5$ and $x = -3$.

6. Evaluate $f(2.5)$ and $f(-3)$ to verify that they are zeros (roots) of f. [Recall $f(x) = (2x - 5)(x + 3)$.]

Algebra II

Module 4: Investigation 5
Zeros (Roots) of Quadratic Functions

7. If the zeros of a function are real numbers, what does this tell us about the function's graph? (For example, what do we know about the graph if $f(2.5) = 0$ and $f(-3) = 0$?)

8. What value of x is equidistant from the zeros (roots) of f? What does this x-value represent for the quadratic function? How do you know?

In Exercises #9-12, find the zeros of each function (*Hint: use the zero-product property*).

9. $f(x) = (x+1)(x+7)$

10. $g(x) = (x-2)(x-5)$

11. $p(x) = -3(x+2)(5x+11)$

12. $f(x) = 6(x-4)^2$

13. What is the line of symmetry for each of the functions in Exercises #9 and #10?

14. What are the coordinates of the vertex for each of the functions in Exercises #9 and #10?

15. Sketch a graph of the functions in Exercises #9 and #10

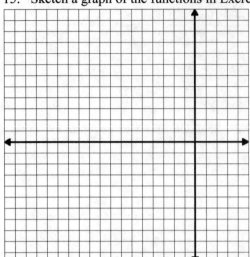
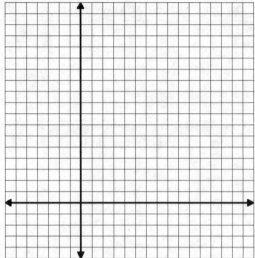

16. Recall that the vertical intercept of a function is the output of the function when the input is 0. That is, the vertical intercept of a function f is $f(0)$.

   a. Find the vertical intercept of f if $f(x) = -2x^2 - 6x + 5$.

   b. Find the vertical intercept of f if $f(x) = (2x - 3)(x + 4)$.

17. A function is defined in standard form as $f(x) = -2x^2 + 11x - 12$ and in factored form as $f(x) = (-2x + 3)(x - 4)$. Find the function's roots, vertical intercept, and vertex, then graph the function.

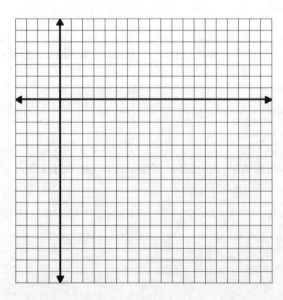

Algebra II

Module 4: Investigation 6
Factoring Quadratic Functions

In Investigation 5 we learned about the factored form of a quadratic function [such as $g(x)=(-2x+3)(x-4)$] and the standard form of a quadratic function [such as $g(x)=-2x^2+11x-12$].

1. What are the advantage(s) of writing a quadratic function in standard form?

2. What are the advantage(s) of writing a quadratic function in factored form?

Since there's an advantage to writing a quadratic function in factored form, it's important that we know how to do this. The key to knowing how to factor is first to think about how to expand from factored form to standard form and imagine reversing the process.

Example 1: Consider the function $f(x)=(2x-5)(x+1)$. Let's expand this to write the formula in standard form. *Pay close attention to the steps in the process.*

$f(x) = (2x-5)(x+2)$

$f(x) = \boxed{(2x-5)}(x+2)$

$f(x) = (2x-5)(x) + (2x-5)(2)$ • distribute $(2x-5)$ over the sum $x+2$

$f(x) = \underbrace{(2x)(x)}_{\substack{\text{product of the}\\\text{first term in}\\\text{each of the}\\\text{factors}}} + \underbrace{(-5)(x)+(2x)(2)}_{\substack{\text{products of the first and}\\\text{second terms in each of}\\\text{the factors}}} + \underbrace{(-5)(2)}_{\substack{\text{product of the}\\\text{second terms in}\\\text{each of the}\\\text{factors}}}$ • distribute two more times

$f(x) = 2x^2 + (-5x) + (4x) + (-10)$

$f(x) = \underbrace{2x^2}_{\substack{\text{product of the}\\\text{first term in}\\\text{each of the}\\\text{factors}}} + \underbrace{(-x)}_{\substack{\text{sum of the products of the}\\\text{first and second terms in}\\\text{each of the factors}}} + \underbrace{(-10)}_{\substack{\text{product of the}\\\text{second terms in}\\\text{each of the}\\\text{factors}}}$

$f(x) = 2x^2 - x - 10$

If we instead began with $f(x)=2x^2-x-10$ and wanted to write the function in factored form, we would need to fill in the parentheses for $f(x)=(\;\;+\;\;)(\;\;+\;\;)$ such that $(\;\;+\;\;)(\;\;+\;\;)=2x^2-x-10$. To make it easier to reference, let's write this as $(a+c)(b+d)=2x^2-x-10$.

We need to find a, b, c, and d such that $ab=2x^2$, $cd=-10$, and $ad+cb=-x$.

- The best place to start is to look at the possible values of a and b. There aren't too many options using integer coefficients: a and b must be $2x$ and x so that $ab=2x^2$. Using this information we have $(2x+c)(x+d)=2x^2-x-10$.

- Next, we observe that $cd=-10$, so c and d must be factors of -10. We have the following possibilities:

  $c=-1, d=10$ $c=-10, d=1$ $c=-5, d=2$ $c=-2, d=5$
  $c=10, d=-1$ $c=1, d=-10$ $c=2, d=-5$ $c=5, d=-2$

Module 4: Investigation 6

© 2014 Carlson, O'Bryan & Joyner

Algebra II

Module 4: Investigation 6
Factoring Quadratic Functions

- Since $ad+cb=-x$, we need to figure out which combination gives us the correct sum.
    - If $c=-1$ and $d=10$, then $ad+cb=(2x)(10)+(-1)(x)=20x-x=19x \neq -x$.
    - If $c=10$ and $d=-1$, then $ad+cb=(2x)(-1)+(10)(x)=-2x+10x=8x \neq -x$.
    - If $c=-10$ and $d=1$, then $ad+cb=(2x)(1)+(-10)(x)=2x-10x=-8x \neq -x$.
    - If $c=1$ and $d=-10$, then $ad+cb=(2x)(-10)+(1)(x)=-20x+x=-19x \neq -x$.
    - If $c=-5$ and $d=2$, then $ad+cb=(2x)(2)+(-5)(x)=4x-5x=-x$ *.

We can stop when we've found the right combination, which is $c=-5$ and $d=2$. So the factored form of $f(x)=2x^2-x-10$ is $f(x)=(a+c)(b+d)=(2x+(-5))(x+2)=\boxed{(2x-5)(x+2)}$.

> **Tip for Success**
> Until you become good at factoring, we recommend that you expand the factored form to make sure you get back the original standard form. If you don't, then you know you made a mistake and you can fix it.

We expanded $f(x)=(2x-5)(x+2)$ earlier in this investigation, so we know our answer is correct. We will do another example and check our answer by expanding the factored form.

Example 2: Let $f(x)=x^2-3x-28$. We want to write this in factored form $f(x)=(a+c)(b+d)$ such that $(a+c)(b+d)=x^2-3x-28$.

- $a=x$ and $b=x$ since $ab=x^2$. So $(x+c)(x+d)=x^2-3x-28$.

- We know that c and d are factors of -28. We have a lot of possibilities, including

    | $c=-4, d=7$ | $c=-7, d=4$ | $c=-1, d=28$ | $c=-28, d=1$ |
    | $c=7, d=-4$ | $c=4, d=-7$ | $c=28, d=-1$ | $c=1, d=-28$ |

- Since $ad+cb=-3x$, we need to figure out which combination gives us the correct sum.
    - If $c=-4$ and $d=7$, then $ad+cb=(x)(-4)+(7)(x)=-4x+7x=3x \neq -3x$.
    - If $c=-7$ and $d=4$, then $ad+cb=(x)(-7)+(4)(x)=-7x+4x=-3x$ *.

The factored form of $f(x)=x^2-3x-28$ is $\boxed{f(x)=(x-7)(x+4)}$.

3. Expand $f(x)=(x-7)(x+4)$ to show that we get $f(x)=x^2-3x-28$.

Algebra II

Module 4: Investigation 6
Factoring Quadratic Functions

A few more abbreviated examples follow.

Example 3: If $f(x) = x^2 + 7x + 6$, then $\boxed{f(x) = (x+1)(x+6)}$ since $(x)(x) = x^2$, $(1)(6) = 6$, and $(x)(6) + (1)(x) = 6x + x = 7x$.

Example 4: If $f(x) = x^2 + 6x + 9$, then $\boxed{f(x) = (x+3)(x+3)}$ (also written $f(x) = (x+3)^2$) since $(x)(x) = x^2$, $(3)(3) = 9$, and $(x)(3) + (3)(x) = 3x + 3x = 6x$.

Example 5: If $f(x) = 3x^2 - 19x - 14$, then $\boxed{f(x) = (3x+2)(x-7)}$ since $(3x)(x) = 3x^2$, $(2)(-7) = -14$, and $(3x)(-7) + (2)(x) = -21x + 2x = -19x$.

Example 6: If $f(x) = x^2 - 16$, we observe that $f(x) = x^2 + 0x - 16$, and then $\boxed{f(x) = (x+4)(x-4)}$ since $(x)(x) = x^2$, $(4)(-4) = -16$, and $(x)(-4) + (4)(x) = -4x + 4x = 0x$.

In Exercises #4-13, write the factored form of the quadratic function.

4. $f(x) = x^2 + 7x + 12$

5. $g(x) = x^2 + 5x - 14$

6. $h(x) = x^2 - 5x - 14$

7. $g(x) = x^2 - 3x + 2$

8. $f(x) = 2x^2 + 9x + 7$

9. $h(x) = -x^2 + 7x - 10$

10. $f(x) = 2x^2 + x - 15$

11. $a(x) = 6x^2 - 7x - 5$

Algebra II

Module 4: Investigation 6
Factoring Quadratic Functions

12. $h(x) = 49x^2 - 81$

13. $m(x) = 25x^2 - 10x + 1$

In Exercises #14-17 you are given quadratic functions written in standard form. For each, do the following.
   a) State the vertical intercept.
   b) Write the function in factored form.
   c) State the zeros of the function.
   d) Find the line of symmetry.
   e) Give the coordinates of the vertex.

14. $f(x) = x^2 + 5x + 6$

15. $f(x) = -x^2 + 6x - 8$

16. $f(x) = 2x^2 - 18x + 40$

17. $f(x) = 4x^2 - 9$

Algebra II

Module 4: Investigation 6
Factoring Quadratic Functions

Not all quadratic expressions are factorable using integers. For example, when attempting to write the function $f(x) = x^2 + 0.5x - 3$ in the factored form $f(x) = (a+c)(b+d)$, we know $a = x$ and $b = x$ since $ab = x^2$. We also know that c and d must be factors of -3. The integer possibilities are

$$c = 3, d = -1 \qquad c = -1, d = 3 \qquad c = -3, d = 1 \qquad c = 1, d = -3$$

However, **none** of these options give us the correct middle term, and guessing the correct non-integer factors would be almost impossible. *When this situation arises, we will use a different method that you will learn about later in this module.*

In Exercises #18-23, rewrite the expression in factored form using integer coefficients if possible. If it's not possible, write *not factorable using integers*.

18. $f(x) = x^2 + 3x + 5$

19. $f(x) = x^2 - 7x + 10$

20. $f(x) = x^2 + 10x + 15$

21. $f(x) = x^2 - x + 1$

22. $f(x) = 2x^2 + x - 1$

23. $f(x) = 3x^2 - 2x + 4$

Algebra II

Module 4: Investigation 7
Solving Quadratic Equations

Factoring and using the zero-product property is useful for solving quadratic equations in addition to finding the zeros of the function.

1. You are given the graph of the function $f(x) = 2x^2 - 7x - 3$. What does it mean to solve the equation $2x^2 - 7x - 3 = 12$? Explain, then represent the meaning on the graph.

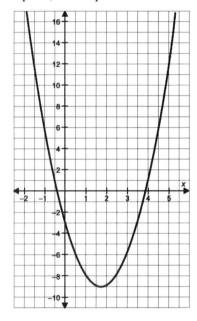

We can attempt to solve the equation using factoring. To begin with, we rewrite the equation so that the value of each side of the equation is 0. Then we factor and use the zero-product property.

$$2x^2 - 7x - 3 = 12$$
$$2x^2 - 7x - 3 - 12 = 0$$
$$2x^2 - 7x - 15 = 0$$
$$(2x + 3)(x - 5) = 0$$

$$\begin{array}{ll} 2x + 3 = 0 & x - 5 = 0 \\ 2x = -3 & x = 5 \\ x = -1.5 & \end{array}$$

2. The solutions to the equation $2x^2 - 7x - 3 = 12$ are $x = -1.5$ and $x = 5$. What do these values represent?

Module 4: Investigation 7
© 2014 Carlson, O'Bryan & Joyner

Algebra II

Module 4: Investigation 7
Solving Quadratic Equations

In Exercises #3-6, use factoring and the zero-product property to solve the given equation, then explain what your solution represents.

3. $x^2 + 12x + 13 = -22$

4. $8 = x^2 - 13x + 30$

5. $2x^2 + 6x - 6 = 2$

6. $6x^2 + 5x - 9 = -10$

7. Suppose the functions $f(x) = x^2 - 3x - 25$ and $g(x) = 3x + 15$ are graphed on the same coordinate plane. Where will the graphs intersect? *(Solve algebraically first, then check your answer by graphing the functions on a calculator.)*

8. Suppose the functions $f(x) = 2x^2 - 4x - 15$ and $g(x) = x^2 - 3x + 5$ are graphed on the same coordinate plane. Where will the graphs intersect? *(Solve algebraically first, then check your answer by graphing the functions on a calculator.)*

Algebra II

Module 4: Investigation 8
Vertex Form and Completing the Square

1. Let $f(x) = 2(x-7)^2 + 3$. A classmate claims that the vertex of the function must be at $(7, 3)$, and it must be the function's minimum. His reasoning follows.
    - When $x = 7$, $(x-7)^2 = 0$, so $2(x-7)^2 = 0$, and so $f(7) = 3$.
    - If $x > 7$, then $(x-7)^2 > 0$, so $2(x-7)^2 > 0$, and so $f(x) > 3$.
    - If $x < 7$, then $(x-7)^2 > 0$, so $2(x-7)^2 > 0$, and so $f(x) > 3$.
    - Therefore, the smallest output value is 3, and this occurs at $x = 7$. So $(7, 3)$ must be the vertex, and it must be a minimum.

    a. Explore his argument. Is he correct?

    b. What are the coordinates for the vertex of each of the following functions? Is each vertex a maximum or minimum?
        i) $f(x) = (x-5)^2 - 4$
        ii) $f(x) = 4(x-1)^2$

        iii) $f(x) = 2(x+6)^2 + 5$
        iv) $f(x) = (x+2)^2 - 8$

2. Let $f(x) = -3(x-2)^2 + 5$ with a vertex of $(2, 5)$. Construct an argument to explain why the vertex must be a maximum.

3. The form $f(x) = a(x-h)^2 + k$ is called the *vertex form for a quadratic function*. This is an appropriate name because the ordered pair (h, k) represents the vertex of the function.
    a. We have now learned three forms for a quadratic function: standard form [$f(x) = ax^2 + bx + c$], factored form [$f(x) = a(x-u)(x-v)$], and now vertex form [$f(x) = a(x-h)^2 + k$]. Each of these forms has an advantage over the other forms. What is the advantage of each form?

    b. Write two different functions that have their vertices at $(-5, -7)$.

Algebra II

Module 4: Investigation 8
Vertex Form and Completing the Square

Vertex Form of a Quadratic Function

The **vertex form** of a quadratic function is [_____] where (h, k) represents the vertex of the function.

- If _____, then the vertex is a minimum
- if _____, then the vertex is a maximum

4. Let $f(x) = 2(x-2)^2 - 8$.

   a. Write f in standard form.

   b. Write f in factored form.

   c. Provide all of the details you can about the function f (such as the vertex, vertical intercept, etc.), then sketch its graph.

   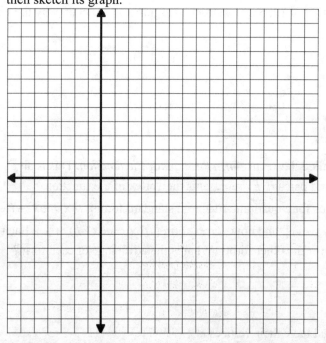

In Exercise #4 you were given the vertex form of the function formula and you had to find the standard form. Going the other direction is also possible. Let's see how we can do this.

Algebra II
Module 4: Investigation 8
Vertex Form and Completing the Square

First, it's important to note that in the vertex form $f(x) = a(x-h)^2 + k$ we have an expression squared: $f(x) = a\boxed{(x-h)^2} + k$. Expanding just this part gives us $(x-h)^2 = x^2 - 2hx + h^2$.

5. a. If $h = 6$, then $(x-h)^2 = (x-6)^2 = (x-6)(x-6) = x^2 - 12x + c$. What is the value of c that completes this statement?

   b. If $h = -4$, then $(x-h)^2 = (x+4)^2 = (x+4)(x+4) = x^2 + bx + 16$. What is the value of b that completes this statement?

If we substitute some different values for h, we get the following:

   I. If $h = 10$, $(x-h)^2 = (x-10)^2 = (x-10)(x-10) = \boxed{x^2 - 20x + 100}$
   II. If $h = -1$, $(x-h)^2 = (x-(-1))^2 = (x+1)^2 = (x+1)(x+1) = \boxed{x^2 + 2x + 1}$
   III. If $h = 4$, $(x-h)^2 = (x-4)^2 = (x-4)(x-4) = \boxed{x^2 - 8x + 16}$
   IV. If $h = -9$, $(x-h)^2 = (x-(-9))^2 = (x+9)^2 = (x+9)(x+9) = \boxed{x^2 + 18x + 81}$

6. For each of the expressions in standard form ($ax^2 + bx + c$), show that $\left(\frac{b}{2}\right)^2 = c$.

   I. $x^2 - 20x + 100$ \hspace{2cm} II. $x^2 + 2x + 1$

   III. $x^2 - 8x + 16$ \hspace{2cm} IV. $x^2 + 18x + 81$

If we have a quadratic expression of the form $ax^2 + bx + c$ where $a = 1$ and $\left(\frac{b}{2}\right)^2 = c$, we can write it in the form $(x-h)(x-h)$ or $(x-h)^2$ [or $(x+h)(x+h)$ or $(x+h)^2$]. Some examples follow.

$$x^2 + 12x + 36 = (x+6)^2 \hspace{1cm} x^2 - 30x + 225 = (x-15)^2 \hspace{1cm} x^2 + 3x + 2.25 = (x+1.5)^2$$

This fact unlocks a technique called *completing the square*. The name comes from the fact that we will provide the necessary value of c for $x^2 + bx +$ ____ that lets us factor the expression as $(x-h)^2$ or $(x+h)^2$. Let's learn about this technique.

Algebra II
Module 4: Investigation 8
Vertex Form and Completing the Square

7. Let $f(x) = x^2 + 14x - 5$. The first step is to group the terms that contain variables:
$$f(x) = (x^2 + 14x) - 5$$

   a. Consider the expression $x^2 + 14x + \underline{}$. What must go in the blank so that we can factor the expression in the form $(x-h)^2$ or $(x+h)^2$?

   b. If we return to the function, we want to add this value in the blank space for
   $f(x) = (x^2 + 14x + \underline{}) - 5$. However, if we simply add something to the function then we change the output values of the function. We don't want to do this! Therefore, anything we add to the function definition we must also subtract.

   Fill in the blanks to write f in vertex form.
   $$f(x) = (x^2 + 14x + \underline{}) - \underline{} - 5$$
   $$f(x) = (x\underline{})^2 \underline{}$$

   c. What are the coordinates of the vertex?

What follows are two more examples that use completing the square. Make sure you understand the steps in each example.

$f(x) = x^2 - 22x + 18$
$f(x) = (x^2 - 22x + \underline{}) - \underline{} + 18$
$f(x) = (x^2 - 22x + 121) - 121 + 18$
$f(x) = (x-11)^2 - 103$

$g(x) = x^2 + 13x$
$g(x) = (x^2 + 13x + \underline{}) - \underline{}$
$g(x) = \left(x^2 + 13x + \left(\frac{13}{2}\right)^2\right) - \left(\frac{13}{2}\right)^2$
$g(x) = \left(x + \frac{13}{2}\right)^2 - \frac{169}{4}$

8. Use the technique from Exercise #7 to write each of the following functions in vertex form, then give the coordinates of the vertex. *Check your answers by graphing the function using a calculator.*

   a. $f(x) = x^2 + 2x + 7$

   b. $g(x) = x^2 - 10x + 3$

   c. $j(x) = x^2 + 6x + 30$

   d. $h(x) = x^2 - 32x + 27$

Algebra II

Module 4: Investigation 8
Vertex Form and Completing the Square

e. $f(x) = x^2 + 9x + 13$

f. $g(x) = x^2 - 5x$

If $a \neq 1$ for $ax^2 + bx + c$, we can still complete the square. We just need to be a little bit more careful.

Let $f(x) = 2x^2 + 16x + 1$. The first step is to again group the terms that contain a variable:
$$f(x) = (2x^2 + 16x) + 1$$

Next, we factor the coefficient in the x^2 term out of the grouped portion:
$$f(x) = 2(x^2 + 8x) + 1$$

This is where we need to be careful. We know that for $x^2 + 8x + \underline{}$, we need to replace the blank space with 64 so that we can write $x^2 + 8x + 16$ as $(x+4)^2$. However, when we examine the process in the context of the entire function, we see that we are adding 16 *inside of a set of parentheses*, and that we are multiplying 2 by *everything inside of the parentheses*. This means that we aren't actually adding 16 to the function definition. We are actually adding 2(16), or 32. Therefore we must subtract 2(16), or 32, not 16.

$$f(x) = 2(x^2 + 8x + \underline{}) - \underline{} + 1$$
$$f(x) = 2(x^2 + 8x + 16) - 2(16) + 1$$
$$f(x) = 2(x^2 + 8x + 16) - 32 + 1$$
$$f(x) = 2(x+4)^2 - 31$$

The vertex form of f is $\boxed{f(x) = 2(x+4)^2 - 31}$. The vertex is located at $\boxed{(-4, -31)}$.

Here are two more examples of completing the square with $a \neq 1$. Make sure you understand the steps in each example.

$f(x) = 4x^2 - 40x + 17$
$f(x) = (4x^2 - 40x) + 17$
$f(x) = 4(x^2 - 10x + \underline{}) - \underline{} + 17$
$f(x) = 4(x^2 - 10x + 25) - 4(25) + 17$
$f(x) = 4(x^2 - 10x + 25) - 100 + 17$
$f(x) = 4(x-5)^2 - 83$

$g(x) = -2x^2 + 14x - 11$
$g(x) = (-2x^2 + 14x) - 11$
$g(x) = -2(x^2 - 7x + \underline{}) - \underline{} - 11$
$g(x) = -2\left(x^2 - 7x + \left(-\frac{7}{2}\right)^2\right) - (-2)\left(-\frac{7}{2}\right)^2 - 11$
$g(x) = -2\left(x^2 - 7x + \frac{49}{4}\right) - (-2)\left(\frac{49}{4}\right) - 11$
$g(x) = -2\left(x^2 - 7x + \frac{49}{4}\right) + \frac{49}{2} - 11$
$g(x) = -2\left(x - \frac{7}{2}\right)^2 + \frac{27}{2}$

Algebra II

Module 4: Investigation 8
Vertex Form and Completing the Square

9. Write each of the following functions in vertex form, then give the coordinates of the vertex. *Check your answers by graphing the function using a calculator.*

   a. $f(x) = 2x^2 + 12x + 23$

   b. $f(x) = 5x^2 + 120x$

   c. $h(x) = -3x^2 - 18x + 12$

   d. $j(x) = 4x^2 - 22x + 19$

10. A local school is planning a community carnival to raise money for charity. Based on their research they believe that they can model the expected attendance by the function $f(x) = -0.5x^2 + 80x - 2350$ where x is the forecasted high temperature for the day in degrees Fahrenheit.

    a. Write the formula in vertex form by completing the square. The first step has been done for you.
    $$f(x) = -0.5x^2 + 80x - 2350$$
    $$f(x) = -0.5(x^2 - 160x \qquad) - 2350$$

    b. Is the vertex a minimum or a maximum?

    c. What are the coordinates of the vertex? What information does this give us about the context?

Algebra II   **Module 4: Investigation 9**
                                             Deriving the Quadratic Formula

In Investigation 6 we learned how to factor quadratic expressions and how to use the zero-product property to find the zeros (or roots) of a quadratic function. However, we also saw that not every quadratic function is factorable over the integers. Therefore, we need an additional method that doesn't rely on factoring. The *quadratic formula* can be used to find the zeros of *any* quadratic function based on the coefficients of the function in standard form (that is, based on a, b, and c in $f(x) = ax^2 + bx + c$). Deriving the formula uses completing the square to solve the equation $ax^2 + bx + c = 0$ for x.

1. When we solve $ax^2 + bx + c = 0$ for x, why does this give us the zeros of f?

Let $f(x) = 3x^2 + 13x + 7$ and $g(x) = ax^2 + bx + c$. We've started the process of finding the zeros using completing the square for these functions. (*Note: We avoid simplifying the expressions that result from completing the square with function f to help you make connections with the process for the generic coefficients in function g.*)

	$3x^2 + 13x + 7 = 0$	$ax^2 + bx + c = 0$
Step 1	$(3x^2 + 13x) + 7 = 0$	$(ax^2 + bx) + c = 0$
Step 2	$3\left(x^2 + \dfrac{13}{3}x + \underline{}\right) - \underline{} + 7 = 0$	$a\left(x^2 + \dfrac{b}{a}x + \underline{}\right) - \underline{} + c = 0$
Step 3	$3\left(x^2 + \dfrac{13}{3}x + \left(\dfrac{13/3}{2}\right)^2\right) - 3\left(\dfrac{13/3}{2}\right)^2 + 7 = 0$	$a\left(x^2 + \dfrac{b}{a}x + \left(\dfrac{b/a}{2}\right)^2\right) - a\left(\dfrac{b/a}{2}\right)^2 + c = 0$

2. In Step 3, why were $\left(\dfrac{13/3}{2}\right)^2$ and $\left(\dfrac{b/a}{2}\right)^2$ added inside of the parentheses?

3. In Step 3, why were $3\left(\dfrac{13/3}{2}\right)^2$ and $a\left(\dfrac{b/a}{2}\right)^2$ subtracted from the left sides of the equations instead of $\left(\dfrac{13/3}{2}\right)^2$ and $\left(\dfrac{b/a}{2}\right)^2$?

Algebra II

Module 4: Investigation 9
Deriving the Quadratic Formula

Step 3 $\quad 3\left(x^2+\dfrac{13}{3}x+\left(\dfrac{13/3}{2}\right)^2\right)-3\left(\dfrac{(13/3)}{2}\right)^2+7=0 \qquad a\left(x^2+\dfrac{b}{a}x+\left(\dfrac{b/a}{2}\right)^2\right)-a\left(\dfrac{b/a}{2}\right)^2+c=0$

Step 4 $\quad 3\left(x^2+\dfrac{13}{3}x+\left(\dfrac{13}{2(3)}\right)^2\right)-3\left(\dfrac{13}{2(3)}\right)^2+7=0 \qquad a\left(x^2+\dfrac{b}{a}x+\left(\dfrac{b}{2a}\right)^2\right)-a\left(\dfrac{b}{2a}\right)^2+c=0$

Step 5 $\quad 3\left(x+\dfrac{13}{2(3)}\right)^2-3\left(\dfrac{(13)^2}{(2(3))^2}\right)+7=0 \qquad a\left(x+\dfrac{b}{2a}\right)^2-a\left(\dfrac{b^2}{(2a)^2}\right)+c=0$

Step 6 $\quad 3\left(x+\dfrac{13}{2(3)}\right)^2=\underline{} \qquad\qquad\qquad a\left(x+\dfrac{b}{2a}\right)^2=\underline{}$

4. What did we do to move from Step 3 to Step 4?

5. What did we do as we moved from Step 4 to Step 5? *(Hint: We made two changes.)*

6. Fill in the blanks to complete Step 6.

Step 7 $\quad \left(x+\dfrac{13}{2(3)}\right)^2=\dfrac{(13)^2}{(2(3))^2}-\dfrac{7}{3} \qquad\qquad \left(x+\dfrac{b}{2a}\right)^2=\dfrac{b^2}{(2a)^2}-\dfrac{c}{a}$

Step 8 $\quad \left(x+\dfrac{13}{2(3)}\right)^2=\dfrac{(13)^2}{2^2(3)^2}-\dfrac{7}{3} \qquad\qquad \left(x+\dfrac{b}{2a}\right)^2=\dfrac{b^2}{2^2a^2}-\dfrac{c}{a}$

Step 9 $\quad \left(x+\dfrac{13}{2(3)}\right)^2=\dfrac{(13)^2}{2^2(3)^2}-\left(\dfrac{7}{3}\right)\cdot\left(\dfrac{2^2(3)}{2^2(3)}\right) \qquad \left(x+\dfrac{b}{2a}\right)^2=\dfrac{b^2}{2^2a^2}-\left(\dfrac{c}{a}\right)\cdot\left(\dfrac{2^2a}{2^2a}\right)$

Step 10 $\quad \left(x+\dfrac{13}{2(3)}\right)^2=\dfrac{(13)^2}{2^2(3)^2}-\dfrac{2^2(7)(3)}{2^2(3)^2} \qquad \left(x+\dfrac{b}{2a}\right)^2=\dfrac{b^2}{2^2a^2}-\dfrac{2^2ac}{2^2a^2}$

7. What was the goal of steps 9 and 10?

Algebra II

Module 4: Investigation 9
Deriving the Quadratic Formula

8. Subtract the fractions on the right sides of the equations to write Step 11.

   Step 11

   Step 12 $x + \dfrac{13}{2(3)} = \pm \sqrt{\dfrac{(13)^2 - 2^2(7)(3)}{2^2(3)^2}}$ $\qquad$ $x + \dfrac{b}{2a} = \pm \sqrt{\dfrac{b^2 - 2^2 ac}{2^2 a^2}}$

   Step 13 $x = -\dfrac{13}{2(3)} \pm \sqrt{\dfrac{(13)^2 - 2^2(7)(3)}{2^2(3)^2}}$ $\qquad$ $x = -\dfrac{b}{2a} \pm \sqrt{\dfrac{b^2 - 2^2 ac}{2^2 a^2}}$

   Step 14

   Step 15 $x = -\dfrac{13}{2(3)} \pm \dfrac{\sqrt{(13)^2 - 4(7)(3)}}{2(3)}$ $\qquad$ $x = -\dfrac{b}{2a} \pm \dfrac{\sqrt{b^2 - 4ac}}{2a}$

9. Fill in Step 14 that links Step 13 and Step 15.

10. Why does the statement $x = -\dfrac{13}{2(3)} \pm \dfrac{\sqrt{(13)^2 - 4(7)(3)}}{2(3)}$ show two unique values of *x*? What do these values of *x* represent?

11. Use a calculator to find the approximate values of *x* represented by $x = -\dfrac{13}{2(3)} \pm \dfrac{\sqrt{(13)^2 - 4(7)(3)}}{2(3)}$. Round to two decimal places.

12. Graph function j using a calculator, then make a rough sketch of the graph below. Locate the points where the graph crosses the horizontal axis and label them with their ordered pairs.

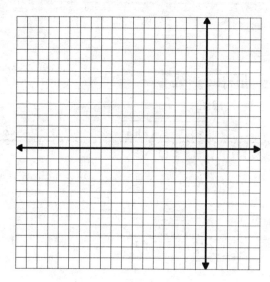

13. Simplify $x = -\dfrac{13}{2(3)} \pm \dfrac{\sqrt{(13)^2 - 4(7)(3)}}{2(3)}$ to show the exact values of the zeros of j. Your final answer will contain a radical.

14. The quadratic formula is defined as $x = -\dfrac{b}{2a} \pm \dfrac{\sqrt{b^2 - 4ac}}{2a}$. When we substitute the values of a, b, and c from the coefficients of a given quadratic function in standard form, what does this formula find for us?

Algebra II

Module 4: Investigation 10
Using the Quadratic Formula

> **The Quadratic Formula**
>
> Let *f* be defined by $f(x) = ax^2 + bx + c$ for real numbers *a*, *b*, and *c*. Then the quadratic formula is
>
> [   ]
>
> The quadratic formula determines …

---

1. Let $f(x) = 2x^2 + 7x + 3$.

    a. Use the quadratic formula to find the roots of *f*.

    b. Find the roots of *f* by factoring and using the zero-product property.

    c. Explain how you can use the graph of *f* to check your answers in parts (a) and (b).

    d. What other method can you use to check your solutions to parts (a) and (b)?

2. Use the quadratic formula to find the roots of each quadratic function. Give exact answers and the decimal equivalents (rounded to two decimal places if necessary).

    a. $f(x) = 2x^2 - 6x + 2.5$

    b. $h(x) = x^2 + 4x - 1$

c. $f(x) = 3x^2 - 5x - 1$ d. $g(x) = -x^2 + 7x - 2$

3. We know that the solutions to $ax^2 + bx + c = 0$ for any quadratic function are given by
   $x = -\frac{b}{2a} + \frac{\sqrt{b^2 - 4ac}}{2a}$ and $x = -\frac{b}{2a} - \frac{\sqrt{b^2 - 4ac}}{2a}$.

   a. This tells us that the zeros (roots) of the function are equidistant from $x = -\frac{b}{2a}$. That is, the roots of the function are $\frac{\sqrt{b^2 - 4ac}}{2a}$ units less than and greater than $-\frac{b}{2a}$. Using what we've learned in previous investigations, what must $x = -\frac{b}{2a}$ represent for a quadratic function?

   b. Find the line of symmetry for each of the following quadratic functions.
      i) $f(x) = x^2 - 6x + 10$ ii) $g(x) = -2x^2 - 12x + 9$ iii) $h(x) = 10x^2 + 5x - 1$

   c. What are the coordinates for the vertex of each function in part (b)?

Algebra II

Module 4: Investigation 10
Using the Quadratic Formula

4. a. Use the quadratic formula to find the roots of $f(x) = -3x^2 + x - 4$.

    b. Describe what went "wrong" during the solution process.

    c. The quadratic formula is supposed to find the zeros of the function. When the zeros are real numbers, they equate to the horizontal intercepts of the function's graph. However, in this case the quadratic formula didn't return any real number answers. Graph function *f* using a calculator and explain what this fact implies about the horizontal intercepts of the graph of *f*.

5. a. Use the quadratic formula to find the roots of $f(x) = 2x^2 - 4x + 2$.

    b. Describe what happened during the solution process.

    c. The quadratic formula is supposed to find the zeros of the function. When the zeros are real numbers, they equate to the horizontal intercepts of the function's graph. However, in this case the quadratic formula only returned one value. Graph function *f* using a calculator and explain what this fact implies about the horizontal intercept(s) of the graph of *f*.

Algebra II

Module 4: Investigation 10
Using the Quadratic Formula

6. Perform the first several steps of using the quadratic formula to find the zeros of the following functions. Complete as many steps as necessary until you can predict the number of horizontal intercepts for the function's graph. (*You do not have to find the intercepts – only how many will exist.*) Use a graphing calculator to check your answers.

   a. $f(x) = x^2 + 7x + 25$ b. $f(x) = x^2 - 2x - 5$ c. $f(x) = 2x^2 + 20x + 50$

7. The expression $b^2 - 4ac$ from the formula is called *the discriminant* because it discriminates (makes a distinction, or highlights the difference) between functions with zero, one, or two horizontal intercepts.

   a. Calculate the value of the discriminant ($b^2 - 4ac$) for the functions from Exercise #6.

   b. Why does the discriminant alone give us enough information to predict the number of horizontal intercepts for the graph of a quadratic function?

The Discriminant of a Quadratic Function

Let *f* be defined by $f(x) = ax^2 + bx + c$ for real numbers *a*, *b*, and *c*. Then the discriminant is the value of ☐

- If _____, then…

- If _____, then…

- If _____, then…

Algebra II

Module 4: Investigation 10
Using the Quadratic Formula

For each of the functions in Exercises #8-9, do the following.
   a. Determine the vertical intercept of the function.
   b. Find the line of symmetry and the coordinates for the function's vertex.
   c. Use the discriminant ($b^2 - 4ac$) to predict the number of horizontal intercepts the function's graph will have.
   d. Use the quadratic formula to find the real-number zeros of the function.
   e. Sketch a graph of each function and label the information you found in parts (a) through (d).

8. $f(x) = x^2 + 10x - 39$
   a. vertical intercept:
   b. axis of symmetry:
      vertex:
   c. number of horizontal intercepts:
   d. horizontal intercept(s):
   e.

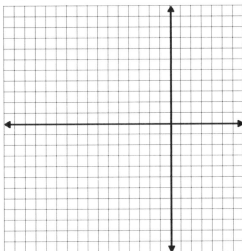

9. $f(x) = -4x^2 + 24x - 36$
   a. vertical intercept:
   b. axis of symmetry:
      vertex:
   c. number of horizontal intercepts:
   d. horizontal intercept(s):
   e.

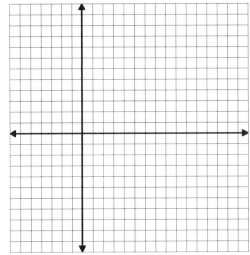

10. In Investigation 8 we examined the following context: *A local school is planning a community carnival to raise money for charity. Based on their research they believe that they can model the expected attendance by the function* $f(x) = -0.5x^2 + 80x - 2350$ *where x is the forecasted high temperature for the day in degrees Fahrenheit.*
   a. Find the roots of *f* using the quadratic formula. Round your answers to the nearest whole number.

   b. What do the roots of *f* tell us about this situation?

Algebra II   **Module 4: Investigation 11**
"Imaginary" Numbers

Have you ever tried to explain negative numbers to a child? For children, numbers are very concrete concepts – they use numbers to count items and (if they're old enough) perhaps to measure something like an object's height (in some units). In these examples negative numbers don't make much sense since they aren't needed for counting and children don't yet have a concept of directional measurement.

Believe it or not, mathematicians didn't always accept negative numbers. Up until about 500 years ago almost every mathematician used geometry to justify their work, and characteristics such as length, area, and volume are represented by positive numbers. So mathematicians threw out any solutions to equations that were negative and believed negative numbers were useless. In this investigation we're going to explore another type of number that comes up when solving quadratic equations. It might be initially strange to think about these numbers, but they have very powerful applications in fields such as science and engineering.

Let's begin by reviewing square roots. The square root of a number x (written as $\sqrt{x}$) is defined to be the positive number that, when multiplied by itself, gives you back the original number x.

1. Complete the following statements.
   a. $\sqrt{9}$ is 3 because _____ .
   b. $\sqrt{64}$ is 8 because _____ .

Non-perfect Squares

For numbers that aren't perfect squares, such as 18, the square root is typically left as $\sqrt{18}$ or simplified as $\sqrt{18} = \sqrt{9 \cdot 2} = \sqrt{9} \cdot \sqrt{2} = 3\sqrt{2}$. These square roots are *irrational*, meaning they can't be written exactly as a rational number or a repeating or terminating decimal. We can *approximate* their value ($\sqrt{18} \approx 4.24$), but in "exact" form the best we can do is say that the square root of 18 is the real number $\sqrt{18}$ (or $3\sqrt{2}$).

What about negative numbers? What's the square root of -9? It can't be 3, since $3 \cdot 3 = 9$, and it can't be -3 since $-3 \cdot -3 = 9$. Therefore, it looks like there isn't a number that, when multiplied by itself, yields -9. If there *were* such a number, however, it would be represented by $\sqrt{-9}$. The number $\sqrt{-9}$ would be the number that, multiplied by itself, yields -9.

2. Solve $x^2 + 35 = 10$ for x to show that the only *possible* solutions are $x = \sqrt{-25}$ and $x = -\sqrt{-25}$.

3. Solve $2x^2 + 14 = -6$ for x to show that the only *possible* solutions are $x = \sqrt{-10}$ and $x = -\sqrt{-10}$.

Algebra II

Module 4: Investigation 11
"Imaginary" Numbers

Since mathematicians initially rejected negative numbers, just imagine what they thought about the *square root* of a negative number! They thought such "numbers" had no place in serious mathematics, and so they called them *imaginary numbers*. The name stuck even though we now know that these so-called "imaginary" numbers are actually very useful. It took mathematicians a long time to accept them, but now they are a critical part of modern mathematics.

> **Imaginary Numbers and Notation**
>
> We use the letter *i* to represent $\sqrt{-1}$. That is, $i = \sqrt{-1}$, and therefore $i^2 = -1$.

With this notation, we can rewrite numbers such as $\sqrt{-25}$, $-\sqrt{-2}$, $\sqrt{-48}$ and $4 - \sqrt{-49}$ as follows.

$\sqrt{-25}$
$\sqrt{25 \cdot -1}$
$\sqrt{25} \cdot \sqrt{-1}$
$5 \cdot i$
$5i$

$-\sqrt{-2}$
$-\sqrt{2 \cdot -1}$
$-\sqrt{2} \cdot \sqrt{-1}$
$-\sqrt{2} \cdot i$
$-\sqrt{2}i$

$\sqrt{-48}$
$\sqrt{48 \cdot -1}$
$\sqrt{16 \cdot 3 \cdot -1}$
$\sqrt{16} \cdot \sqrt{3} \cdot \sqrt{-1}$
$4 \cdot \sqrt{3} \cdot i$
$4\sqrt{3}i$

$4 - \sqrt{-49}$
$4 - \sqrt{49 \cdot -1}$
$4 - \sqrt{49} \cdot \sqrt{-1}$
$4 - 7 \cdot i$
$4 - 7i$

4. Rewrite each of the following expressions using *i*.

   a. $\sqrt{-4}$
   b. $-\sqrt{-121}$
   c. $\sqrt{-13}$
   d. $\sqrt{-41}$

   e. $-\sqrt{-32}$
   f. $\sqrt{-99}$
   g. $5 + \sqrt{-3}$
   h. $12 - \sqrt{-50}$

Algebra II
Module 4: Investigation 11
"Imaginary" Numbers

5. If $i = \sqrt{-1}$ and $i^2 = -1$, then what's the value of each of the following? (*Simplify your answers if possible.*)

   a. i^3 b. i^4 c. i^5

   d. i^6 e. i^{21} f. i^{46}

Complex Numbers

A *complex number* **a + bi** is a number made up of both a real number portion (*a*) and an imaginary number portion (*b*).
- If $a = 0$, then the number is called *purely imaginary*.
- If $b = 0$, then the number is a *real number*.

Examples: $3 + 2i$ $5 - 7i$ $7i$ -14

Complex numbers have similar properties to expressions involving only real numbers. We can add them, subtract them, multiply them, etc. by mostly just following the processes we already know. When the operations are complete, we make sure to collect the real numbers together and the imaginary numbers together so that the final answer is written as a single complex number. Remember that $i^2 = -1$.

$(3 + 5i) + (6 - 13i)$
$3 + 5i + 6 - 13i$
$3 + 6 + 5i - 13i$
$9 - 8i$

$(1 + 3i)(7 - i)$
$(1 + 3i)(7) + (1 + 3i)(-i)$
$(1)(7) + (3i)(7) + (1)(-i) + (3i)(-i)$
$7 + 21i - i - 3i^2$
$7 + 20i - 3i^2$
$7 + 20i - 3(-1)$
$7 + 20i + 3$
$10 + 20i$

$(4 - 3i)^2$
$(4 - 3i)(4 - 3i)$
$(4 - 3i)(4) + (4 - 3i)(-3i)$
$(4)(4) + (-3i)(4) + (4)(-3i) + (-3i)(-3i)$
$16 - 12i - 12i + 9i^2$
$16 - 24i + 9i^2$
$16 - 24i + 9(-1)$
$16 - 24i - 9$
$7 - 24i$

Notice that when we perform operations with complex numbers we end up with complex numbers as the result.

Algebra II

Module 4: Investigation 11
"Imaginary" Numbers

6. Perform the indicated operations. Write your final answer as a complex number $a + bi$ in simplest form.

   a. $2i(9 + 4i)$ b. $(-4 + i) + (19 - 6i)$

   c. $3(7 - 5i) - 8(6 + 3i)$ d. $(1 + 4i)(5 - 2i)$

   e. $(3 + 5i)^2$ f. $(a + bi)(c + di)$

Algebra II
Module 4: Investigation 12
The Quadratic Formula, Complex Roots, and Conjugates

Recall that in this module we learned that the quadratic formula $\left(x = -\frac{b}{2a} \pm \frac{\sqrt{b^2-4ac}}{2a}\right)$ tells us the zeros of a quadratic function (the values of x such that $f(x) = 0$).

1. We saw that some quadratic functions have exactly two real number zeros, such as $f(x) = 4x^2 + 2x - 2$. Use the quadratic formula to find the two real zeros for f.

2. We saw that some quadratic functions have only one real number zero, such as $g(x) = 3x^2 - 12x + 12$. Use the quadratic formula to find the one real number zero for g.

Algebra II
Module 4: Investigation 12
The Quadratic Formula, Complex Roots, and Conjugates

3. We know that some quadratic functions don't have real number zeros.
    a. Find the roots of $h(x) = x^2 + 6x + 10$. Write your final answers as complex numbers (*in the form $a + bi$*).

    b. Can the roots of h be represented on the graph of h? Explain.

4. Rewrite the following numbers using i.
    a. $\sqrt{-72}$
    b. $3\sqrt{-39}$
    c. $-2\sqrt{-32}$
    d. $\sqrt{3} \cdot \sqrt{-27}$

5. We found that the roots of $h(x) = x^2 + 6x + 10$ are $x = -3 + i$ and $x = -3 - i$, so we expect $h(-3+i) = 0$ and $h(-3-i) = 0$. We can confirm that $-3+i$ is a root of h by evaluating $h(-3+i)$.

$$h(x) = x^2 + 6x + 10$$
$$h(-3+i) = (-3+i)^2 + 6(-3+i) + 10$$
$$= (-3+i)(-3+i) + 6(-3+i) + 10$$
$$= 9 - 6i + i^2 - 18 + 6i + 10$$
$$= 1 + i^2$$
$$= 1 + (-1)$$
$$= 0$$

Prove that $-3-i$ is also a root of h by evaluating $h(-3-i)$ to show that $h(-3-i) = 0$. (*Use the space above*).

Algebra II

Module 4: Investigation 12
The Quadratic Formula, Complex Roots, and Conjugates

In Exercises #6-11, find the zeros (roots) of the following quadratic functions. For those with complex roots, make sure to write the roots as complex numbers (in the form $a + bi$).

6. $f(x) = x^2 + x + 2.5$

7. $f(x) = x^2 + 5x + 6$

8. $f(x) = x^2 + 5x + 9$

9. $f(x) = 2x^2 - 5x + 2$

10. $f(x) = 4x^2 + 2x - 12$

11. $f(x) = -3x^2 + 3x - 4$

You may have observed that when the discriminant is negative ($b^2 - 4ac < 0$) for a quadratic function with real-number coefficients, we get two complex roots, and that they will always be in a pair like $3 + 2i$ and $3 - 2i$. These are called *conjugate pairs*.

Algebra II

Module 4: Investigation 12
The Quadratic Formula, Complex Roots, and Conjugates

> **Conjugate Pairs**
> If we know that a complex number $a + bi$ is a root of a quadratic function, then the complex number $a - bi$ must also be a root, and we call the two numbers *a conjugate pair*.

12. For each of the following complex numbers, write the other number in the conjugate pair.
    a. $7 - 10i$
    b. $11 + 2i$
    c. $-2 - 2i$
    d. $-i$

There are two interesting facts about conjugate pairs. We already know the first fact – if one of them is a root of a quadratic function, then the other one is also a root. Let's see if we can discover the second interesting fact by doing an exploration.

For Exercises #13-18, find the product. Write your final answer as a complex number (in the form $a + bi$).

13. $(1 + 2i)(1 - 2i)$

14. $(3 - 4i)(2 + 5i)$

15. $(7 - i)(6 + 2i)$

16. $(8 - i)(8 + i)$

17. $(4 + 3i)(5 - 3i)$

18. $(a + bi)(a - bi)$

19. What appears to be true about *conjugate* pairs that isn't true about *all* pairs of complex numbers?

Algebra II

Module 4: Homework
Quadratic Functions

INVESTIGATION 1: INTRODUCTION TO QUADRATIC FUNCTIONS

1. Gina wanted to explore the idea of *speeding up* with her students. She walked across the front of the room, traveling 1 foot over the 1st second, 3 feet over the 2nd second, 5 feet over the 3rd second, 7 feet over the 4th second and 9 feet over the 5th second.

   a. Complete the table to show how far Gina traveled over each of the following intervals.

interval	distance Gina travels during the interval (feet)
from 0 to 1 second	
from 1 to 2 seconds	
from 2 to 3 seconds	
from 3 to 4 seconds	
from 4 to 5 seconds	

   b. Do you notice a pattern in how *the distance Gina travels during the interval* changes from one interval of time to the next? If so, describe the pattern in your own words.

   c. Explain how you know that Gina's speed is increasing. (Increasing speed is called *acceleration*).

   d. Assuming Gina continues the same pattern of acceleration, complete the table showing *the total distance Gina traveled d* (measured in feet) in terms of *the elapsed time t since she started walking* (measured in seconds).

number of seconds since Gina began walking (t)	total distance (in feet) Gina has traveled (d)	distance Gina travels (in feet) during the interval (Δd)	change in Δd
0		1	
1		3	
2		5	
3		7	
4		9	
5			
6			
7			

   e. How are *the number of seconds since Gina began walking (t)* and *total distance (in feet) Gina has traveled (d)* related? Give the formula for *d* in terms of *t*.

   f. Construct a graph of *Gina's total distance traveled* (in feet) in terms of *the number of seconds since she started walking* for integer values of *t*.

   g. On your graph, show the **change** in *the number of seconds since Gina began walking* from $t = 0$ to $t = 1$, then show the corresponding **change** in *the total distance Gina has traveled*. Repeat this for the intervals from $t = 1$ to $t = 2$, $t = 2$ to $t = 3$, and so on.

   h. Discuss how the graph shows that Gina's speed was not constant.

Algebra II — Module 4: Homework — Quadratic Functions

2. Ed, an avid golfer, is 140 yards away from the pin. He selects an 8-iron from his golf bag and hits the ball. The given table shows the ball's height in yards at various seconds after the ball was hit.

Number of seconds since the ball was hit (t)	Height of the ball in yards (h)	Change in height of the ball in yards Δh	
0	0		
1	38.5		
2	63		
3	73.5		
4	70		
5	52.5		
6	21		

a. Describe how the ball's height (in yards) varies with the time (in seconds) since the ball was hit. Be specific in your description.
b. Fill in the values of the 3rd column of the table by determining the change in height of the ball for each one-second interval from 0 to 6 seconds.
c. Describe how Δh changes with the change in time (in seconds) since the ball was hit.
d. Add a fourth column onto the table that gives the change in Δh.
e. Describe how the changes in Δh change with the elapsed time (in seconds) since the ball was hit and explain how we can use this information to conclude that the function that describes the height of the ball in terms of the number of seconds since it was hit is a quadratic function.
f. A graph of the height of the ball in yards in terms of the number of seconds after the ball was hit is provided below. On this graph, show the *change in the number of seconds elapsed since the ball was hit* from $t = 0$ to $t = 1$, then show the corresponding *change in the height of the ball over that interval*. Repeat this for the intervals from $t = 1$ to $t = 2$, $t = 2$ to $t = 3$, $t = 3$ to $t = 4$. How does the graph support your response to part (c)?

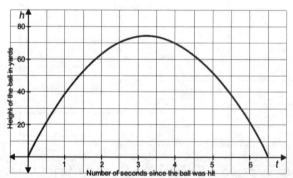

g. The height of the ball in yards is a function of the time elapsed since the ball was hit and is defined by the formula $g(t) = -7t^2 + 45.5t$. Using this formula, and your responses to the previous parts of this task,
   i) Construct a table that gives the height of the ball (in yards) $g(t)$ for half-second intervals from $t = 0$ to $t = 3.5$ since the ball was hit.
   ii) On your table from (i), display: (1) the change in time for half-second intervals from $t = 0$ to $t = 3.5$ seconds since the ball was hit, (2) Δh for each half-second interval from $t = 0$ to $t = 3.5$ seconds since the ball was hit, and (3) the change in Δh for each half-second from $t = 0$ to $t = 3.5$ seconds since the ball was hit.
h. Is the height of the ball in yards a quadratic function of the number of seconds since the ball was hit? Justify your response.

Algebra II
Module 4: Homework
Quadratic Functions

3. Determine whether each of the tables represent input (x) and output (y) quantities of a quadratic function by finding the Δy and the change in Δy. Explain your reasoning.

a.

x	y	Δy	change in Δy
0	3		
2	6.5		
4	13.5		
6	24		
8	38		
10	55.5		

b.

x	y	Δy	change in Δy
0	30		
3	36		
6	36		
9	30		
12	18		
15	12		

4. Determine whether each of the tables represent input (x) and output (y) quantities of a quadratic function by finding the Δy and the change in Δy. Explain your reasoning.

a.

x	y	Δy	change in Δy
0	−3		
1	−6		
2	−7		
3	−6		
4	−3		
5	2		

b.

x	y	Δy	change in Δy
2	2		
4	−10		
6	−38		
8	−80		
10	−140		
12	−220		

5. Kim, a champion road cyclist, has ridden to the top of a mountain and is beginning her descent down a long, straight road. The following table represents Kim's speed in miles per hour as a function of the number of seconds since she began her descent.

Number of seconds since Kim began her descent (t)	Speed in miles per hour (s)
0	10
1	10.25
2	11
3	12.25
4	14
5	16.25
6	19
7	22.25
8	26
9	30.25

Determine if Kim's speed on her descent in miles per hour and the number of seconds since she began her descent is quadratic relation. Justify your response by building a table to illustrate how the number of seconds since she began her descent changes with Δs, and the change in Δs.

Algebra II

Module 4: Homework
Quadratic Functions

6. An automobile company has created a prototype for a new racecar and performs a test race to see how their new car performs. The following table represents the total distance the car has traveled in yards as a function of the number of seconds since the race began.

Number of seconds since the race began (t)	Total distance traveled in yards (d)
0	0
1	1.5
2	6.25
3	14.25
4	29.5
5	54.25

   Determine whether the total distance traveled in yards in terms of the number of seconds since the race began is a quadratic relationship. Justify your response by building a table to illustrate how the number of seconds since the race began changes with Δd, and the change in Δd.

7. Complete the following sentences:
   a. f is a quadratic function of x if as x varies…
   b. g is a linear function of x if as x varies…
   c. h is a constant function of x if as x varies…

8. Consider the functions $f(x) = 2$, $g(x) = 2x$, and $h(x) = 2x^2$. For each of these functions describe how the output changes as x varies by constant amounts and use that to justify whether the function is quadratic, linear, or constant.

9. Use the given formulas to complete the tables. Which of these functions are quadratic? Use the values in the table to justify your response.

   a. $y = x^2 - 3x$

x	y	Δy	change in Δy
0			
3			
6			
9			
12			
15			

   b. $y = \frac{1}{4}x^2$

x	y	Δy	change in Δy
1			
3			
5			
7			
9			
11			

10. Use the given formulas to complete the tables. Which of these functions are quadratic? Use the values in the table to justify your response.

    a. $y = -3(x-1)^2$

x	y	Δy	change in Δy
0			
1			
2			
3			
4			
5			

    b. $y = x(x+4)^2$

x	y	Δy	change in Δy
-2			
-1			
0			
1			
2			
3			

Algebra II

Module 4: Homework
Quadratic Functions

c. $y = (x+2)^2 - 3$

x	y	Δy	change in Δy
-8			
-6			
-4			
-2			
0			
2			

11. The graph of a quadratic function is called a parabola. Examine the patterns of change of the output while considering equal increments of change of the input for each of the following graphs. Which, if any, could represent the graph of a quadratic function?

a.

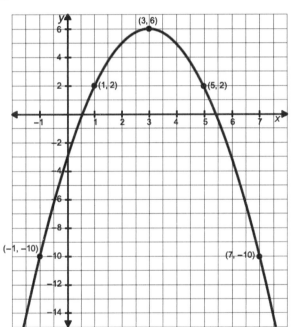

b.

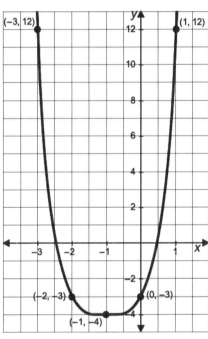

12. The graph of a quadratic function is called a parabola. Examine the patterns of change of the output while considering equal increments of change of the input for each of the following graphs. Which, if any, could represent the graph of a quadratic function?

    a.
    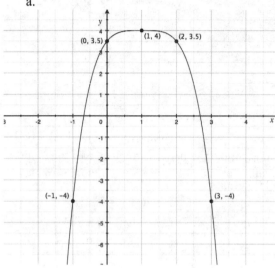

    b.
    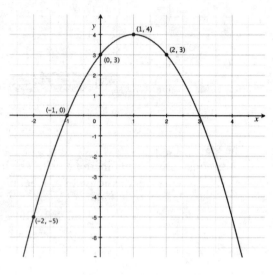

13. The graph of a quadratic function is called a parabola. Examine the patterns of change of the output while considering equal increments of change of the input for each of the following graphs. Which, if any, could represent the graph of a quadratic function?

    a.
    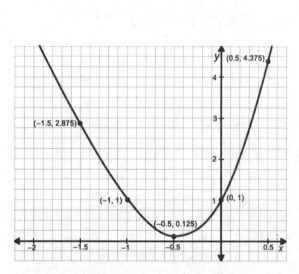

    b.
    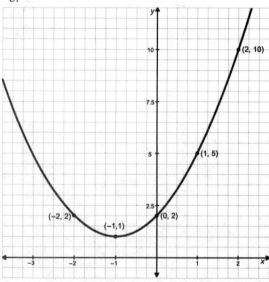

14. Examine the table and explain why it does not represent a quadratic function of y with respect to x. (Hint: Examine Δy and the change in Δy).

x	0	1	2	3	4	5
y	0	1	16	81	256	625

Algebra II

Module 4: Homework
Quadratic Functions

15. Examine the table and explain why it does not represent a quadratic function of y with respect to x. (Hint: Examine Δy and the change in Δy).

x	−4	−2	0	2	4	6
y	14	8	3	7	11	17

16. Janessa conjectured the following characterization of quadratic functions: "A function f is quadratic if Δf defines a linear function where $\Delta f = f(x+1) - f(x)$." This task will examine Janessa's conjecture.

   a. Let $f(x) = ax^2 + bx + c$ where a, b, and c are constants. Construct a table that gives the values of $f(x)$ for $x = 0$, $x = 1$, $x = 2$, $x = 3$, and $x = 4$ as well as the corresponding values of $\Delta f(x)$.

   b. What does it mean for a function to be linear? How are linear functions expressed symbolically?

   c. Suppose $f(x) = ax^2 + bx + c$ where a, b, and c are constants. Express $\Delta f(x)$ in terms of the constants a, b, and c. Is $\Delta f(x)$ linear? Explain.

   d. According to Janessa's characterization of quadratic functions, is f quadratic? Explain.

INVESTIGATION 2: EXPLORING CONCAVITY

17. Which of the following tables could represent ordered pairs from a quadratic function?

   a.

x	y	Δy	
0	1		
1	3		
2	9		
3	19		
4	33		

   b.

x	y	Δy	
2	−4		
4	−16		
6	−36		
8	−64		
10	−100		

   c.

x	y	Δy	
−25	2		
−20	4		
−15	6		
−10	8		
−5	10		

   d.

x	y	Δy	
4	5		
8	10		
12	20		
16	40		
20	80		

   e.

x	y	Δy	
−3	−1		
0	−10		
3	−1		
6	26		
9	71		

   f. Pick two of the tables (in (a) through (e) above) that you believe represent values from a quadratic function. Add one more row to each of these tables and predict the y value that corresponds to the new x-value in your table.

Algebra II
Module 4: Homework
Quadratic Functions

18. Consider the following tables representing the relationship between x and y. Find the missing values so the function that describes how x and y change together is quadratic. Explain how you determined the values that you inserted.

x	0.5	1	1.5	2	2.5	3	3.5	4
y	0.5	2		8	12.5	18		32

x	1	3	5		9	11	13	15
y	0	8	24	48	80	120		224

19. Consider the following tables representing the relationship between x and y. Find the missing values so the function that describes how x and y change together is quadratic. Explain how you determined the values that you inserted.

x	−3	−2	−1	0	1		3	4
y	4	−1	−4		−4	−1	4	

x	0		6	9	12	15	18	21
y	−1	−4	−25		−121	−196		−400

20. The following table of values shows a function that is concave up (has positive concavity). Determine Δy and the change in Δy for the given values of x. Explain in your own words why this function is concave up. (Note: It may help you to re-write the table vertically).

x	−5	−4	−3	−2	−1	0	1	2	3	4	5
y	11.5	7	3.5	1	−0.5	−1	−0.5	1	3.5	7	11.5

21. The following table of values shows a function that is concave up (has positive concavity). Determine Δy and the change in Δy for the given values of x. Explain in your own words why this function is concave up. (Note: It may help you to re-write the table vertically).

x	−8	−6	−4	−2	0	2	4	6	8	10	12
y	81	49	25	9	1	1	9	25	49	81	121

22. The following table of values shows a function that is concave down (has negative concavity). Determine Δy and the change in Δy for the given values of x. Explain in your own words why this function is concave down. (Note: It may help you to re-write the table vertically).

x	−2.5	−2	−1.5	−1	−0.5	0	0.5	1	1.5	2	2.5
y	−16	−11	−7	−4	−2	−1	−1	−2	−4	−7	−11

23. The following table of values shows a function that is concave down (has negative concavity). Determine Δy and the change in Δy for the given values of x. Explain in your own words why this function is concave down. (Note: It may help you to re-write the table vertically).

x	−5	−4	−3	−2	−1	0	1	2	3	4	5
y	−26	−15	−6	1	6	9	10	9	6	1	−6

24. For the following graphs of quadratic functions, determine whether the function is concave up or concave down. Explain your reasoning.
   a.
   b.

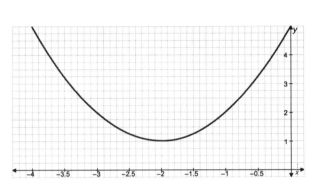

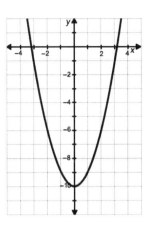

25. For the following graphs of quadratic functions, determine the intervals where the function is concave up (if any) or concave down (if any). Explain your reasoning.
   a.
   b.

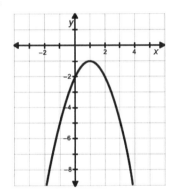

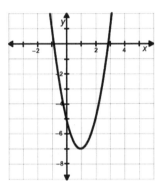

For each given quadratic function in Exercises #26-27, do the following.
   i) Find the missing output values, then determine the values of Δy and the *changes* in Δy.
   ii) Identify the x-value where the function changes from increasing to decreasing or from decreasing to increasing.
   iii) Does the function have positive concavity or negative concavity?
   iv) Sketch the function's graph.

26. $y = -4(x+2)^2 + 8$

x	y	Δy	
-3	4	1.75	
-2.75	5.75	1.25	-0.5
-2.5	7	0.75	-0.5
-2.25	7.75	0.25	-0.5
-2	8	-0.25	-0.5
-1.75	7.75	-0.75	-0.5
-1.5	7	-1.25	-0.5
-1.25	5.75		

27. $y = 3(x-4)^2 + 6$

x	y	Δy	
1	33	-27	
4	6	27	54
7	33	81	54
10	114	135	54
13	249	189	54
16	438	243	54
19	681	297	54
22	978		

Algebra II   **Module 4: Homework**
                                                                                                  Quadratic Functions

28. For each of the tables below, do the following:
    i. Determine whether the table represents values from a quadratic function by determining the Δy and the change in Δy over each x-interval.
    ii. For those functions that define a quadratic relationship, state the interval(s) over which the function is increasing (if any), then state the interval(s) over which the function is decreasing (if any).
    iii. For those functions that define a quadratic relationship, state the interval(s) over which the function has positive concavity (if any), then state the interval(s) over which the function has negative concavity (if any).
    iv. Sketch the graph of each quadratic function.

    a.
x	0	1	2	3	4
y	2	3	6	11	18

    b.
x	−2	−1	0	1	2
y	10	6	5	6	10

    c.
x	0	1	2	3	4
y	12.5	24.5	40.5	60.5	84.5

29. For each of the tables below, do the following:
    i. Determine whether the table represents values from a quadratic function by determining the Δy and the change in Δy over each x-interval.
    ii. For those functions that define a quadratic relationship, state the interval(s) over which the function is increasing (if any), then state the interval(s) over which the function is decreasing (if any).
    iii. For those functions that define a quadratic relationship, state the interval(s) over which the function has positive concavity (if any), then state the interval(s) over which the function has negative concavity (if any).
    iv. Sketch the graph of each quadratic function.

    a.
x	−6	−5	−4	−3	−2
y	40	26	16	8	2

    b.
x	4	8	12	16	20
y	3.2	51.2	259.2	819.2	2000

    c.
x	−6	−2	2	6	10
y	108	12	12	108	300

30. Upon examining a variety of quadratic functions graphically, Maria conjectures, "All quadratic functions are either concave up or concave down, but never both." Examine Maria's conjecture.
    a. What does it mean for a function to be quadratic?
    b. What does it mean for a function to be concave up over a given interval of the input variable?
    c. What does it mean for a function to be concave down over a given interval of the input variable?
    d. Use your responses to parts (a)–(c) to justify Maria's conjecture.

31. What type of functions are neither concave up nor concave down? Justify your response using the definition of concavity and support your response with a graphical representation.

Algebra II

Module 4: Homework
Quadratic Functions

Investigation 3: Properties of Quadratic Functions

For each quadratic function in #32-33, do the following. (Parts (i)–(iii) can be done in the form of a table).
  i. Compute the value of *y* for each whole number value of *x* from $x = -3$ to $x = 3$.
  ii. Determine the value of Δy over each 1-unit interval of *x* from $x = -3$ to $x = 3$.
  iii. Determine the changes in Δy over each 1-unit interval of *x* from $x = -3$ to $x = 3$.
  iv. Determine the vertex of each function and whether it is a maximum or a minimum. Explain.
  v. Determine the line of symmetry for each function. Justify your answer.
  vi. Determine whether the function is concave up or concave down. Explain.

32. a. $y = -x^2 + 2x + 6$ b. $y = 3(x+1)^2$ c. $y = -2(x-1)(x+2)$

33. a. $y = x^2 + 13$ b. $y = -2(x+3)^2 - 4$ c. $y = 3(x+5)^2 + 7$

34. A partial table of values is given below for two quadratic functions. The function *f* represented by the first table has a vertex at $(2, -5)$. The function *g* represented by the next table has a vertex at $(3, -2)$.

x	-2	-1	0	1	2	3
y = f(x)	-53		-17			-8

x	-1	1			7	9
y = g(x)			-2	3		43

  a. Complete the table of values for each quadratic function.
  b. Determine the line of symmetry for each function. Show work to support your answer.
  c. Determine the concavity of each quadratic function. Show work to support your answer.
  d. Determine if the vertex of each function is a maximum or a minimum. Show work to support your answer.

35. A partial table of values is given below for two quadratic functions. The function *f* represented by the first table has a vertex at $(1, 0)$. The function *g* represented by the next table has a vertex at $(-2, 2)$.

x	-4	-2		2	4	6
y = f(x)	-25		-1		-9	-25

x	-4	-3	-2		0	1
y = g(x)	6			3		11

  a. Complete the table of values for each quadratic function.
  b. Determine the line of symmetry for each function. Show work to support your answer.
  c. Determine the concavity of each quadratic function. Show work to support your answer.
  d. Determine if the vertex of each function is a maximum or a minimum. Show work to support your answer.

36. After identifying several axes of symmetry by examining tables of values, Natalie notices that the line of symmetry is always given by $x = -\frac{b}{2a}$ for quadratic functions of the form $f(x) = ax^2 + bx + c$ where *a*, *b*, and *c* are constants. In this task, we examine Natalie's conjecture.
  a. Suppose *f* is a quadratic function and its line of symmetry is given by $x = k$ where *k* is a constant. What can you say about the values $f(x + h)$ and $f(x - h)$ where *h* is a constant? Explain.

(continues....)

b. Suppose $f(x) = ax^2 + bx + c$ where a, b, and c are constants. Use your response in part (a) to derive a formula for the line of symmetry (i.e., use your response in part (a) to set up an equation involving k and solve).

c. Was Natalie's conjecture correct? How can you be sure?

INVESTIGATION 4: MODELING QUADRATIC RELATIONSHIPS

37. If an object starts at rest and accelerates at a rate of a feet per second each second then the distance the object travels (in feet) after t seconds is given by the quadratic function $f(t) = \frac{1}{2}at^2$.

    a. Suppose a car accelerates at a rate of 18 feet per second each second over a 6 second period of time.
        i. Explain what this statement means.
        ii. Write the formula used to calculate the car's distance traveled (in feet) after accelerating for t seconds.
        iii. How far did the car travel during the 6 seconds it spent accelerating?
    b. Suppose another car travels at a constant speed for the same 6-second period. How fast must this car be traveling if it travels the same distance as the accelerating car?

38. A farmer uses a stream as the south barrier to corral his herd of sheep. He has 300 feet of fencing available to construct the remaining three sides of the corral.
    a. Draw the stream and then illustrate three possible configurations for the dimensions of the corral.
    b. If the length on the north side of the corral increases from 20ft to 40ft, how do the combined lengths of the west and east sides of the corral change?
    c. If the length on the north side of the corral increases from 20ft to 40ft, how does the length of the west side of the corral change?
    d. If the length on the north side of the corral increases from 20ft to 40ft, how does the area of the corral change?
    e. If the length on the east side of the corral increases from 15ft to 25ft, how does the length of the north side of the corral change?
    f. If the length on the east side of the corral increases from 15ft to 25ft, how does the area of the corral change?
    g. Determine the length of the east side of the corral, the total amount of fencing used, and the area of the corral for three different lengths of the north side of the corral.
    h. How is the length of the east side of the corral related to the length of the north side of the corral?
    i. Define the area of the pasture as a function of the length of the north side of the corral.
    j. How long should the farmer build the north side and east side of the corral so that its area is as large as possible? How do you know this is the largest area possible?

39. A company makes bolts, screws, and washers. The washers are made with holes of different sizes to fit the various size bolts and screws, as illustrated below. (*For this exercise, recall that the area of a circle A is given by* $A = \pi r^2$ *where r is the circle's radius.*)

    a. Represent the outer radius of the washer for varying sizes of the hole x.
    b. What is the area of the hole in the washer if the radius of the hole is 0.6 cm?
    c. What is the area of the washer's top surface if the radius of the hole is 0.6 cm?
    d. Write a formula that calculates the area of the washer's top surface if the radius of the circular hole is x cm.

Algebra II

Module 4: Homework
Quadratic Functions

40. *Pipes-R-Us* manufactures water pipes of varying thicknesses to accommodate for the amount of water pressure inside the pipe. The company is in the planning stages of determining how much PVC plastic is needed to manufacture a batch of pipes that have a 1-inch inner diameter. The pipes will have varying thicknesses.
    a. Draw a diagram of a pipe. Identify and label all relevant quantities.
    b. Describe how to determine the amount of PVC plastic needed to create a pipe of a specified thickness and length.
    c. For shipping purposes, the pipes are cut at a length of 10 feet. Determine the amount of PVC plastic needed for pipes with an inner diameter of 1 inch and a thickness of 0.25, 0.5, and x inches. (Express your solution in cubic inches. Note that the length of the pipe is given in feet).
    d. Define a function g to determine the amount of PVC plastic needed for one pipe (in cubic inches) in terms of the thickness of the pipe (in inches) x if the length of the pipe is 10 feet. Is g quadratic? Explain.
    e. Graph the function g. What is the vertex of the graph and what, if anything, does it represent in the context of this situation? What, if anything, do the intercepts represent?

41. Suppose a baseball outfielder catches a ball and throws it back to the infield releasing it from his hand 5.5 feet above ground. The baseball's height above the ground h (in feet) after t seconds since it was released can be modeled by the function $f(t) = -16t^2 + 31.5t + 5.5$ where $h = f(t)$.
    a. What is the value of $f(2)$ and what does this value represent.
    b. What is $f(0)$ and what does the point $(0, f(0))$ represent?
    c. If $x = 0.984375$ is the axes of symmetry for the quadratic function f, determine the maximum value of f and explain what, if anything, it represents in the context of this question.

42. Over the years, the owner of a store has found that the number of scented candles she can sell in a month depends on the price according to the formula $c = 200 - 10p$ where c represents the number of candles sold in a month, and p is the price of a candle in dollars. From experience, the owner knows that charging $20 or more for a candle results in no candle sales.
    a. List three possible candle prices and the anticipated number of candles a month the owner expects to sell at that price.
    b. For the three possible prices you gave in part (a), what are the expected amounts of money the owner will generate from sales (referred to as the revenue) that month?
    c. If the price of a candle increases from $6.00 to $6.75, how does the monthly revenue change?
    d. If the price of a candle increases from $14.00 to $15.25, how does the monthly revenue change?
    e. Define a function f to determine the monthly revenue R from scented candle sales in terms of the price p of the candle. Is f quadratic? Explain.
    f. Sketch a graph of the function f. Identify quantities and label your axis. If the function is quadratic, what is the vertex of the graph? What, if anything, does the vertex convey about this situation?
    g. At what price (or prices) should the owner sell the candles if she needs a revenue of $750 per month for her candle sells?

43. A farmer owns a fruit orchard on which 40 trees per acre were planted. Although the trees are bearing fruit, the orchard is overcrowded. Prior experimentation has shown that for every tree per acre removed, the amount of fruit harvested per tree increases by three-fourths of a bushel. With 40 trees per acre, the amount of fruit harvested per tree is 10 bushels.
    a. If four trees per acre are removed, what is the number of bushels of fruit the farmer expects to have per tree? What would the expected harvest be for the entire acre?

(continues…)

b. Select four different amounts of trees removed per acre and find the number of trees remaining, the harvest per tree (in bushels), and the total harvest per acre (in bushels).
c. Write algebraic expressions for the number of trees per acre and the amount of harvest per tree (in bushels). Let *x* represent the number of trees per acre that are removed.
d. Write a function that determines the total harvest per acre if *x* trees are removed per acre. Is the function quadratic? Explain.
e. Draw a graph that describes the total harvest of fruit per acre, in bushels, as a function of the number of trees that were removed per acre. Label your axes and define the relevant quantities.
f. Based on the graph, what is the vertex of the function? In the context of the problem, what, if anything, is the meaning of the vertex? Is such a situation possible? Is the information useful to the farmer?
g. Considering your responses to parts (e) and (f), what is the maximum harvest per acre that can be actually achieved by removing trees? How many trees per acre should be removed to achieve this harvest?

INVESTIGATION 5: ZEROS (ROOTS) OF QUADRATICS

44. Determine the zeroes (roots), if any, of the quadratic functions graphed below.
   a.   b.   c.

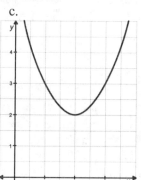

45. Determine the zeroes (roots), if any, of the functions graphed below.
   a.   b.   c.

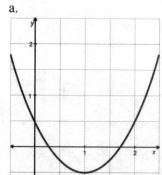
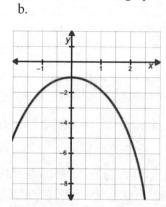
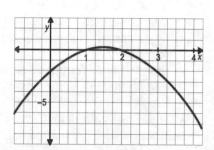

46. For the following quadratic functions written in factored form, determine the standard form by expanding the product of factors.
   a. $f(x) = (x+3)(x-2)$
   b. $g(x) = (2x-3)(x+4)$
   c. $h(x) = (-2x+4)(x+1)$
   d. $p(x) = 3(x+3)(2x-1)$

Algebra II
Module 4: Homework
Quadratic Functions

47. For the following quadratic functions written in factored form, determine the standard form by expanding the product of factors.
    a. $r(x) = -2(5x+3)(6x-5)$
    b. $s(x) = (2x-4)^2$
    c. $b(x) = (3x-7)(2x+11)$
    d. $t(x) = (2x+8)(-x+4)$

48. For each of the functions, determine the zeroes (roots) of the function.
    a. $f(x) = (x+3)(x-2)$
    b. $g(x) = (2x-3)(x+4)$
    c. $h(x) = (-2x+4)(x+1)$
    d. $p(x) = 3(x+3)(2x-1)$

49. For each of the functions, determine the roots of the function.
    a. $r(x) = -2(5x+3)(6x-5)$
    b. $s(x) = (2x-4)^2$
    c. $b(x) = (3x-7)(2x+11)$
    d. $t(x) = (2x+8)(-x+4)$

50. Consider your solutions to Exercises #44 and #45. How many roots can a quadratic equation have? Be sure to justify your answer.

51. Sketch a graph of a function that has no zeroes, one zero, and two zeroes.

52. For each of the functions,
    i. Find the zeros of the quadratic function.
    ii. Determine the line of symmetry of the function.
    iii. Determine the coordinates of the vertex of the function.
    iv. Sketch a graph of the function.
    a. $f(x) = (x+3)(x-2)$
    b. $g(x) = (2x-3)(x+4)$
    c. $h(x) = (-2x+4)(x+1)$

53. For each of the functions,
    i. Find the zeros of the quadratic function.
    ii. Determine the line of symmetry of the function.
    iii. Determine the coordinates of the vertex of the function.
    iv. Sketch a graph of the function.
    a. $p(x) = 3(x+3)(2x-1)$
    b. $r(x) = -2(5x+3)(6x-5)$
    c. $s(x) = (2x-4)^2$
    d. $b(x) = (3x-7)(2x+11)$

54. A function g is given in standard form as $g(x) = 24x^2 + 30x + 9$ and in factored form as $g(x) = 3(2x+1)(4x+3)$.
    a. Determine the zeros of the function. Indicate which form of the function you used to determine this value.
    b. Determine the vertical intercept of the function. Indicate which form of the function you used to determine this value.
    c. Determine the coordinates of the vertex and indicate which form of the function you used to determine this value.
    d. Sketch a graph of this function and indicate which form of the function you used to sketch this graph.

55. A function g is given in standard form as $g(x) = -2x^2 - 7x + 15$ and in factored form as $g(x) = -(x+5)(2x-3)$.
   a. Determine the zeros of the function. Indicate which form of the function you used to determine this value.
   b. Determine the vertical intercept of the function. Indicate which form of the function you used to determine this value.
   c. Determine the coordinates of the vertex and indicate which form of the function you used to determine this value.
   d. Sketch a graph of this function and indicate which form of the function you used to sketch this graph.

56. Sketch the graph of a quadratic function with roots at $x = 4$ and $x = -3.5$.

57. Sketch the graph of a quadratic function with only one roots at $x = 1$.

58. We are given the graph of f, whose factored form looks like $f(x) = a(x-u)(x-v)$.

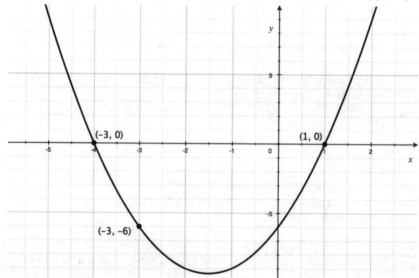

   a. What are the values of u and v for this function?
   b. Changing the value of a can create many different functions with the same roots. To find the value of a that will produce the given graph we can substitute a known ordered pair for the values of x and y and solve the resulting equation for a. Given that $f(-3) = -6$, find the value of a that will generate the graph of f.
   c. Write the formula for function f.

Algebra II

Module 4: Homework
Quadratic Functions

59. We are given the graph of g, whose factored form looks like $g(x) = a(x - u)(x - v)$.

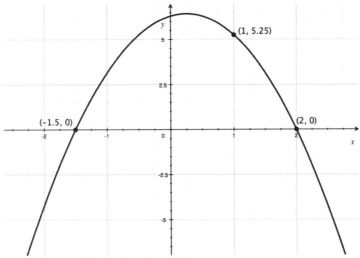

a. What are the values of u and v for this function?
b. Given that $g(1) = 5.25$, find a value of a that will generate the graph of g.
c. Write the formula for function g.
d. Why can't we use one of the horizontal intercepts in part (b) to find the value of a? For example, why can't we use the fact that $g(2) = 0$ to find the value of a?

60. We are given the graph of g, whose factored form looks like $g(x) = a(x - u)(x - v)$.

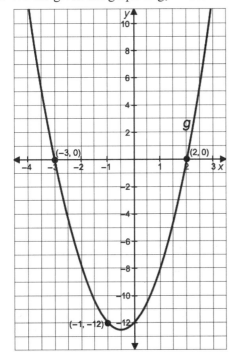

a. What are the values of u and v for this function?

b. Given that $g(-1) = -12$, find a value of a that will generate the graph of g.

c. Write the formula for function g.

Algebra II

Module 4: Homework
Quadratic Functions

61. We are given the graph of h, whose factored form looks like $h(x) = a(x-u)(x-v)$.

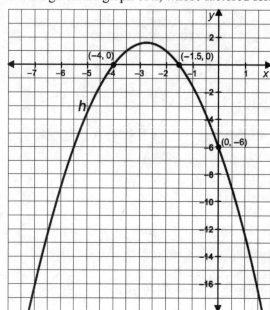

a. What are the values of u and v for this function?

b. Given that $h(0) = -6$, find a value of a that will generate the graph of h.

c. Write the formula for function h.

INVESTIGATION 6: FACTORING QUADRATIC FUNCTIONS

In Exercises #62–77, write the factored form of the quadratic function. If it is not possible say, "not factorable using integers."

62. $h(x) = x^2 + 4x - 12$
63. $p(x) = x^2 + x - 3$
64. $g(x) = 2x^2 + 2x - 4$

65. $c(x) = -x^2 - 13x - 40$
66. $k(x) = 3x^2 + 16x + 21$
67. $f(x) = 3x^2 + 12x - 14$

68. $b(x) = 5x^2 + 3x + 2$
69. $b(x) = 3x^2 + 12x - 15$
70. $t(x) = 6x^2 + 31x + 35$

71. $f(x) = 8x^2 + 32x + 30$
72. $h(x) = 4x^2 - 3x + 2$
73. $p(x) = -3x^2 - x + 10$

74. $g(x) = -6x^2 - 21x - 15$
75. $j(x) = -6x^2 + 5x + 6$
76. $s(x) = x^2 - 25$

77. $m(x) = 3x^2 - 3$

78. Consider the function $p(x) = x^2 - x - 6$.
   a. State the vertical intercept.
   b. Write the function in factored form.
   c. State the zeros of the function.
   d. Find the line of symmetry.
   e. Give the coordinates of the vertex.

79. Consider the function $p(x) = x^2 - 5x - 6$.
   a. State the vertical intercept.
   b. Write the function in factored form.
   c. State the zeros of the function.
   d. Find the line of symmetry.
   e. Give the coordinates of the vertex.

Algebra II

Module 4: Homework
Quadratic Functions

80. Consider the function $p(x) = x^2 + 11x + 28$.
    a. State the vertical intercept.
    b. Write the function in factored form.
    c. State the zeros of the function.
    d. Find the line of symmetry.
    e. Give the coordinates of the vertex.

INVESTIGATION 7: SOLVING QUADRATIC EQUATIONS

81. Consider the function $f(x) = x^2 + 2x + 5$.
    a. What does it mean to solve $x^2 + 2x + 5 = 8$? Sketch a graph to support your explanation.
    b. Determine the value(s) of x that satisfy $x^2 + 2x + 5 = 8$.
    c. What do the value(s) you found in part (b) represent?

82. Consider the function $f(x) = 2x^2 - x - 8$.
    a. What does it mean to solve $2x^2 - x - 8 = 7$? Sketch a graph to support your explanation.
    b. Determine the value(s) of x that satisfy $2x^2 - x - 8 = 7$.
    c. What do the value(s) you found in part (b) represent?

83. Use factoring and the zero-product property to solve the equation $x^2 - 2x - 8 = 16$, then explain what your solution represents.

84. Use factoring and the zero-product property to solve the equation $x^2 + 10 = -13x - 26$, then explain what your solution represents.

85. Given the functions f and g defined by the formulas $f(x) = 2x^2 - 4x - 15$ and $g(x) = x^2 - 3x + 5$, use algebraic methods to determine the point(s) where f and g intersect. Check your answer by graphing f and g on the same axes.

86. Given the functions g and h defined by the formulas $g(x) = 5x^2 + 15x + 3$ and $h(x) = -x^2 + 4x - 1$, use algebraic methods to determine where the graphs intersect. Graph g and h on the same axes and label the point(s) of intersection.

87. Given $f(x) = x^2 + 3x + 3$ and $g(x) = 2x + 9$, use algebraic methods to determine where the graphs intersect. Graph f and g on the same axes and label the point(s) of intersection.

88. Given $f(x) = 2x^2 - 4x - 10$ and $g(x) = 3x + 5$, use algebraic methods to determine where the graphs intersect. Graph f and g on the same axes and label the point(s) of intersection.

INVESTIGATION 8: VERTEX FORM AND COMPLETING THE SQUARE

89. A quadratic function f has a vertex at $(-3, 3)$.
    a. Explain what it means for a function to have a vertex at this point.
    b. How can you determine if this vertex is a maximum or minimum? Explain.
    c. Explain how the values of $f(-4)$ and $f(-2)$ compare to the value of $f(-3)$.
    d. Compare the values of $f(-4)$ and $f(-2)$.

Algebra II   Module 4: Homework
Quadratic Functions

90. A quadratic function f has a vertex at (6, –2).
   a. Explain what it means for a function to have a vertex at this point.
   b. How can you determine if this vertex is a maximum or minimum? Explain.
   c. Explain how the values of $f(2)$ and $f(10)$ compare to the value of $f(6)$.
   d. Compare the values $f(2)$ and $f(10)$.

For Exercises #91–96, determine the vertex for the each function and determine if the vertex is a maximum or minimum. Show your work.

91. $f(x) = 2(x-3)^2 + 7$ 92. $f(x) = (x-1)^2$ 93. $f(x) = 3(x+5)^2 + 5$

94. $f(x) = -3(x-2)^2 + 6$ 95. $f(x) = 5(x+1)^2 - 16$ 96. $f(x) = -7x^2 - 2$

97. Define two different functions, f and g, in vertex form that have their maximum at (2, –5). Explain how you determined these functions.

98. Define two different functions, h and j, in standard form that has their minimum at (3, 6). Explain how you determined these functions.

99. A function k is defined by $k(x) = (x-2)^2 - 4$.
   a. Write k in standard form.
   b. Write k in factored form.
   c. What is the vertex of k?
   d. What is the y-intercept of k?

100. A function h is defined by $h(x) = -(x+3)^2 + 1$.
   a. Write h in standard form.
   b. Write h in factored form.
   c. What is the vertex of h?
   d. What is the y-intercept of h?

101. A function $g(x)$ is given in standard form as $g(x) = 2x^2 + 12x + 16$,
   a. Rewrite the function g in both factored and vertex form.
   b. Determine the roots of the function using whichever form is most helpful.
   c. Determine the vertical intercept of the function.
   d. Determine the vertex and indicate which form of the function you used to determine this value.
   e. Sketch a graph of this function and indicate which form of the function you used to determine this graph.

102. A function $h(x)$ is given in vertex form as $h(x) = -(x-2)^2 + 1$,
   a. Rewrite the function h in both factored and standard form.
   b. Determine the roots of the function using whichever form is most helpful.
   c. Determine the vertical intercept of the function.
   d. Determine the vertex of the function.
   e. Sketch a graph of this function.

For Exercises #103–106 determine the value of c so that you can factor the expression in the form $(x-h)^2$ or $(x+h)^2$.

103. $f(x) = x^2 + 16x + c$ 104. $g(x) = x^2 + 4x + c$ 105. $p(x) = x^2 - 8x + c$ 106. $j(x) = x^2 - 18x + c$

Algebra II

Module 4: Homework
Quadratic Functions

107. Let $f(x) = x^2 + 10x - 3$.
   a. Write f in vertex form.
   b. What is the vertex of f?

108. Let $g(x) = x^2 - 8x + 2$.
   a. Write g in vertex form.
   b. What is the vertex of g?

109. Let $h(x) = 2x^2 + 6x - 1$.
   a. Write h in vertex form.
   b. What is the vertex of h?

110. Let $k(x) = 3x^2 + 12x - 7$.
   a. Write k in vertex form.
   b. What is the vertex of k?

111. Let $m(x) = -4x^2 + 20x + 9$.
   a. Write m in vertex form.
   b. What is the vertex of m?

112. A scientist claims that the number of molecules of hydrogen in a flask containing a particular mixture can be predicted by the quadratic formula $f(x) = -x^2 + 12x + 36$, where x represents the number of minutes that the flask has been placed over a flame.
   a. Write the function f in vertex form.
   b. What is the vertex of f?
   c. Is the vertex of f a maximum or a minimum?
   d. Interpret the meaning of the vertex in the context of this situation.

113. The shape of the cables that hold up a suspension bridge (think of the Golden Gate Bridge) can be modeled by a quadratic formula. A civil engineer is building a new bridge with two towers and claims that the height of the suspension cables between the towers can be modeled by
   $f(x) = 0.5x^2 - 20x + 250$ where x represents the horizontal distance from the first tower in yards, and the height of the cable is measured from the water in yards.
   a. Write the function f in vertex form.
   b. What is the vertex of f?
   c. Is the vertex of f a maximum or a minimum?
   d. Interpret the meaning of the vertex in the context of this situation.

INVESTIGATION 9: DERIVING THE QUADRATIC FORMULA

No homework is associated with Investigation 9: Deriving the Quadratic Formula.

INVESTIGATION 10: THE QUADRATIC FORMULA

114. Let $g(x) = 12x^2 - x - 6$.
   a. Use the Quadratic Formula to find the roots of g.
   b. Find the roots of g by factoring and using the zero-product property.
   c. Sketch a graph of g and explain how you can use it to check your answers in parts (a) and (b).
   d. What other method can you use to check your solutions to parts (a) and (b)?

Algebra II

Module 4: Homework
Quadratic Functions

115. Let $h(x) = 8 + 10x + 2x^2$.
   a. Use the Quadratic Formula to find the roots of h.
   b. Find the roots of h by factoring and using the zero-product property.
   c. Sketch a graph of h and explain how you can use it to check your answers in parts (a) and (b).
   d. What other method can you use to check your solutions to parts (a) and (b)?

116. Let $k(x) = -x - x^2 + 6$.
   a. Use the Quadratic Formula to find the roots of k.
   b. Find the roots of k by factoring and using the zero-product property.
   c. Sketch a graph of k and explain how you can use it to check your answers in parts (a) and (b).
   d. What other method can you use to check your solutions to parts (a) and (b)?

117. Let $m(x) = 2x^2 - x - 3$.
   a. Use the Quadratic Formula to find the roots of m.
   b. Find the roots of m by factoring and using the zero-product property.
   c. Sketch a graph of m and explain how you can use it to check your answers in parts (a) and (b).
   d. What other method can you use to check your solutions to parts (a) and (b)?

For the functions in Exercises #118–121 do the following:
   a. Determine the vertical intercept of the function.
   b. Find the line of symmetry and the coordinates for the function's vertex.
   c. Use the discriminant ($b^2 - 4ac$) to predict the number of horizontal intercepts the function's graph will have.
   d. Use the quadratic formula to find the real-number zeros of the function. Provide both exact answers (if they exist), and approximate rounded decimals for graphing the zeros.
   e. Sketch a graph of each function below using the information produced in your answers to parts (a) through (d).

118. $f(x) = x^2 + 12x - 40$

119. $f(x) = -x^2 - \pi x + \sqrt{7}$

120. $f(x) = 0.1x^2 + 0.2x + 0.3$

121. $f(x) = 24x - 9x^2 - 16$

122. A ball was thrown up the air and the height of the ball, measured in feet above the ground, was modeled by the function $h(t) = -1.6t^2 + 10.5t + 5.7$, where t represents the number of seconds since the ball was thrown.
   a. After how many seconds since the ball was thrown did it reach its original height for the second time?
   b. Based on the model, after how many seconds will the ball hit the ground?
   c. Based on the model, after how many seconds will the ball be at its maximum height? What is the maximum height of the ball? Explain how you determined these values.
   d. Sketch a graph of h. Label your axes.
   e. Does the ball ever reach a height of 20 feet? If yes, when does this happen? If this doesn't happen, explain why.

123. A biologist monitors the way a fish population in a local lake rises and falls. The population is modeled by the function $f(t) = 2000 + 7000t - 250t^2$, where t is measured in years since January 1, 1990 when the lake was stocked with fish.
   a. In what month and year will the fish population be the same as it was on January 1, 1990?

(continues…)

Algebra II

Module 4: Homework
Quadratic Functions

   b. Based on the model, in what month and year will all the fish in the lake have died?
   c. Based on the model, in what month and year will the fish population be maximized? What is the maximum number of fish? Explain how you determined these values.
   d. Sketch a graph of f. Label your axes.
   e. Does the population of fish in the lake reach 75,000? If yes, in which year (or years) does this happen? If this doesn't happen, explain why.

INVESTIGATION 11: "IMAGINARY" NUMBERS

124. Show that the only possible solutions to the equation $x^2 - 7 = -18$ are $x = \sqrt{-11}$ and $x = -\sqrt{-11}$.

125. Show that the only possible solutions to the equation $-2x^2 - 5 = 19$ are $x = \sqrt{-12}$ and $x = -\sqrt{-12}$.

126. Rewrite each of the following expressions using i where $i = \sqrt{-1}$.
   a. $\sqrt{-9}$
   b. $-\sqrt{-144}$
   c. $\sqrt{-23}$
   d. $\sqrt{-73}$
   e. $-\sqrt{-24}$
   f. $\sqrt{-80}$
   g. $5 + \sqrt{-8}$
   h. $-7 - \sqrt{-60}$

127. Rewrite each of the following expressions using i where $i = \sqrt{-1}$.
   a. $\sqrt{-16}$
   b. $\sqrt{-81}$
   c. $\sqrt{-22}$
   d. $2 - \sqrt{-11}$
   e. $\sqrt{-25}$
   f. $\sqrt{-28}$
   g. $3 + \sqrt{-9}$
   h. $1 - \sqrt{-44}$

128. Perform the indicated operations. Write your final answer as a complex number $a + bi$ in simplest form.
   a. $5i(13 - 2i)$
   b. $7\sqrt{5}i(3i - \sqrt{5})$
   c. $(1 - 2i) - 2(3i)$ d.
   d. $(-\sqrt{5} + 5i) + (6 - \sqrt{3}i)$
   e. $6(\frac{7}{4} - \frac{5}{3}i) - \frac{1}{2}(4 + 5i)$
   f. $(1 + \sqrt{2}i)(4 - \sqrt{3}i)$
   g. $(2 - 3i)^2$
   h. $(2 - 3i)(9i + 8)$
   i. $\sqrt{2}i(2 - 3i)$
   j. $(a + bi) - (c + di)$

129. Perform the indicated operations. Write your final answer as a complex number $a + bi$ in simplest form.
   a. $3i(2 + 3i)$
   b. $-2i(1 - i)$
   c. $\left(-\frac{6}{5} + 5i\right) + \left(\frac{9}{2} - \frac{16}{3}i\right)$
   d. $(5 + i)(2 - 2i)$
   e. $(-1 - i)(6 + 3i)$
   f. $(3 - 2i)^2$
   g. $(a + bi) + (c + di)$
   h. $ai(b + ci)$

Exercises #130–135 represents a factored form for a quadratic function with "imaginary" numbers for zeros. State the zeros of the function, then write the function in standard form.

130. $g(x) = 5(x + 9i)(x - 9i)$

131. $j(x) = 3(2x - \sqrt{-11})(2x + \sqrt{-11})$

132. $p(x) = (x - 5\sqrt{3}i)(x + 5\sqrt{3}i)$

133. $q(x) = (3x - 12i)(3x + 12i)$

134. $f(x) = (x - \sqrt{2}i)(x + \sqrt{2}i)$

135. $r(x) = (4x - 2i\sqrt{2})(4x + 2i\sqrt{2})$

Algebra II

Module 4: Homework
Quadratic Functions

INVESTIGATION 12: THE QUADRATIC FORMULA, COMPLEX ROOTS, AND CONJUGATES

In Exercises #136-138,
(i) prove that the given numbers are roots of each quadratic function and
(ii) write the quadratic function in factored form.

136. $x = 1+i$ and $x = 1-i$ are roots for $f(x) = x^2 - 2x + 2$

137. $x = 2+5i$ and $x = 2-5i$ are roots for $g(x) = x^2 - 4x + 29$

138. $x = -4+3i$ and $x = -4-3i$ are roots for $h(x) = x^2 + 8x + 25$

In Exercises #139–144 find the zeros (roots) of the following quadratic functions. For those with imaginary roots, make sure to write the roots as complex numbers (that is, in the form $a + bi$).

139. $f(x) = x^2 - \frac{3}{2}x + \frac{9}{16}$

140. $g(x) = 5x^2 + 3x - 1$

141. $h(x) = 2x^2 - 2x + 1$

142. $p(x) = 4x^2 - 14x + 12$

143. $q(x) = 4x^2 + 53$

144. $w(x) = -2x^2 - 3x - 2$

145. Determine if the conjugate pair $14 + i$ and $14 - i$ are roots of the function f defined by $f(x) = 2x^2 - 56x + 394$.

146. Determine if $-\sqrt{2}i$ is a root of the function g defined by $g(x) = x^2 + 8$.

For the following complex numbers in Exercises #147–150, write the other number in the conjugate pair.

147. $14 + i$

148. $-\sqrt{2}i$

149. $-3 - 5i$

150. $-13 + \sqrt{7}i$

151. Evaluate the following products.
  a. $(\sqrt{7} + 2i)(\sqrt{5} - 3i)$
  b. $(\sqrt{7} + 2i)(\sqrt{7} - 2i)$
  c. $(\frac{2}{5} + 4i)(\frac{2}{5} - 2i)$
  d. Compare the products $(a + bi)(c - di)$, $(a + bi)(a - di)$, and $(a + bi)(a - bi)$. What do you notice about the answer? Is it an imaginary number? How does this explain the nature of your answers in parts (a)–(c)?

152. Evaluate the following products.
  a. $(\sqrt{7} + 2i)(\sqrt{7} - 3i)$
  b. $(\frac{2}{5} + 4i)(\frac{3}{7} - 2i)$
  c. $(\frac{2}{5} + 4i)(\frac{2}{5} - 4i)$
  d. Compare the products $(a + bi)(c - di)$, $(a + bi)(a - di)$, and $(a + bi)(a - bi)$. What do you notice about the answer? Is it an imaginary number? How does this explain the nature of your answers in parts (a)–(c)?

Algebra II

Module 5: Investigation 1
The Box Problem: Context, Table, Formula, and Graph

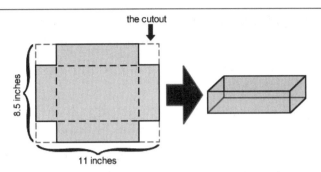

1. a. Create an open box (without a top) by: i) cutting four equal-sized square corners from an 8.5 by 11-inch sheet of paper and ii) folding up the sides and taping them together at the edges. Is it necessary that the cutout be square? Explain.

    b. List the relevant quantities for determining the volume of a box made by cutting four equal-sized square corners from a sheet of paper and folding up the sides.

    c. Use a ruler to measure the length of the side of your square cutout (in inches). Then calculate the volume of your box.

    Cutout length:_____   Box's height:_____

    Length of the box's base: _____   Width of the box's base: _____

    Volume:_____

    d. Describe your method for computing a box's volume when all you know is the length of the side of the square cutout, and you are unable to use a ruler to measure other dimensions.

2. Explain how the length of the side of the cutout and the volume of the box changed together.

3. How are you thinking about the volume? What does it mean for the volume of the box to be 25 cubic inches?

Algebra II

Module 5: Investigation 1
The Box Problem: Context, Table, Formula, and Graph

4. In your groups answer the following:
   a. How are the length of the side of the cutout and the width of the box changing together?

   b. Define a function *f* to relate the *width of the box* and the *length of the side of the square cutout*. Be sure to define your variables.

   c. Describe how the length of the side of the cutout and the length of the base of the box change together.

   d. Define a function *g* to relate the *length of the box* and the *length of the side of the square cutout*. Be sure to define your variables.

   e. Define a function *h* to relate the *area of the box's base* to the *length of the side of the square cutout*. Be sure to define your variables.

   f. Define a function *k* to relate the *volume of the box* to the *length of the side of the square cutout*.

   g. Let *x* represent the length of the side of the square cutout; let *w* represent the width of the box's base; let *l* represent the length of the box's base. Complete the following table.

Quantity	Smallest possible value	Largest possible value
x		
w		
l		

Module 5: Investigation 1

© 2014 Carlson, O'Bryan & Joyner

Algebra II

Module 5: Investigation 1
The Box Problem: Context, Table, Formula, and Graph

5. a. Using your graphing calculator, create a graph that represents the volume of the box V (measured in cubic inches) in terms of the length of the side of the square cutout x (measured in inches). (*When determining the window setting on your calculator, consider the possible values of x and the possible values of V.*) Draw the graph on the given axes.

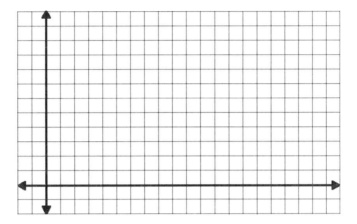

b. Label two points on the graph and describe what each point conveys about the box.

   Point 1:_____   Point 2:_____

c. Identify the point on the graph that corresponds to the dimensions of the box you created.

d. As x (the length of the side of the cutout) increases from 0.5 to 0.75 inches, how does the volume of the box change?

e. As x (the length of the side of the cutout) increases from 2.1 to 2.7 inches, how does the volume of the box change?

f. Represent a change of cutout length from 2 inches to 3 inches on your graph, then represent the corresponding change in volume.

g. Estimate the interval of values for the length of the side of the cutout x for which the volume of the box decreases.

Algebra II

Module 5: Investigation 1
The Box Problem: Context, Table, Formula, and Graph

6. Using your graphing calculator, determine the following.
   a. An approximate value for the maximum volume of the box. Label the point on the graph you created in Exercise #5 that corresponds to the approximate maximum value that you determined.

   b. The dimensions of the box with the maximum volume you estimated in part (a).

7. Using your graphing calculator, determine the following.
   a. The length of the side of the cutout when the volume of the box is 25 cubic inches. [This is equivalent to saying "Solve for x when $V = 25$."]

   b. How the length of the side of the cutout x changes as the volume decreases from 50 cubic inches to 25 cubic inches.

Algebra II

Module 5: Investigation 4
Polynomial Functions and End Behavior

Investigation 2: *The Bottle Problem: From Bottle to Graph* and Investigation 3: *Co-varying Distance & Time* are available online along with associated homework at www.rationalreasoning.net.

Polynomial Functions

A ***polynomial function*** is a function of the form $p(x) = a_n x^n + a_{n-1} x^{x-1} + \ldots + a_1 x^1 + a_0$ where $a_n, a_{n-1}, \ldots a_0$ are real numbers and n is a non-negative integer. $a_n, a_{n-1}, \ldots a_0$ are called ***coefficients*** and $a_n x^n$ is called the ***leading term*** of the function (the term with the largest exponent on the variable).

When the polynomial function is written so that the terms with larger exponents on the variable come before terms with smaller exponents, we say the polynomial is written in ***standard form*** (such as $f(x) = 3x^5 - 5x^4 + 12x + 10$). The leading term $a_n x^n$ is the first term when the polynomial is written in standard form.

In this course we've seen many examples of polynomial functions:
- Quadratic functions, such as $f(x) = 2x^2 - 5x + 3$, are a type of polynomial function.
- Linear functions, such as $g(x) = \frac{5}{3}x - 7$, are also a type of polynomial function.

We often classify polynomial functions by the largest exponent on the variable in the formula.

Degree

The degree of a polynomial function is the largest exponent n on the variable in the formula, which will also be the exponent on the variable in the leading term.

> ***Ex:*** All quadratic functions are degree 2 polynomials because the leading term will be in the form $a_2 x^2$ where a_2 is some real number and 2 is the exponent on x in the leading term. A linear function is a degree 1 polynomial.

1. Determine if each of the following is a polynomial function. If so, list the degree of the polynomial.

   a. $f(x) = 2x^3 + 5x - 5$

   b. $g(x) = -5x^2 + 2x - 6x^4 + 3$

   c. $h(x) = 3x^3 - 2x^{-2} + 5x - 18$

   d. $j(x) = \frac{1}{5}x^3 - 6$

   e. $k(x) = 25$

   f. $m(x) = -4x^{1/2} - 5x + 11$

   g. $n(x) = x^4$

Algebra II

Module 5: Investigation 4
Polynomial Functions and End Behavior

2. a. For each of the given functions, do the following:
   i. Determine the degree of the polynomial.
   ii. Complete the tables.
   iii. Graph the function on the given set of axes.

$f(x) = x^3$ $g(x) = x^7$ $h(x) = x^4$ $j(x) = x^{12}$

x	f(x)
2	
10	
100	
-2	
-10	
-100	

x	g(x)
2	
10	
100	
-2	
-10	
-100	

x	h(x)
2	
10	
100	
-2	
-10	
-100	

x	j(x)
2	
10	
100	
-2	
-10	
-100	

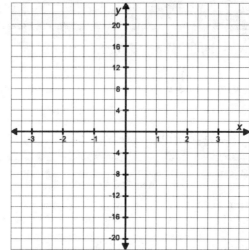

b. As the value of x increases without bound (gets larger and larger, often notated by $x \to +\infty$), how does the value of
   i. $f(x)$ change?

   ii. $g(x)$ change?

   iii. $h(x)$ change?

   iv. $j(x)$ change?

c. As the value of x decreases without bound (gets more and more negative or smaller and smaller, often notated by $x \to -\infty$), how does the value of
   i. $f(x)$ change?

   ii. $g(x)$ change?

   iii. $h(x)$ change?

   iv. $j(x)$ change?

Algebra II

Module 5: Investigation 4
Polynomial Functions and End Behavior

> **End Behavior of a Function**
> How a function behaves as the input quantity increases or decreases without bound is called its **end behavior**. End behavior is also sometimes referred to as *long-run behavior*.

d. How are the functions *f* and *g* alike and how are they different?

e. How are the functions *h* and *j* alike and how are they different?

f. How are the functions *f* and *h* alike and how are they different?

g. What general statements do you think you can make about how the exponent on a power function impacts the behavior of the function?

h. If all of the function outputs were multiplied by –3, how is the end-behavior of each function affected?

There are two common ways to determine the end behavior of a polynomial function. The first method is to use a table or graph to examine what happens to the output values of a function as *x* increases and decreases without bound.

A second method uses the formula of the polynomial function. For example, let $p(x) = x + 5$. When the magnitude of *x* gets very large, the "+ 5" portion of the function rule is a proportionally smaller and smaller part of determining the final value of the expression $x + 5$.

For example,
- $p(20) = 25$,
- $p(1,000) = 1,005$, and
- $p(100,000,000) = 100,000,005$.

We can say that as *x* increases without bound, $p(x) \to x$ (and likewise as *x* decreases without bound, $p(x) \to x$) because the relative size of $x + 5$ and *x* are nearly identical.

Algebra II

Module 5: Investigation 4
Polynomial Functions and End Behavior

3. a. As x increases or decreases without bound, what happens to the value of the expression $x^2 - 4$? Why does this happen?

   b. As x increases or decreases without bound, what happens to the value of the expression $x^3 + 6x^2$? Why does this happen?

   c. As x increases or decreases without bound, what happens to the value of the expression $5x^2 - 32x$? Why does this happen?

   d. As x increases or decreases without bound, what happens to the value of the expression $-7x^4 + 4x^3 - x^2 + 3x + 1$? Why does this happen?

An important conclusion from Exercise #3 is that the end behavior of a polynomial function can be determined by examining the leading term of the function and analyzing its end behavior since, in the long run, a polynomial function behaves likes its leading term.

4. For each of the polynomial functions below, identify the leading term of the function and describe the end behavior of the function based on your analysis of the leading term. (*You may need to write the function in standard form first*). Graph each of the functions on a calculator to confirm your answer.
    a. $f(x) = 5x^3 + 3x^2 - 16x + 8$
    b. $g(x) = 9x^2 - 4x - 2x^4$

    c. $h(x) = x^2(x+3)^2(3x-4)(x+1)$
    d. $k(x) = -7(x-5)(x^2+1)$

Algebra II

Module 5: Investigation 5
Zeros (Roots) of Polynomial Functions

1. As we determined for quadratic functions in Module 4, the roots of any polynomial function *f* are the values of the input quantity *x* such that $f(x) = 0$. Consequently, if these *roots* are real numbers then they are the *x*-value(s) where the graph touches or crosses the *x*-axis (e.g., $x = c$). Remember that *roots* are also referred to as *zeros* – the terms are interchangeable.

   a. Suppose a candle that was originally 19 inches long was burning at a rate of 2.5 inches per hour from one end. We defined the function $f(x) = 19 - 2.5x$ to represent the length of the candle (in inches) in terms of the number of hours *x* the candle had been burning. If we want to determine how many hours the candle had been burning when it was 0 inches long, we substituted 0 for $f(x)$ and solve the resulting equation for *x*.

   $$0 = 19 - 2.5x$$
   $$2.5x = 19$$
   $$x = 7.6$$

   i. Graph *f*, label the root on the graph, and explain what $x = 7.6$ represents in the context of the problem.

   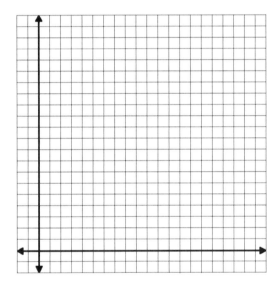

   ii. Justify your answer for each of the true/false questions

   T F   7.6 is a root of *f*

   T F $f(0) = 7.6$ is a zero of *f*

   T F $x = 7.6$ is a zero of *f*

Algebra II

Module 5: Investigation 5
Zeros (Roots) of Polynomial Functions

b. In the "Box Problem" from Investigation #1 we created the function $v(x) = (11-2x)(8.5-2x)x$, which determined the volume of an open-top box with respect to the length of the side of the square cutout. Recall that this is written in *factored form* and that we can multiply the terms together to re-write the function in *standard form*. Doing so results in $v(x) = 4x^3 - 39x^2 + 93.5x$, revealing that v is a degree 3 (or cubic) polynomial function.

   i. Which form of the function will make it easiest to solve the equation $v(x) = 0$? Explain.

   ii. Determine the roots of v.

   iii. What do the roots correspond to in the context of the problem?

2. Determine the roots of the following polynomial functions algebraically. Factor if necessary. Confirm your solutions either by graphing the functions or inputting the value of the root(s) to the function to show the output will be zero.

   a. $f(x) = 2(x+3)(3x-5)$ b. $g(x) = x(2x-6)(x-8)$ c. $h(x) = x^2 - x - 30$

   d. $j(x) = 2x^3 - 14x^2 + 24x$ e. $k(x) = -6x(3x+2)^2(x^2-4)$ f. $m(x) = 3x^2(x^2-5)$

   g. $n(x) = 2x^2 + 3x - 9$

3. Suppose that the roots of a polynomial function f are $x = 2$, $x = -3$, and $x = 5$.
   a. What factors must be included in the formula that defines f? Explain.

Algebra II

Module 5: Investigation 5
Zeros (Roots) of Polynomial Functions

b. Define three different functions that have roots of $x = 2$, $x = -3$, and $x = 5$. Explain how these functions differ from the function f where $f(x) = (x-2)(x+3)(x-5)$.

c. Given that $g(x) = 2(x-2)(x+3)(x-5)$ and $h(x) = -2(x-2)(x+3)(x-5)$,
   i. How do the formulas of functions g and h differ?

   ii. How does this difference affect the outputs and graph of the two functions?

   iii. Graph the two functions on your calculator to confirm your hypothesis.

   iv. Describe how the sign of the leading coefficient affects the sign and behavior of the function.

4. Given that f is defined by $f(x) = (x-2)^2(x+3)$,
   a. Determine the zeros of the function (Recall that these are values of x such that $f(x) = 0$).

   b. Determine the sign of f (is it positive or negative?) on each of the given intervals.
      i) $x < -3$

      ii) $-3 < x < 2$

      iii) $x > 2$

   c. Determine the y-intercept of f.

Algebra II

Module 5: Investigation 5
Zeros (Roots) of Polynomial Functions

d. Plot one point from each of the intervals in part (b), then draw a rough sketch of the remaining graph of f.

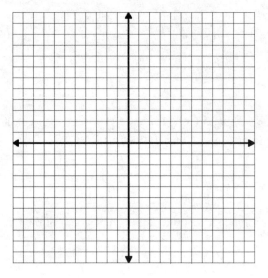

e. Explain why f changes signs at $x = -3$.

f. Explain why f does not change signs at $x = 2$.

g. What general statement can you make about the relationship between the *power of the factor* of a polynomial function, and the sign of the function on the intervals immediately to the left and right of the root that is associated with a factor?

Algebra II

Module 5: Investigation 5
Zeros (Roots) of Polynomial Functions

Multiplicity and Repeated Roots

In the factored form of a function definition, the number of times a factor appears is its ***multiplicity***. Roots associated with factors that have a multiplicity greater than one are called ***repeated roots***.

Ex: Let $f(x) = 2x(x+4)(x+4)(x-3)$, which we can also write as $f(x) = 2x(x+4)^2(x-3)$. The factor $(x+4)$ has multiplicity 2, and the associated root $x = -4$ is a repeated root.

---

5. Suppose a function has zeros at $x = -1$ and $x = 5$. Define a function in factored form such that
   a. The function changes signs at both $x = -1$ and $x = 5$.

   b. The function changes signs at *only* $x = 5$.

   c. The function never changes signs (does not change at either $x = -1$ or $x = 5$).

6. Let $f(x) = -\frac{1}{2}(x-3)^2(x+1)^3$.
   a. Identify the leading term of *f* (in standard form) and use it to determine the end-behavior of the function.

   b. Determine the zeros of *f* and state the multiplicity of each zero.

Algebra II

Module 5: Investigation 5
Zeros (Roots) of Polynomial Functions

c. Determine the sign of f (whether f is positive or negative) on each interval of the domain where the function may change signs (*the intervals that are separated by the zeros*). Explain why the function cannot change signs within the intervals.

d. Evaluate $f(0)$.

e. Use the information you gathered from parts (a) through (d) to sketch a graph of f on the given axes.

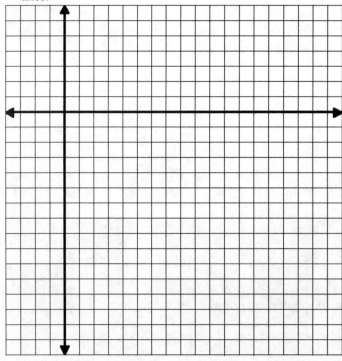

Algebra II   **Module 5: Investigation 6**
                                                Determining the Zeros of a Polynomial Function Algebraically

Prior to the existence of graphing calculators, mathematicians had to find the zeros of polynomial functions algebraically (and by hand). They used known theorems and the coefficients of a polynomial to create a list of possible zeros and then tested these possible zeros to determine whether or not they were actual roots. In this investigation, we'll examine some of these techniques while applying what we know about polynomial functions and their zeros.

1. Consider all 1st degree polynomial functions.
   a. What kinds of functions are 1st degree polynomial functions?

   b. What is the maximum number of real zeros a 1st degree polynomial function can have? What is the minimum number of real zeros it can have?

   c. Sketch the graph of a 1st degree polynomial.

2. Consider all 2nd degree polynomial functions.
   a. What kinds of functions are 2nd degree polynomial functions?

   b. What is the maximum number of real zeros a 2nd degree polynomial function can have? What is the minimum number of real zeros it can have?

   c. Sketch graphs of 2nd degree polynomials with each of the possible number of zeros based on your answer to part (b).

Algebra II

Module 5: Investigation 6
Determining the Zeros of a Polynomial Function Algebraically

3. Consider all 3rd degree polynomial functions.
   a. What kinds of functions are 3rd degree polynomial functions?

   b. What is the maximum number of real zeros a 3rd degree polynomial function can have? What is the minimum number of real zeros it can have?

   c. Sketch graphs of 3rd degree polynomials with each of the possible number of zeros based on your answer to part (b).

4. Look at your answers from Exercises #1–3. Did you notice a relationship between the maximum number of real zeros of a polynomial function and the degree of the function? What was this relationship?

The relationship you observed in Exercise #4 helped mathematicians to narrow down the possibilities while trying to identify the zeros of a polynomial function. But what else do we know about the zeros of polynomial functions? Suppose that we know that a zero for a polynomial function *f* exists at $x = 2$. This means then that the formula for *f must* have a factor of $(x-2)$ when written in factored form.

For example, let $f(x) = x^2 - x - 2$. If we know $x = 2$ is a zero, we know that the factored formula will have a factor of $(x-2)$. Since this is a quadratic function, it's fairly easy to determine the other factor by using the techniques discussed in Module 4. The other factor is $(x+1)$ and the function can be rewritten as $f(x) = (x-2)(x+1)$. This also reveals that $x = -1$ is a zero.

Polynomial functions with a degree greater than two, however, are harder to factor than quadratic functions. For example, $g(x) = x^3 + 2x^2 - 5x - 6$ is not easy to factor, and this makes it harder to find the zeros of *g*. But suppose you knew that *g* has a zero at $x = -1$. Then we know one factor of *g must* be $(x+1)$. We don't know what the other factors are, but we can use *long division of polynomials* to help us find them.

Algebra II
Module 5: Investigation 6
Determining the Zeros of a Polynomial Function Algebraically

Think about how you learned to solve $1015 \div 7$ without using a calculator. You used long division, and it looked something like this.

$$\begin{array}{r} 145 \\ 7{\overline{\smash{)}1015}} \\ \underline{-7} \\ 31 \\ \underline{-28} \\ 35 \\ \underline{-35} \\ 0 \end{array}$$

Notice that at the last step we end up with 0, which we call the remainder. When the remainder is 0 it means that $1015 \div 7 = 145$ exactly, or that $7 \cdot 145 = 1015$. The number 7 is a *factor* of 1015. Had the remainder not been equal to 0, then 7 would not be a factor of 1015.

Similar reasoning can be used to factor polynomial functions (although the division is slightly more complicated). To write $g(x) = x^3 + 2x^2 - 5x - 6$ in factored form knowing that $(x+1)$ is a factor, as in the previous example, we do the following.

$$\begin{array}{r} x^2 + x - 6 \\ x+1{\overline{\smash{)}x^3 + 2x^2 - 5x - 6}} \\ \underline{-\left(x^3 + 1x^2\right)} \\ x^2 - 5x \\ \underline{-\left(x^2 + 1x\right)} \\ -6x - 6 \\ \underline{-(-6x - 6)} \\ 0 \end{array}$$

Since the remainder of the long division is 0, it confirms that $(x+1)$ is a factor of $x^3 + 2x^2 - 5x - 6$.

We can now re-write $g(x) = x^3 + 2x^2 - 5x - 6$ as $g(x) = (x+1)(x^2 + x - 6)$. Since we know how to factor $(x^2 + x - 6)$, we can go further and say that $g(x) = (x+1)(x+3)(x-2)$. We can now see that the g has three real zeros, namely $x = -1$, $x = -3$, and $x = 2$.

Factors and Zeros: *The Remainder Theorem*

$(x - a)$ is a factor of g and $x = a$ is a zero of g if and only if the remainder is 0 when $g(x)$ is divided by $(x - a)$.

Algebra II

Module 5: Investigation 6
Determining the Zeros of a Polynomial Function Algebraically

5. Confirm that the given *x*-value is a factor of the function.
    a. $x = 3$
    $h(x) = x^3 + 2x^2 - 11x - 12$

    b. $x = 5$
    $k(x) = x^3 + 0x^2 - 19x - 30$

6. For each of the given *x*-values and functions, do the following.
    i. Determine whether the *x*-value is a zero of the function by writing the corresponding factor, then using long division to determine the remainder after division.
    ii. If the *x*-value is a zero of the function, re-write the function in factored form (as much as possible using your knowledge of factoring).
    If the *x*-value is not a zero of the function, determine the value of $p(x)$ for the given value of *x*. Compare this value to the remainder you found in part (a).

    a. $x = 2$
    $p(x) = x^3 - 3x^2 - 3x - 20$

    b. $x = 6$
    $p(x) = x^3 - 7x^2 + 0x + 36$

c. $x = -3$
    $p(x) = x^3 + 3x^2 - x - 3$

d. $x = -2$
    $p(x) = x^3 - 4.5x^2 - x - 12$

7. Let $f(x) = 3x^3 + 8x^2 - 30x$ and $g(x) = 2x^3 + 10x^2 - x + 42$. There are three solutions to this system, one of which is (7, 1211). Find the other two solutions algebraically and then verify your answers by graphing the system.

Algebra II
Module 5: Investigation 6
Determining the Zeros of a Polynomial Function Algebraically

8. Let $f(x) = -2x^3 + 3x^2 - 5x + 10$ and $g(x) = -3x^3 + 3x^2 + 2x + 4$. There are three solutions to this system, one of which is $(-3, 106)$. Find the other two solutions algebraically and then verify your answers by graphing the system.

Algebra II

Module 5: Investigation 7
Seeing Structure in Polynomial Expressions

You may recall that the difference of two perfect square terms will factor according to a special pattern. For example, $y^2 - x^2 = (y-x)(y+x)$. We can sometimes use this pattern to help us factor higher-order polynomial expressions. First, observe that x^6 can be thought of as a perfect square because $(x^3)^2 = x^6$. Likewise, $4y^{10}$ can be thought of as a perfect square because $(2y^5)^2 = 4y^{10}$.

Now, consider the expression $x^6 - 4y^{10}$. From our previous conclusions we can factor this expression as a difference of two perfect squares as shown.

$$x^6 - 4y^{10}$$
$$(x^3)^2 - (2y^5)^2$$
$$(x^3 - 2y^5)(x^3 + 2y^5)$$

Here are two more examples for you to consider. Look them over carefully and make sure you can follow the reasoning.

$$a^4 - 144 \qquad\qquad 100x^8 - 25y^{12}$$
$$(a^2)^2 - (12)^2 \qquad\qquad (10x^4)^2 - (5y^6)^2$$
$$(a^2 - 12)(a^2 + 12) \qquad\qquad (10x^4 - 5y^6)(10x^4 + 5y^6)$$

1. Factor the following expressions.
   a. $y^4 - 64$

   b. $36y^2 - 81x^2$

   c. $y^6 - 25$

   d. $49y^4 - 9x^{10}$

Algebra II **Module 5: Investigation 7**
Seeing Structure in Polynomial Expressions

2. Recall that the roots of a function f are the x-values where $f(x) = 0$. For each given function, do the following.
   i) Factor the expression that defines the function output.
   ii) Determine the real roots and non-real roots (if any) of each function.

   a. $f(x) = x^4 - 16$

   b. $g(x) = 4x^4 - 36$

3. For each given function, do the following.
   i. Factor the expression that defines the function's outputs. *Hint: We can think of an expression like* $x^4 + 5x^2 + 6$ *as* $(x^2)^2 + 5(x^2) + 6$.
   ii. Determine the real roots and non-real roots (if any) of each function.

   a. $f(x) = x^4 - x^2 - 20$

   b. $h(x) = x^4 + 29x^2 + 100$

Algebra II

Module 5: Investigation 8
Polynomial Translations

1. A local school is planning a community carnival to raise money for charity. Based on their research they believe that they can model the expected attendance by the function $f(x) = -0.5x^2 + 80x - 2350$ where x is the forecasted high temperature for the day in degrees Fahrenheit.

   a. Suppose the carnival will have 50 volunteers helping at booths, taking tickets, and selling refreshments. What is the total number of people (including attendees and volunteers) that will be at the carnival if the forecasted high temperature is 60°F? How did you determine this value?

   b. Draw the graph of f by constructing a table of values or by using a graphing calculator. Use this graph to create the graph of the total number of people (including attendees and volunteers) that will be at the carnival as a function of the forecasted high temperature (in degrees Fahrenheit). [Call this new function g.]

   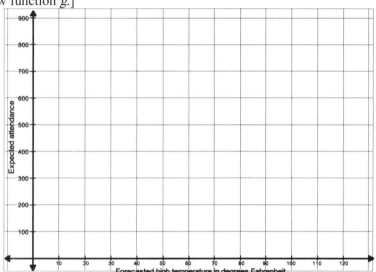

   c. What is the formula that defines g? How did you determine it?

2. Suppose f and g are related such that $g(x) = f(x) + 6$.
   a. For any given input value, how will the output values of f and g compare?

   b. How will the graphs of f and g be related?

c. Repeat parts (a) and (b) if instead $g(x) = f(x) - 5$.

> **Vertical Translations (Shifts)**
> When two functions are related such that for any input value the output values are always separated by a constant difference, we say that their graphs are related by a *vertical translation*.

3. Each of the following pairs of functions are related by a vertical translation. Use a graphing calculator to graph each pair of functions and explore their relationship.

   $f(x) = x^2$ $f(x) = 2x^3$ $f(x) = -3x^2 + x$ $f(x) = 4x^5 + 5$
   $g(x) = x^2 + 4$ $g(x) = 2x^3 - 8$ $g(x) = -3x^2 + x + 1$ $g(x) = 4x^5 + 7$

4. The graph of f is given below. Draw the graph of g if $g(x) = f(x) - 3$.

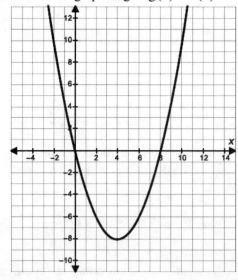

5. A table of values for h is given below. Create a table of values for function f if $f(x) = h(x) + 2.4$.

x	$h(x)$
−1.5	12.5
−0.9	2.89
−0.7	1.16
1.6	−1.51
3.5	−50

6. Let $f(x) = 2x^2 - 5x + 6$ and $g(x) = x^5 + 3x^2 - 1$. What are the formulas defining functions p and q if $p(x) = f(x) + 8$ and $q(x) = g(x) - 1$?

7. The following graph models the student enrollment at a certain high school as a function of the number of years since 1990. [Call this function f.]

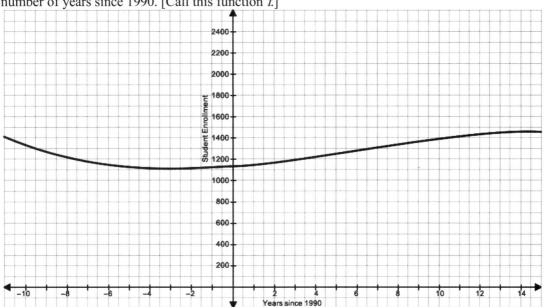

a. Draw a new graph representing function g, the student enrollment at the high school as a function of the number of years since 1980.

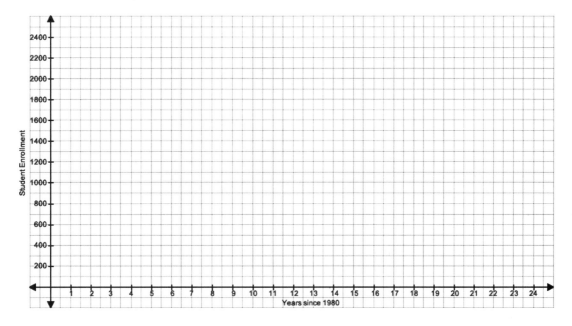

b. Which of the following statements is true? Defend your choice.
$$g(1) = f(11) \quad \text{or} \quad g(1) = f(-9)$$

c. What is the relationship between the graphs of f and g?

d. If x represents the number of years since 1980, which of the following statements is true? Defend your choice. $\quad g(x) = f(x+10) \quad \text{or} \quad g(x) = f(x-10)$

e. Draw a new graph representing function h, the student enrollment at the high school as a function of the number of years since 1995.

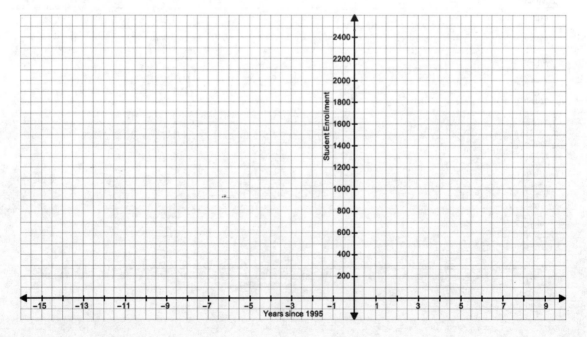

Algebra II

Module 5: Investigation 8
Polynomial Translations

f. Which of the following statements is true? Defend your choice.
$$h(7) = f(12) \quad \text{or} \quad h(7) = f(2)$$

g. What is the relationship between the graphs of *f* and *h*?

h. If *t* represents the number of years since 1995, which of the following statements is true? Defend your choice. $\quad h(t) = f(t+5) \quad \text{or} \quad h(t) = f(t-5)$

Horizontal Translation (Shift)

When two functions are related such that the same output values occur in both functions when the input values are separated by a constant difference, we say that their graphs are related by a *horizontal translation.*

---

8. Consider the graph of *f*.

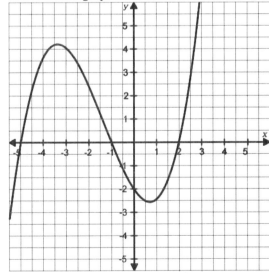

a. Draw the graph of *g* given that $g(x) = f(x-2)$.

b. Draw the graph of *h* given that $h(x) = f(x+1) + 2$.

Algebra II

Module 5: Investigation 8
Polynomial Translations

9. The table below shows ordered pairs for function f.

x	$f(x)$
−2	16
−1	14
0	8
1	4
2	8
3	26

    a. Create a table of values for g if $g(x) = f(x+3)$.

    b. Create a table of values for h if $h(x) = f(x-4) + 6$.

10. Given that $f(x) = 4x^5 - 3x^3 + x$ and $g(x) = -5x^2 + 7$, answer the following questions.

    a. Evaluate $h(2)$ if $h(x) = f(x-5)$.

    b. Evaluate $j(2)$ if $j(x) = f(x+1)$.

    c. Write the formula defining p if $p(x) = f(x-1)$.

    d. Write the formula defining q if $q(x) = g(x+4) + 2$.

Algebra II

Module 5: Investigation 9
Polynomial Reflections

1. The given table shows ordered pairs for function f.

x	f(x)
−2	16
−1	14
0	8
1	4
2	8
3	26

   a. What is the value of $g(2)$ given that $g(x) = -f(x)$?

   b. What is the value of $g(-1)$ given that $g(x) = -f(x)$?

   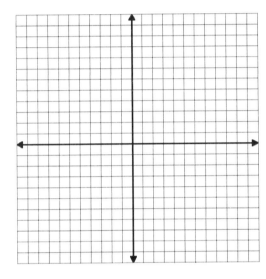

   c. Plot the points for f on the given axes, then sketch a best-fit curve through the points.

   d. On the same axes, draw the graph of g given that $g(x) = -f(x)$.

   e. What is the value of $h(2)$ given that $h(x) = f(-x)$?

   f. What is the value of $h(-1)$ given that $h(x) = f(-x)$?

   g. On the axes from part (c), draw the graph of h given that $h(x) = f(-x)$.

Reflections

When two functions are related such that for any input value the output values have the same magnitude but opposite signs (such as $g(x) = -f(x)$), we say that their graphs are related by a _vertical reflection_.

When two functions are related such that the same output value occurs for input values that have the same magnitude but opposite signs (such as $h(x) = f(-x)$), we say that their graphs are related by a _horizontal reflection_.

2. The graph of f is given.

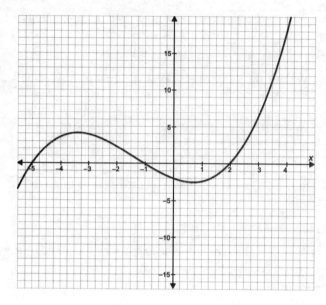

a. On the same axes, draw the graph of g given that $g(x) = -f(x)$.

b. On the same axes, draw the graph of h given that $h(x) = f(-x)$.

3. Given that $f(x) = 3x^4 + x^2$, $g(x) = x^3 + 4x^2 - 8$, and $h(x) = \frac{1}{2}x + 3$, answer the following questions.

   a. Evaluate $j(3)$ given that $j(x) = -f(x)$.

   b. Evaluate $k(-1)$ given that $k(x) = f(-x)$.

   c. Write the formula defining n given that $n(x) = -h(x)$.

   d. Write the formula defining p given that $p(x) = g(-x)$.

   e. Write the formula defining q given that $q(x) = -f(x+1) + 5$.

Algebra II

Module 5: Investigation 10
Combining Polynomial Functions: Adding and Subtracting

1. The following tables represent the populations of two cities (City A and City B) between 1950 and 2000.

   function f

input	output
years since 1950	population of City A (in millions)
0	1
10	0.8
20	0.8
30	1
40	1.4
50	2

   function g

input	output
years since 1950	population of City B (in millions)
0	3.6
10	4.25
20	4.8
30	5.25
40	5.6
50	5.85

   a. Plot the values from the table on the given axes, then draw the approximate best-fit curve for functions f and g.

   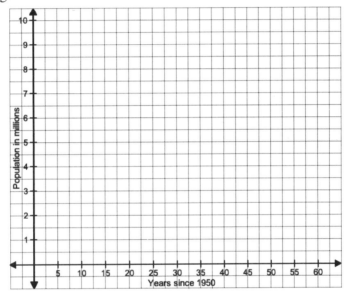

   b. Function h models the combined population of the two cities (in millions) as a function of the number of years since 1950.
      i. Explain how to determine the value of $h(40)$.

      ii. Graph h. Use the axes in part (a).

Module 5: Investigation 10

c. If t is the number of years since 1950, then $f(t) = 0.001t^2 - 0.03t + 1$ and $g(t) = -0.0005t^2 + 0.07t + 3.6$.

   i. Write the formula that defines h.

   ii. Evaluate $h(17)$ and explain its meaning.

d. Function j models the difference in population between the two cities (in millions) as a function of the number of years since 1950. Explain how we can construct the graph of j.

You are probably very familiar with how the graphs of simple polynomial functions look, such as $f(x) = 2x^2$ or $f(x) = -x^4$. However, more complicated polynomial functions like $f(x) = -3x^3 + 10x^2 + 5x - 8$, are much harder to imagine.

One way to think about how these functions are built is to think about them as being created by the sum of two or more functions. For example, we can break apart $f(x) = x^2 + 2x$ by letting $g(x) = x^2$ and $h(x) = 2x$, and then thinking of f as $f(x) = g(x) + h(x)$. When it comes to generating the graph of f, we can plot points by evaluating $g(x)$ and $h(x)$ and using these results to determine $f(x)$.

2. Consider $f(x) = x^2 + 2x$.

   a. Complete the tables showing the values of x^2 and $2x$ for the given values of x.

x	x^2
−2	
−1.5	
−1	
−0.5	
0	
0.5	
1	
1.5	
2	

x	$2x$
−2	
−1.5	
−1	
−0.5	
0	
0.5	
1	
1.5	
2	

Algebra II

Module 5: Investigation 10
Combining Polynomial Functions: Adding and Subtracting

b. Graph $y = x^2$ and $y = 2x$.

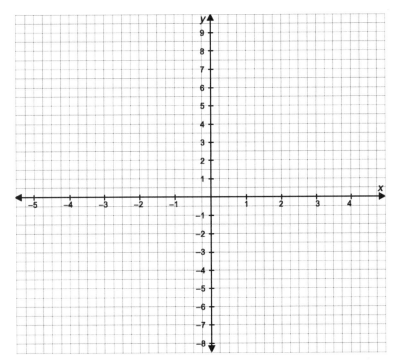

c. Graph f by using the fact that for every value of x, $f(x)$ is the sum of x^2 and $2x$. (Use the axes from part (b).)

3. Graph $f(x) = 2x^3 - 4x$ given the graphs of $y = 2x^3$ and $y = -4x$.

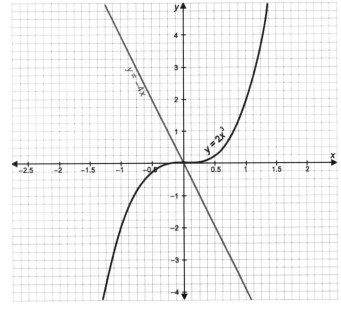

4. Graph $f(x) = -2x^2 + 3x + 1$ given the graphs of $y = -2x^2$, $y = 3x$, and $y = 1$.

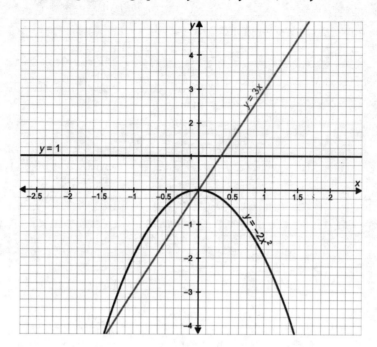

5. Let $f(x) = 3x^4 + 7x$, $g(x) = -5x^3 + 4x + 1$, and $h(x) = x^2 - 10x + 5$.
   a. What is the formula for j given that $j(x) = g(x) + h(x)$?

   b. What is the formula for n given that $n(x) = h(x) - f(x)$?

   c. If $p(x) = g(x) + f(x)$, evaluate $p(3.4)$.

   d. If $q(x) = f(x) - g(x) + h(x)$, evaluate $q(-1.5)$.

Algebra II

Module 5: Investigation 11
Combining Polynomial Functions: Composing and Inverting

Recall that a function's inverse represents the same relationship between two quantities but with the input and output quantities reversed. That is, if $y = f(x)$ (with an input of x and an output of y), then $x = f^{-1}(y)$ (with an input of y and an output of x).

For Exercises #1-3, do the following.
    a. Graph the function.
    b. Graph the function's inverse relation.
    c. Answer any questions provided.

1. $f(x) = 0.5x^3$

   a. 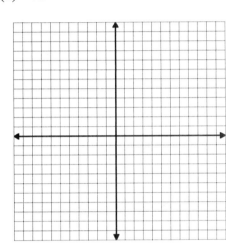   b.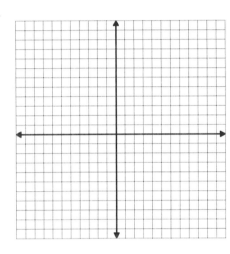

   c. i. Is the inverse relation a function? Defend your answer.

      ii. Write a formula to define f^{-1}.

2. $f(x) = x^2$

   a. 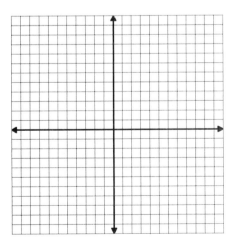   b.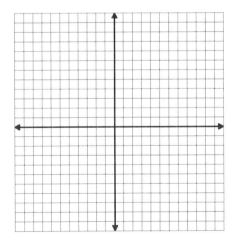

c. i. Is the inverse relation a function? Explain your reasoning.

ii. If we restrict the domain of f, we can limit its ordered pairs so that the inverse will be a function. How should we restrict the domain of f so that f^{-1} is a function?

3. $f(x) = \frac{2}{3}x - 4$
   a.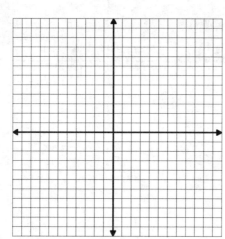
   b.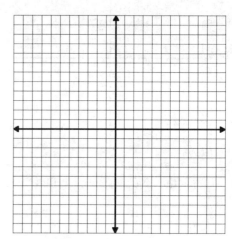

   c. Write a formula to define f^{-1}.

For Exercises #4-7, do the following.
   a. Use a calculator or other graphing software to graph the function. Using this graph as a guide, restrict the domain of the function (if necessary) so that its inverse relation will be a function.
   b. Write a formula to define the inverse function.

4. $f(x) = 2x^2 - 4$

5. $f(x) = 4(x-2)^3$

Algebra II
Module 5: Investigation 11
Combining Polynomial Functions: Composing and Inverting

6. $f(x) = \frac{3}{2}x + \frac{7}{2}$

7. $f(x) = (x+5)^4 - 2$

In a previous module we explored the composition of inverse functions. Specifically, we found that, if f and f^{-1} are inverses, then $f^{-1}(f(m)) = m$ and $f(f^{-1}(n)) = n$. This is because inverse functions "undo" each other.

8. Compose each pair of inverse functions from Exercises #4-7 in both possible ways ($f^{-1}(f(m))$ and $f(f^{-1}(n))$.

9. Use the graphs of the polynomial functions *f* and *g* to evaluate each expression.

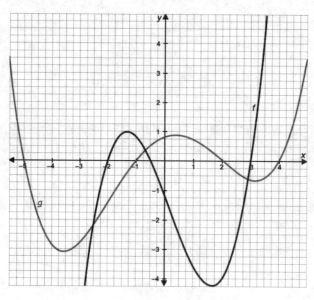

a. $g(f(-2))$
b. $g(f(1))$
c. $f(g(3))$
d. $f(g(0))$

10. Given that $f(x) = -4x^2 + x$, $g(x) = x^3 + 2x - 1$, and $h(x) = 0.5x + 2$, answer the following.
    a. Evaluate $g(f(1))$.

    b. Evaluate $f(h(-5))$.

    c. Find the formula to define *p* if $p(x) = h(g(x))$.

    d. Find the formula to define *q* if $q(x) = f(h(x))$.

Algebra II

Module 5: Homework
Polynomial Functions

INVESTIGATION 1: THE BOX PROBLEM: CONTEXT, TABLE, FORMULA, AND GRAPH

1. A box designer has been charged with the task of determining the volume of various boxes that can be constructed by cutting four equal-sized square corners from a 14-inch by 17-inch sheet of cardboard and folding up the sides.
   a. What quantities vary in this situation? What quantities remain constant?
   b. Create an illustration to represent the situation and label the relevant quantities.
   c. What are the dimensions of the box (length, width and height) if the cutout is 0.5 inches? 1 inch? 2 inches?
   d. Let x represent the length of the side of the square cutout; let w represent the width of the box's base; let l represent the length of the box's base. Complete the following table assuming each quantity is measured in a number of inches.

Quantity	Smallest possible value	Largest possible value
x		
w		
l		

   e. Define a function to relate the *length of the box* and the *length of the side of the square cutout*. Be sure to define your variables.
   f. Define a function to relate the volume of the box to the length of the side of the square cutout.
   g. Use your graphing calculator to approximate the maximum volume of the box rounded to the nearest tenth of a cubic inch. Explain how you know that the value you obtained is the maximum value of the box rounded to the nearest tenth.

2. An open box is constructed by cutting four equal-sized square corners from a 10-inch by 13-inch sheet of cardboard and folding up the sides. The graph below represents the volume of a box (measured in cubic inches) in relation to the length of the side of the square cutout (measured in inches).

   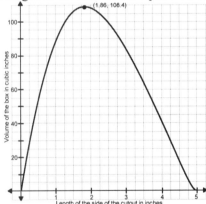

   a. Define a function to relate the volume of the box to the length of the side of the square cutout.
   b. Determine the value of the volume when the length of the side of the square cutout is:
      i) 0 inches   ii) 1 inch   iii) 2.5 inches   iv) 5 inches
   c. Use the graph above to approximate the value(s) of the length of the side of the square cutout when the volume is:
      i) 0 cubic inches   ii) 100 cubic inches   iii) 108.4 cubic inches
   d. As the length of the side of the square cutout increases from 0 to 1 inch, by how much does the volume change? Illustrate this change on the graph.
   e. As the length of the side of the cutout increases from 4 to 5 inches, by how much does the volume change? Illustrate this change on the graph.

   (continues...)

f. Use the given graph to determine the dimensions of two boxes with volume 65 cubic inches.
g. Use the given graph to determine for what values of the length of the side of the square cutout the volume is increasing.
h. Use the given graph to determine for what values of the length of the side of the square cutout the volume is decreasing.

3. A box designer has been charged with the task of determining the surface area of various open boxes (no lids) that can be constructed by cutting four equal-sized square corners from an 8-inch by 11.5-inch sheet of cardboard and folding up the sides. (Note: The surface area is the total area of the box's sides and bottom.)
    a. What quantities vary in this situation? What quantities remain constant?
    b. Create an illustration to represent the situation and label the relevant quantities.
    c. What are the dimensions of the box (length, width and height) if the cutout is 0.5 inches? 1 inch? 2 inches?
    d. What is the surface area of the box when the cutout is 0.5 inches? 1 inch? 2 inches?
    e. Define a function for the following relationships:
        i) the length of the base of the box as a function of the length of the side of the square cutout
        ii) the width of the box as a function of the length of the side of the square cutout
    f. Define a function s that relates the total surface area (measured in square inches) of the open box to the size of the square cutout x (measured in inches).
    g. There are two ways to think of determining a box's surface area: by adding up the area of each piece of the box; and by subtracting the area of the four cutouts from the area of the initial sheet of cardboard.
        (1) Which method did you use in defining your function in part (f)?
        (2) Define a function to compute the box's surface area by the other method. Compare the two formulas. Are they equivalent (that is, do they always produce the same result)? Show this algebraically.

4. Wynndows, Inc. manufactures specialty windows. One of their styles is in the shape of a fanned semicircle, as shown below. When a customer orders this window, she specifies the length of the base of the window. The cost of materials for manufacturing the window includes the cost of the glass panes together with the cost of the framing material.

    a. Determine the length of the outer frame (not including the inside fans) for a window with a base of 4 feet.
    b. Define a function to determine the total length of the framing material used in making a window, including the two inner supports, based on the length of the base. Define your variables.
    c. If the cost of the framing material is $12 per linear foot, what is the cost of the material for a window frame with a base-length of 6 feet?
    d. Suppose that your budget limits you to spending $500 on your window frames and supports. What is the longest base length that you can afford?
    e. How much does the cost for framing material change if the base of the window changes from 4 feet to 5 feet? How much does the cost for framing material change if the base of the window changes from 8 feet to 9 feet?
    f. Does it cost more to add one foot to the base of a smaller or larger window? Explain whether your answer will hold for any change of one foot in the base length of a window.

Algebra II

Module 5: Homework
Polynomial Functions

INVESTIGATION 2: THE BOTTLE PROBLEM: FROM BOTTLE TO GRAPH

> Investigation 2: *The Bottle Problem: From Bottle to Graph* and Investigation 3: *Co-varying Distance & Time* are available online along with associated homework at www.rationalreasoning.net.

INVESTIGATION 3: CO-VARYING DISTANCE & TIME

INVESTIGATION 4: POLYNOMIAL FUNCTIONS AND END BEHAVIOR

24. Determine if each of the following is a polynomial function. If so, list the degree of the function.
    a. $a(x) = 2x^2 + 5$
    b. $b(x) = 3x^3 + 2x^{-2} + 1$
    c. $c(x) = -.5$
    d. $d(x) = -x$

25. Determine if each of the following is a polynomial function. If so, list the degree of the function.
    a. $a(x) = \dfrac{1}{x^2}$
    b. $b(x) = 3x - 4.5x^3 + 1$
    c. $c(x) = 0$
    d. $d(x) = -8x^3 + 2x^{-3} + 10$

26. For each of the polynomial functions given, identify the leading term of the function (multiply factors if necessary) and describe the end behavior of the function based on your analysis of the leading term.
    a. $f(x) = 16x^3 - 4x^6 + 26x^2 + 8x$
    b. $g(x) = 9x^9 - 3 + 8x^4 + 12x^5$
    c. $k(x) = 5 + 5x^3 - x^8 + 2x^2$
    d. $k(x) = -4(x+7)(x^2-4)$

27. For each of the polynomial functions given, identify the leading term of the function (multiply factors if necessary) and describe the end behavior of the function based on your analysis of the leading term.
    a. $f(x) = 13x - 2x^5 - 2$
    b. $g(x) = x^4 + 2.3x^3 + 12$
    c. $h(x) = -12 + 3x^7 + 4x^6$
    d. $h(x) = x(x-2)^2(2x+3)(x-1)$

INVESTIGATION 5: ZEROS (ROOTS) OF POLYNOMIAL FUNCTIONS

28. Determine the zeros of the following polynomial functions given in factored form.
    a. $g(x) = -4x(x-6)(2x+9)$
    b. $h(x) = x^2(x+8)^2(x^3-12)$
    c. $k(x) = -(5x+1)(x-2)(3x+8)$
    d. $m(x) = (2-x)(x-1.5)(x^2-1)$
    e. Construct a graph of *m* and label the zeros of *m* on its graph.

29. Determine the zeroes of the following polynomials given in factored form.
   a. $a(x) = -(x-2)(x+2.5)$
   b. $b(x) = x(x-1)(2x-1)$
   c. $c(x) = -x(x^2-4)(2x+6)$
   d. $d(x) = (-3x-15)(x+2)(2x^2-32)$
   e. Construct a graph of c and label the zeroes of c on its graph.

30. Determine the exact zeros of the following polynomial functions algebraically. Show work.
   a. $f(x) = x^2 + 4x - 12$
   b. $g(x) = 2x^2 + 10x$
   c. $h(x) = -x^4 + x^2$
   d. Construct a graph of h and label the zeroes of the function on the graph.

31. Determine the exact zeros of the following polynomial functions algebraically. Show work.
   a. $k(x) = 3x^2 + 6x - 72$
   b. $m(x) = x^3 - 10x^2 + 21x$
   c. $n(x) = 2x^2 - 5x - 3$
   d. Construct a graph of m and label the zeroes of the function on the graph.

32. Suppose that the roots of a polynomial function f are $x = 6$, $x = -1.5$, and $x = -4$. Define three different polynomial functions that have these roots.

33. Suppose that the roots of a polynomial function g are $x = 3$, $x = -1$, $x = 0$, and $x = -0.5$. Define three different polynomial functions that have these roots.

34. Given the graph of a polynomial function g whose graph is given:

   a. What are the roots (zeros) of g?
   b. Approximate $g(0)$.
   c. On what interval(s) is the function g concave up? (See Module 4 Investigation 2 for review of concavity if needed)
   d. On what intervals is the function g concave down?
   e. As x increases without bound, how does $g(x)$ change?
   f. As x decreases without bound, how does $g(x)$ change?

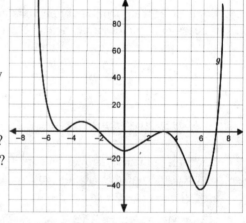

35. Given the graph of a polynomial function h whose graph is given:

   a. What are the roots (zeros) of h?
   b. Find $h(0)$.
   c. On what interval(s) is the function h concave up? (See Module 4 Investigation 2 for review of concavity if needed)
   d. On what intervals is the function h concave down?
   e. As x increases without bound, how does $h(x)$ change?

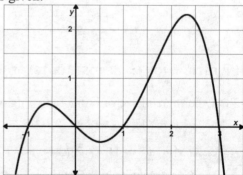

Algebra II   Module 5: Homework
                                                      Polynomial Functions

36. Suppose a function has zeros at $x=-3$, $x=0$, and $x=2$. Define a function in factored form so that
    a. The function never changes signs.
    b. The function only changes signs at $x=0$.
    c. The function changes signs at all three zeros ($x=-3$, $x=0$, and $x=2$).

37. Suppose a function has zeros at $x=-5$, $x=-1$, and $x=4$. Define a function in factored form so that
    a. The function changes signs at all three zeros ($x=-5$, $x=-1$, and $x=4$).
    b. The function never changes signs.
    c. The function changes signs only at $x=-1$ and $x=4$.

38. Given the function $f(x) = -2(x+4)^2(x-1)$
    a. Identify the leading term of f (in standard form) and use it to determine the end-behavior of the function.
    b. Determine the roots of the function. Describe whether or not the function changes signs at each of the roots.
    c. Determine the sign of f (whether f is positive or negative) on each interval of the domain where the function may change signs.
    d. Evaluate $f(0)$.
    e. Use your responses in parts (a) – (d) to sketch a graph of f.

39. Given the function $g(x) = x^2(x-2)^3(2x+5)^2$
    a. Identify the leading term of g (in standard form) and use it to determine the end-behavior of the function.
    b. Determine the roots of the function. Describe whether or not the function changes signs at each of the roots.
    c. Determine the sign of g (whether g is positive or negative) on each interval of the domain where the function may change signs.
    d. Evaluate $g(0)$. What does this value represent?
    e. Use the information you gathered from parts (a) – (d) to sketch a graph of the function.

For Exercises #40-41 use the following graphs.
    i.   ii.

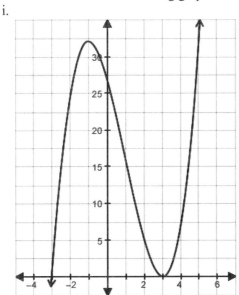
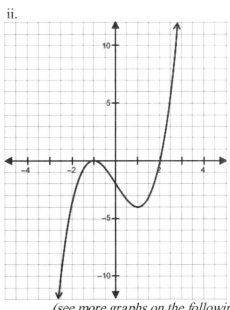

(see more graphs on the following page)

iii.
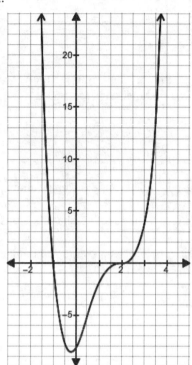

iv.
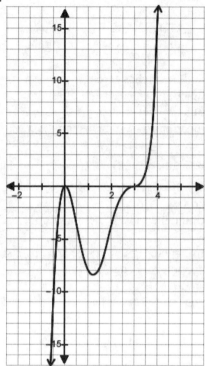

v.
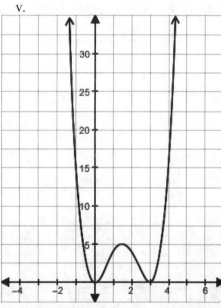

vi.
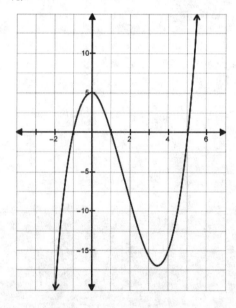

40. Without the use of a calculator, match each of the following functions with its graph above and provide a short justification for your choice.

$$f(x) = (x+3)(x-3)^2$$
$$g(x) = x^2(x-3)^3$$
$$h(x) = x^2(x-3)^2$$

Algebra II

Module 5: Homework
Polynomial Functions

41. Without the use of a calculator, match each of the following functions with its graph above and provide a short justification for your choice.

$$j(x) = (x+1)^2(x-2)$$
$$k(x) = (x+1)(x-2)^3$$
$$m(x) = (x+1)(x-1)(x-5)$$

INVESTIGATION 6: DETERMINING THE ZEROS OF A POLYNOMIAL FUNCTION ALGEBRAICALLY

42. Sketch an example graph of a function for each of the possible number of zeros that a fourth degree polynomial function can have.

43. Sketch an example graph of a function for each of the possible number of zeros that a fifth degree polynomial function can have.

44. Complete the following calculations:
   a. $x-5 \overline{) 2x^3 - 11x^2 - x + 30}$
   b. $x-1 \overline{) 2x^3 - 15x^2 + 24x - 11}$
   c. $x+6 \overline{) 2x^3 + 5x^2 - 64x - 132}$

45. Complete the following calculations:
   a. $x+3 \overline{) x^3 + 2x^2 - 3x}$
   b. $x+1 \overline{) x^3 + x^2 - 4x - 4}$
   c. $x-2 \overline{) 3x^3 - 3x^2 - 6x}$

46. For each of the given *x*-values and functions,
   i. Determine whether the *x*-value is a zero of the function by writing the corresponding factor, then using long division to determine the remainder of the division.
   ii. If the *x*-value is a zero of the function, re-write the function in factored form (as much as possible using your knowledge of factoring).
   If the *x*-value is not a zero of the function, determine the value of $p(x)$ for the given value of *x*. Compare this value to the remainder you found in part (i).

   a.
   $x = -3$
   $p(x) = x^3 - 2x^2 - 5x + 6$

   b.
   $x = -1$
   $p(x) = x^3 - 11x^2 + 4x + 60$

   c.
   $x = 2$
   $p(x) = x^3 - 15x^2 + 56x - 60$

47. For each of the given *x*-values and functions,
   i. Determine whether the *x*-value is a zero of the function by writing the corresponding factor, then using long division to determine the remainder of the division.
   ii. If the *x*-value is a zero of the function, re-write the function in factored form (as much as possible using your knowledge of factoring).
   If the *x*-value is not a zero of the function, determine the value of $p(x)$ for the given value of *x*. Compare this value to the remainder you found in part (a).

   a.
   $x = -5$
   $p(x) = x^3 + 6x^2 + 3x - 10$

   b.
   $x = 4$
   $p(x) = x^3 - 3x^2 - 33x + 35$

   c.
   $x = -2$
   $p(x) = x^3 + 9x^2 + 6x - 16$

48. Let $f(x) = 2x^3 + 7x^2 + x + 2$ and $g(x) = 3x^3 + 10x^2 - 5x - 6$. There are three solutions to this system, one of which is $(-1, 6)$. Find the other two solutions algebraically and then verify your answers by graphing the system.

Algebra II

Module 5: Homework
Polynomial Functions

49. Let $f(x) = 3x^3 + 6x^2 + 2x + 12$ and $g(x) = 4x^3 + 11x^2 - 2x - 8$. There are three solutions to this system, one of which is (2, 64). Find the other two solutions algebraically and then verify your answers by graphing the system.

50. Let $f(x) = x^3 - 4x^2 - 15x - 42$ and $g(x) = -x^3 - 7x^2 + 8x - 30$. There are three solutions to this system, one of which is (−4, −110). Find the other two solutions algebraically and then verify your answers by graphing the system.

51. Let $f(x) = 2x^3 - 10x^2 + 10x - 2$ and $g(x) = -x^3 + 4x^2 + 17x - 12$. There are three solutions to this system, one of which is (5, 48). Find the other two solutions algebraically and then verify your answers by graphing the system.

INVESTIGATION 7: SEEING STRUCTURE IN POLYNOMIAL EXPRESSIONS

52. Factor the following expressions.
    a. $x^4 - 4y^2$
    b. $b^4 - 36$
    c. $4x^6 - 16x^4$
    d. $225s^8 - 400t^2$

53. Factor the following expressions.
    a. $9c^4 - d^4$
    b. $16a^6 - 9b^{10}$
    c. $100r^4 - 64r^2w^3$
    d. $121k^6 - 81m^8$

54. For each of the following functions,
    i. Factor the expression that defines the function's outputs.
    ii. Determine the real roots and non-real roots (if any) of each function.
    a. $f(x) = 2x^4 - 8x^2$
    b. $g(x) = 9x^5 - x$
    c. $h(x) = x^6 - 36$

55. For each of the following functions,
    i. Factor the expression that defines the function's outputs.
    ii. Determine the real roots and non-real roots (if any) of each function.
    a. $f(x) = 2x^4 - 18x^2$
    b. $g(x) = 8x^5 - 2x$
    c. $h(x) = 16x^8 - 9$

56. For each of the following
    i. Factor the expression that defines the function's outputs.
    ii. Determine the real roots and non-real roots (if any) of each function.
    a. $f(x) = x^4 - 2x^2 - 8$
    b. $g(x) = x^4 + 4x^2 - 45$
    c. $h(x) = x^4 + 16x^2 + 63$

57. For each of the following
    i. Factor the expression that defines the function's outputs.
    ii. Determine the real roots and non-real roots (if any) of each function.
    a. $f(x) = x^4 + x^2 - 12$
    b. $g(x) = x^4 + 9x^2 + 16$
    c. $h(x) = x^4 - 10x^2 + 9$

INVESTIGATION 8: POLYNOMIAL TRANSLATIONS

58. Suppose f and g are related such that $g(x) = f(x) - 7$.
    a. For any given input value, how will the output values of f and g compare?
    b. How will the graphs of f and g be related?

Algebra II

Module 5: Homework
Polynomial Functions

59. Suppose f and g are related such that $g(x) = f(x) + 3$.
    a. For any given input value, how will the output values of f and g compare?
    b. How will the graphs of f and g be related?

60. The graph of f is given. Draw the graph of g if $g(x) = f(x) + 4$.

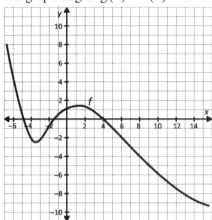

61. The graph of f is given. Draw the graph of g if $g(x) = f(x) - 2$.

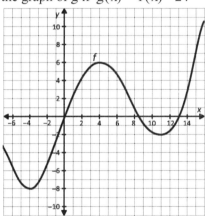

62. Tables of values for f, g, and h are given.

x	f(x)
−3.2	17.48
−1	−1
−0.1	−2.98
1.5	1.5
4	29

x	g(x)
−5	−65
−1.3	27.13
−1	31
1.4	42.52
3.5	24.25

x	h(x)
−2.7	−5.10
−1.7	0.87
0	0
4.7	148
5.1	184.67

a. Evaluate $a(1.5)$ if $a(x) = f(x) + 3$.
b. Evaluate $b(1.4)$ if $b(x) = g(x) - 6$.
c. Evaluate $c(-2.7)$ if $c(x) = h(x) - 10.2$.
d. Evaluate $d(-0.1)$ if $d(x) = f(x) + 3.4$.
e. Create a table of values for function p if $p(x) = g(x) - 7.5$.
f. Create a table of values for function q if $q(x) = h(x) + 1.375$.

Algebra II

Module 5: Homework
Polynomial Functions

63. Tables of values for *f*, *g*, and *h* are given.

x	f(x)
–4.5	6.75
–0.2	–0.56
0	0
0.4	1.36
3.2	19.84

x	g(x)
–2	0
–0.5	0.37
0.25	0.14
1.3	5.58
2.7	34.26

x	h(x)
–3.6	–38.50
–2.1	–5.61
0.1	–2.94
1.5	–3.77
2.9	13.05

   a. Evaluate $a(0.4)$ if $a(x) = f(x) + 7$.
   b. Evaluate $b(2.7)$ if $b(x) = g(x) - 1.5$.
   c. Evaluate $c(-3.6)$ if $c(x) = h(x) + 8.3$.
   d. Evaluate $d(-0.5)$ if $d(x) = g(x) - 7.7$.
   e. Create a table of values for function p if $p(x) = f(x) + 2.3$.
   f. Create a table of values for function q if $q(x) = h(x) - 8.75$.

64. Let $f(x) = -x^2 + 7x + 2$ and $g(x) = 4x^3 - 5x^2 + 6x$. What are the formulas defining functions p and q if $p(x) = f(x) - 3.5$ and $q(x) = g(x) + 7.25$?

65. Let $f(x) = -2x^3 + 3.6x^2 + 9.52$ and $g(x) = x^2 - 12.1x - 9.8$. What are the formulas defining functions p and q if $p(x) = f(x) + 2.9$ and $q(x) = g(x) - 9.11$?

66. Consider the graphs of *f* and *g*.

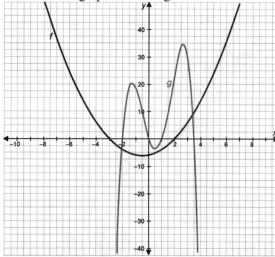

   a. Draw the graph of h given that $h(x) = f(x-1)$.
   b. Draw the graph of j given that $j(x) = g(x+4)$.
   c. Draw the graph of p given that $p(x) = f(x+3) + 2$.
   d. Draw the graph of q given that $q(x) = g(x-2) - 6$.

67. Consider the graph of f.

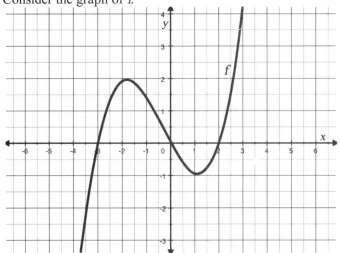

a. Draw the graph of a given that $a(x) = f(x+2)$.
b. Draw the graph of b given that $b(x) = f(x-3)$.
c. Draw the graph of c given that $c(x) = f(x-2)+4$.
d. Draw the graph of d given that $d(x) = f(x+2)-4$.

68. The tables below shows ordered pairs for function f and g.

x	f(x)
−2	20
−1	22
0	18
1	14
2	16
3	30

x	g(x)
−5	−21
−2	0
0	4
1	3
4	−12
6	−32

a. Create a table of values for h if $h(x) = f(x-2)$.
b. Create a table of values for j if $j(x) = g(x+1)$.
c. Create a table of values for p if $p(x) = f(x+3)-5$.
d. Create a table of values for q if $q(x) = g(x-1)+7$.

69. The tables below shows ordered pairs for function f and g.

x	f(x)
−3	19
−1	16
0	10
1	17
3	20
5	36

x	g(x)
−6	−13
−4	−20
0	−11
2	−2
4	1
6	9

a. Create a table of values for h if $h(x) = f(x+4)$.
b. Create a table of values for j if $j(x) = g(x-5)$.
c. Create a table of values for p if $p(x) = f(x-1)+3$.
d. Create a table of values for q if $q(x) = g(x+2)-4$.

Algebra II

Module 5: Homework
Polynomial Functions

70. Given that $f(x) = 2x^3 + x - 10$ and $g(x) = 3x^4 - x^2 + 5x$, answer the following questions.
    a. Evaluate $h(1)$ if $h(x) = f(x+5)$.
    b. Evaluate $j(-2)$ if $j(x) = g(x+1)$.
    c. Write the formula defining p if $p(x) = f(x-3)$.
    d. Write the formula defining q if $q(x) = g(x+1) + 2$.

71. Given that $f(x) = x^4 + x^2 + 2$ and $g(x) = -x^3 + 7x^2 - 3$, answer the following questions.
    a. Evaluate $h(8)$ if $h(x) = f(x-7)$.
    b. Evaluate $j(-4)$ if $j(x) = g(x+1)$.
    c. Write the formula defining p if $p(x) = f(x+2)$.
    d. Write the formula defining q if $q(x) = g(x-1) - 1$.

72. The following graph shows a sprinter's distance from the starting line (in meters) with respect to the number of seconds since the runner left the starting line. Call this function f.

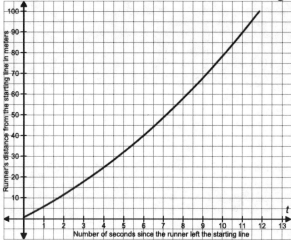

    The race begins when the starter fires the starting pistol. Suppose that the sprinter paused before beginning to run such that he didn't leave the starting line until half a second after the gun was fired.
    a. Draw the graph of the relationship between the sprinter's distance from the starting line (in meters) with respect to the number of seconds since the starting gun fired. Call this function g.
    b. What is the relationship between the graphs of f and g?
    c. If t is the number of seconds since the starting gun was fired, which of the following is true? Defend your choice.
    $g(t) = f(t + 0.5)$ or $g(t) = f(t - 0.5)$

73. The following graph represents the average cost of a CD (in dollars) with respect to the number of years since 1990. Call this function f.

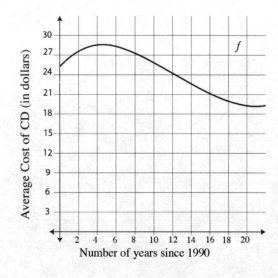

    a. Imagine instead, that we would like to graph the average cost of a CD (in dollars) with respect to the number of years since 1995. Call this function g. Draw the graph of g.
    b. What is the relationship between the graphs of f and g?
    c. If t is the number of years since 1995, which of the following is true? Defend your choice.
    $g(t) = f(t+5)$ or $g(t) = f(t-5)$

Algebra II

Module 5: Homework
Polynomial Functions

INVESTIGATION 9: POLYNOMIAL REFLECTIONS

74. The graphs of *f* and *g* are given.

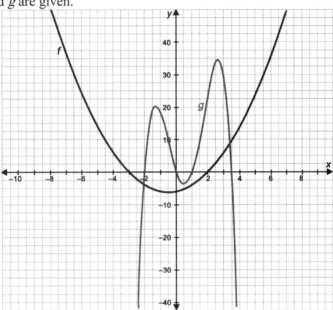

   a. Draw the graph of *p* given that $p(x) = -f(x)$.
   b. Draw the graph of *q* given that $q(x) = g(-x)$.
   c. Draw the graph of *r* given that $r(x) = -g(-x)$.
   d. Draw the graph of *w* given that $w(x) = -f(x+2)$.

75. Consider the graph of *f*.

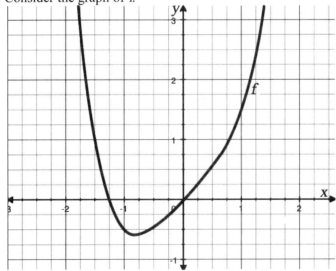

   a. Draw the graph of *h* given that $h(x) = -f(x)$.
   b. Draw the graph of *j* given that $j(x) = f(-x)$.

76. Given that $f(x) = -2x^2 - 3x + 4$, $g(x) = x+3$, and $h(x) = 2x^3 + 3x$, answer the following questions.
    a. Evaluate $j(7)$ given that $j(x) = g(-x)$.
    b. Evaluate $k(-5)$ given that $k(x) = -f(x)$.
    c. Write the formula defining n given that $n(x) = -f(-x)$.
    d. Write the formula defining p given that $p(x) = -g(x) + 2$.
    e. Write the formula defining q given that $q(x) = h(-x) - 3$.

77. Given that $f(x) = x^3 - 2x^2 + 6$, $g(x) = 3x^2 + 2x$, and $h(x) = -x^3 + 2x^2$, answer the following questions.
    a. Evaluate $j(-1)$ given that $j(x) = g(-x)$.
    b. Evaluate $k(2)$ given that $k(x) = -h(x)$.
    c. Write the formula defining n given that $n(x) = -f(-x)$.
    d. Write the formula defining p given that $p(x) = g(-x) + 4$.
    e. Write the formula defining q given that $q(x) = -h(-x) - 1$.

INVESTIGATION 10: COMBINING POLYNOMIAL FUNCTIONS: ADDING AND SUBTRACTING

78. Chad owns two houses – one is his primary residence and the other house is a rental property (Chad owns the house and other people pay him rent to live there). The following graph shows function f (the value of Chad's primary residence (in thousands of dollars) with respect to the number of years since 2000) and function g (the value of Chad's rental property (in thousands of dollars) with respect to the number of years since 2000).

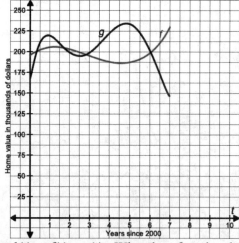

   a. Function h is defined by $h(t) = f(t) + g(t)$. What does function h represent?
   b. Graph function h.
   c. Function j is defined by $j(t) = f(t) - g(t)$. What does function j represent?
   d. Graph function j.
   e. If $f(t) \approx 0.4t^5 - 7.7t^4 + 51.9t^3 - 144.5t^2 + 154.1t + 165.4$ and $g(t) \approx t^3 - 8.7t^2 + 16.7t + 197$, write the formulas defining h and j.

Algebra II

Module 5: Homework
Polynomial Functions

79. Let *f* be the function that represents the number of trucks owned in Small, Texas in terms of the number of years since 2000. Let *g* be the function that represents the number of cars owned in Small, Texas in terms of the number of years since 1990. The graphs of these two functions are shown below.

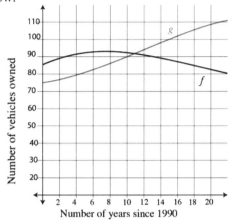

a. Function *h* is defined by $h(t) = f(t) + g(t)$. What does function *h* represent?
b. Function *j* is defined by $j(t) = f(t) - g(t)$. What does function *j* represent?

80. Graph $f(x) = -x^2 + 3x$ given the graphs of $y = -x^2$ and $y = 3x$.

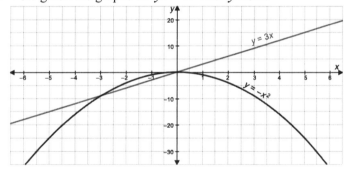

81. Graph $f(x) = x^3 + 2x^2 + 5$ given the graphs of $y = x^3$, $y = 2x^2$, and $y = 5$.

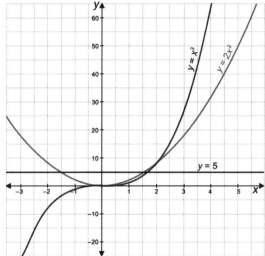

Algebra II

Module 5: Homework
Polynomial Functions

82. Let $f(x) = -2x^3 + 4$, $g(x) = x^2 - 6$, and $h(x) = x^3 - 8x^2 - x$.
    a. What is the formula for j given that $j(x) = f(x) + h(x)$?
    b. What is the formula for n given that $n(x) = f(x) - g(x) + h(x)$?
    c. If $p(x) = g(x) + f(x)$, evaluate $p(-2)$.
    d. If $q(x) = f(x) - g(x)$, evaluate $q(3.2)$.

83. Let $f(x) = 3x^2 + x - 7$, $g(x) = -x^3 + 1$, and $h(x) = 5x^3 - 2x^2$.
    a. What is the formula for j given that $j(x) = f(x) + g(x)$?
    b. What is the formula for n given that $n(x) = f(x) - g(x) + h(x)$?
    c. If $p(x) = f(x) - h(x)$, evaluate $p(2)$.
    d. If $q(x) = g(x) + h(x)$, evaluate $q(-0.5)$.

INVESTIGATION 11: COMBINING POLYNOMIAL FUNCTIONS: COMPOSING AND INVERTING

For Exercises #84-89, do the following.
  a. Use a calculator or other graphing software to graph the function. Using this graph as a guide, restrict the domain of the function (if necessary) so that its inverse will be a function.
  b. Write a formula to define the inverse function.
  c. Compose the inverse functions in both possible ways ($f^{-1}(f(m))$ and $f(f^{-1}(n))$) to verify that they are inverses.

84. $f(x) = x^2$

85. $f(x) = 3x^3 + 1$

86. $f(x) = 4x - 6$

87. $f(x) = (x+2)^2 + 1$

86. $f(x) = -2(x-7)^2$

89. $f(x) = x^4 + 2x + 1$

90. Use the graphs of the polynomial functions f and g to evaluate each expression.

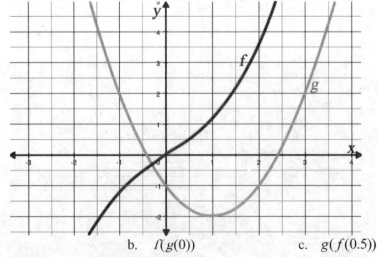

    a. $g(f(0))$ b. $f(g(0))$ c. $g(f(0.5))$
    d. $f(g(3))$ e. $g(f(1.4))$ f. $f(g(2.4))$

Algebra II

Module 5: Homework
Polynomial Functions

91. Use the graphs of the polynomial functions *f* and *g* to evaluate each expression.

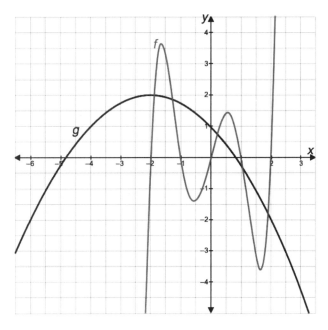

a. $g(f(-1))$
b. $g(f(1.5))$
c. $g(f(-0.5))$
d. $f(g(-4))$
e. $f(g(0))$
f. $f(g(1.2))$

92. Given that $f(x) = 2x^3 - 5x$, $g(x) = x^3 + x^2 + x + 1$, and $h(x) = x - 1$, answer the following questions.
   a. Evaluate $f(h(6))$.
   b. Evaluate $g(f(-2))$.
   c. Find the formula to define function p if $p(x) = g(h(x))$.
   a. Find the formula to define function q if $q(x) = h(f(x))$.

93. Given that $f(x) = -x^2 + 2x$, $g(x) = 3x^3 - x^2 - 2x$, and $h(x) = x^2 + 8$, answer the following questions.
   a. Evaluate $f(g(2))$.
   b. Evaluate $h(g(-0.5))$.
   c. Find the formula to define function p if $p(x) = f(h(x))$.
   d. Find the formula to define function q if $q(x) = h(f(x))$.

Algebra II

Module 6: Investigation 1
Vertical Asymptotes of Rational Functions

1. The Leadville 100 MTB is a marathon mountain biking race held each year. The organizers need to determine the amount of time it may take various bikers to finish the 100-mile race.

    a. *Complete the following statement:* A biker with a faster average speed over the entire race will take _____ time to complete the race than a biker with a slower average speed over the entire race.

    b. Write a function formula that determines the amount of time t (in hours) to complete the 100-mile bike race given a biker's average speed s (in miles per hour) over the course of the race.

    c. Graph f and explain what a point on the graph represents in this context.

    d. *Complete the following statement:* As the average speed of a biker for the 100-mile race decreases, the amount of time that it takes to complete the race _____ .

    e. Estimate the practical domain of f. In other words, what values make sense for a biker's average speed during the race? (*Hint: 15.5 mph was the fastest average speed for the 100-mile race last year.*)

    f. Describe how the amount of time it takes a biker to complete the race co-varies with the biker's average speed for the race.

    g. Use the graph (and the practical domain you chose in part (e)) to estimate the practical range, and then explain what this range represents.

Algebra II

Module 6: Investigation 1
Vertical Asymptotes of Rational Functions

It's common to imagine the values of a quantity increasing or decreasing towards a certain number. For example, imagine x increasing toward a value of $x = 4$. Then we imagine x taking on values such as $x = 3.5$, $x = 3.9$, $x = 3.98$, $x = 3.99999$, and so on. Mathematicians have developed notation for this idea. We represent "x approaches a value of $x = 4$ from values less than 4" (or "x approaches a value of $x = 4$ from the left on the number line") by writing $x \to 4^-$.

Likewise, $x \to 4^+$ represents the idea "x approaches a value of $x = 4$ from values greater than 4" (or "x approaches a value of $x = 4$ from the right on the number line"). So when we see $x \to 4^+$ we are imagining x taking on values such as $x = 4.5$, $x = 4.1$, $x = 4.06$, $x = 4.000001$ and so on.

"Approaches" Notation

We write $x \to a^-$ to represent the statement "x approaches a value of $x = a$ from values less than a" (or "from the left on the number line").

We write $x \to a^+$ to represent the statement "x approaches a value of $x = a$ from values greater than a" (or "from the right on the number line").

---

2. Refer back to the context in Exercise #1.
   a. What does it mean to say "as $s \to 12^-$"?

   b. What does it mean to say "as $s \to 8^+$"?

3. Given the function g, defined by $g(x) = \dfrac{1}{x}$,
   a. Complete the following table of values. Show the calculations that produced your answers.

x	$g(x)$	x	$g(x)$
−5		0.001	
−2		0.01	
−1		0.1	
−0.5		0.5	
−0.1		1	
0.01		2	
−0.001		5	

Module 6: Investigation 1

© 2014 Carlson, O'Bryan & Joyner

Algebra II

Module 6: Investigation 1
Vertical Asymptotes of Rational Functions

b. What do you notice about the value of $g(x)$ as $x \to 0^-$ (as x gets closer and closer to 0 from the negative side, or from values less than 0)? Why does this happen?

c. What do you notice about the value of $g(x)$ as $x \to 0^+$ (as x gets closer and closer to 0 from the positive side, or from values greater than 0)? Why does this happen?

d. Graph g and explain how the graph supports your conclusions in parts (b) and (c).

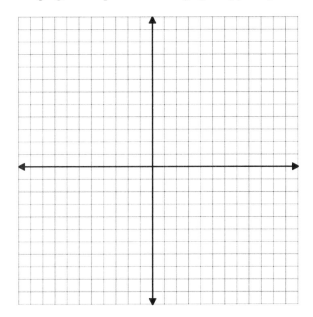

e. What do you notice at $x = 0$?

We say that g has a **vertical asymptote** at $x = 0$.

Vertical Asymptote

Let a be a real number. If $g(x)$ increases or decreases without bound as $x \to a^-$, and if $g(x)$ increases or decreases without bound as $x \to a^+$, then g has a **vertical asymptote** at $x = a$.

It's common to draw a dashed vertical line at $x = a$ to help visualize the behavior of the function for input values near $x = a$.

Algebra II

Module 6: Investigation 1
Vertical Asymptotes of Rational Functions

4. Given that $f(x) = \dfrac{1}{x-2}$, answer the questions below.

   a. Predict where you think a vertical asymptote will exist. Explain your reasoning.

x	f(x)
1.9	
1.99	
1.999	
2.001	
2.01	
2.1	

   b. Complete the table of values for f. Show the calculations that produced your answers.

   c. What do you notice about the values of $f(x)$ as $x \to 2^-$ (as x approaches 2 from the left)? Why does this happen?

   d. What do you notice about the values of $f(x)$ as $x \to 2^+$ (as x approaches 2 from the right)? Why does this happen?

   e. Use a calculator to graph f. Explain how the graph supports your conclusions in parts (c) and (d).

Algebra II

Module 6: Investigation 1
Vertical Asymptotes of Rational Functions

5. For each of the following functions, predict where a vertical asymptote will exist and then test your prediction using a graphing calculator.

   a. $f(x) = \dfrac{3}{x-5}$

   b. $g(x) = \dfrac{x}{3x+6}$

   c. $h(x) = \dfrac{8x^2 - 24}{2(x+1)(x-1)}$

   d. $m(x) = \dfrac{x+3}{x^2 + 6x + 12}$

   e. $n(x) = \dfrac{x^5 + 2}{x^2 + 4}$

   f. $p(x) = \dfrac{3(x-3)}{x-3}$

6. Write an explanation describing how to determine the vertical asymptote(s) of a rational function and *why* this approach works.

Algebra II

Module 6: Investigation 2
End Behavior of Rational Functions

1. Let $g(x) = \dfrac{1}{x-3}$.

   a. Determine the vertical asymptote of the function. Explain why the vertical asymptote exists.

   b. We see that g is not defined at $x = 3$. What is the domain of g?

   c. Complete the following table of values. Show the calculations that produced your answers.

x	g(x)	x	g(x)
−10,000		1	
−1,000		10	
−100		100	
−100		1000	
−1		10,000	

   d. As x increases without bound (as x gets larger and larger, written as $x \to +\infty$), what happens to the value of $g(x)$?

   e. As x decreases without bound (gets smaller and smaller, written as $x \to -\infty$), what happens to the value of $g(x)$?

   f. Graph g. Use a dashed line to represent the vertical asymptote.

   g. What do you notice about the value of $g(x)$ as x increases without bound and as x decreases without bound?

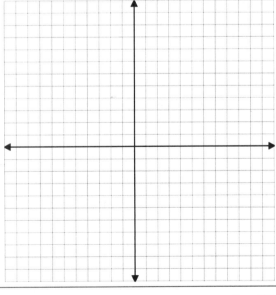

Module 6: Investigation 2

© 2014 Carlson, O'Bryan & Joyner

Algebra II

Module 6: Investigation 2
End Behavior of Rational Functions

> **Horizontal Asymptote**
>
> Let a be a real number. As x increases or decreases without bound [as $x \to \pm\infty$], if $g(x)$ approaches a [$g(x) \to a$] then g has a **horizontal asymptote** at $y = a$.
>
> It's common to draw a dashed horizontal line at $y = a$ to help visualize the end behavior of the function when a horizontal asymptote exists.

In Exercise #1, a horizontal asymptote occurred at $y = 0$ because as x increases and decreases without bound, $g(x) \to 0$.

2. Let $f(x) = \dfrac{2x}{x+1}$.

   a. Complete the following table of values.

x	$2x$	$x+1$	$f(x) = \dfrac{2x}{x+1}$
$-1{,}000{,}000$			
$-10{,}000$			
-100			
100			
$10{,}000$			
$1{,}000{,}000$			

   b. What do you notice as x increases without bound? Why does this happen?

   c. What do you notice as x decreases without bound? Why does this happen?

Algebra II

Module 6: Investigation 2
End Behavior of Rational Functions

d. Use a calculator to graph f. Explain how the graph supports your conclusions in parts (b) and (c).

There are two ways to determine horizontal asymptotes. The first method is to use a table or graph to examine what happens to the output value of a function as the input value increases and decreases without bound.

A second method uses the formula of the function and compares the long-run behavior of the polynomial functions that make up the numerator and denominator. For example, consider the polynomial function $p(x) = x + 4$. When the magnitude of x gets very large, the "+ 4" portion of the expression $x + 4$ is a proportionally smaller and smaller part of the final value of the expression. For example,

- $p(20) = 24$ (*the "+ 4" makes up 16.7% of the final value*)
- $p(1,000) = 1,004$ (*the "+ 4" makes up 0.04% of the final value*)
- $p(100,000,000) = 1,000,004$ (*the "+ 4" makes up 0.000004% of the final value*)

We can say that as $x \to \infty$, $p(x) \to x$ (and likewise as $x \to -\infty$, $p(x) \to x$) because the relative size of $x + 4$ and x are nearly identical.

---

3. a. As x increases or decreases without bound, what happens to the value of the expression $x^2 - 15$?

   b. As x increases or decreases without bound, what happens to the value of the expression $x^3 - 6x^2$?

   c. As x increases or decreases without bound, what happens to the value of the expression $-3x^2 - 19x$?

   d. As x increases or decreases without bound, what happens to the value of the expression $6x^4 - 2x^3 + 14x^2 + 35x - 13$?

Algebra II

Module 6: Investigation 2
End Behavior of Rational Functions

Let's apply the ideas from Exercise #3. Consider the function $f(x) = \dfrac{12x-9}{2x^2 - x}$.

- As x increases or decreases without bound, $(12x-9) \to 12x$ and $(2x^2 - x) \to 2x^2$.

- Therefore, $\left(f(x) = \dfrac{12x-9}{2x^2 - x}\right) \to \dfrac{12x}{2x^2} = \dfrac{6}{x}$ as x increases or decreases without bound.

- If $f(x) \to \dfrac{6}{x}$ as x increases or decreases without bound, then f must have a horizontal asymptote at $y = 0$ since $g(x) = \dfrac{6}{x}$ has a horizontal asymptote at $y = 0$.

4. Use this reasoning to find the horizontal asymptotes for the following functions.

   a. $f(x) = \dfrac{7}{x+12}$

   b. $g(x) = \dfrac{x}{2x+14}$

   c. $h(x) = \dfrac{3x^2 - 21}{(x-3)(x+1)}$

   d. $m(x) = \dfrac{x-8}{x^2 + 4x - 5}$

   e. $n(x) = \dfrac{x^3 + 1}{x^2 + 1}$

Generalizing the reasoning from Exercise #4, any polynomial expression

$$p(x) = a_n x^n + a_{n-1} x^{n-1} + a_{n-2} x^{n-2} + \ldots + a_1 x + a_0$$

will behave like $y = a_n x^n$ as x increases or decreases without bound. Therefore, a rational function

$$r(x) = \dfrac{p(x)}{q(x)} = \dfrac{a_n x^n + a_{n-1} x^{n-1} + a_{n-2} x^{n-2} + \ldots + a_1 x + a_0}{b_m x^m + b_{m-1} x^{m-1} + b_{m-2} x^{m-2} + \ldots + b_1 x + b_0}$$

will behave like $y = \dfrac{a_n x^n}{b_m x^m}$ as x increases or decreases without bound.

Algebra II

Module 6: Investigation 2
End Behavior of Rational Functions

5. Take a moment to reflect on the work we've done so far.
   a. What conditions are necessary for a rational function to have a horizontal asymptote at $y = 0$?

   b. What conditions are necessary for a rational function to have a horizontal asymptote at $y = a$ where $a \neq 0$? How can you determine the exact location of the horizontal asymptote?

   c. What conditions are necessary for a rational function *not* to have a horizontal asymptote? Explain why a horizontal asymptote does not exist in the situation.

6. Let $f(x) = \dfrac{x-2}{(x-2)(x+4)}$.
   a. What vertical asymptote(s) do you expect the function to have?

   b. What horizontal asymptote(s) do you expect the function to have?

   c. Graph *f* using a calculator.

   d. Does the graph surprise you based on your answers to parts (a) and (b)? Explain.

Algebra II

Module 6: Investigation 2
End Behavior of Rational Functions

7. Your answers to Exercise #6 revealed that there was a *removable discontinuity* (or *hole*) in the graph of f at $x = 2$ instead of a vertical asymptote.
    a. Look at the formula defining the function. Why do you think a hole exists at $x = 2$ instead of a vertical asymptote?

    b. Given a rational function of the form $r(x) = \dfrac{p(x)}{q(x)}$ where $p(x)$ and $q(x)$ are polynomial functions and $q(a) = 0$, describe the situation where the graph of r has
        i. a hole at $x = a$.

        ii. a vertical asymptote at $x = a$.

Algebra II
Module 6: Investigation 3
Graphing Rational Functions

In the first two investigations we learned how to use the general form of a rational function

$$r(x) = \frac{p(x)}{q(x)} = \frac{a_n x^n + a_{n-1} x^{n-1} + a_{n-2} x^{n-2} + \ldots + a_1 x + a_0}{b_m x^m + b_{m-1} x^{m-1} + b_{m-2} x^{m-2} + \ldots + b_1 x + b_0}$$ (where $q(x) \neq 0$) to reason about the behaviors of a rational function.

We learned that r has
  a. a vertical asymptote at $x = a$ if $q(a) = 0$ [unless $p(a) = 0$ also].
  b. a hole in the graph at $x = a$ if $p(a) = 0$ and $q(a) = 0$.
  c. a horizontal asymptote at $y = 0$ if the degree of p is less than the degree of q.
  d. a horizontal asymptote at $y = \dfrac{a_m}{b_n}$ if the degree of p is equal to the degree of q.

---

1. For each of the given functions, determine the following information.
   i. all vertical asymptotes of the function (if any exist)
   ii. all holes in the graph (if any exist)
   iii. all horizontal asymptotes (if any exist)

   a. $f(x) = \dfrac{4x}{x+5}$
   b. $g(x) = \dfrac{2x^3 + 2}{(x-2)(x+3)}$
   c. $h(x) = \dfrac{4x+4}{(x+1)(x+2)}$

2. Recall that the vertical intercept of a function can be determined by inputting a value of 0 into the function and determining the associated output. Find the vertical intercept for each of the following rational functions.

   a. $f(x) = \dfrac{4x}{x+5}$
   b. $g(x) = \dfrac{2x^3 + 2}{(x-2)(x+3)}$
   c. $h(x) = \dfrac{4x+4}{(x+1)(x+2)}$

Algebra II
Module 6: Investigation 3
Graphing Rational Functions

3. Recall that a root of a function is a value of *x* that corresponds to where the output of the function is equal to 0.
   a. The roots of a rational function occur at the roots of the numerator that are not also a root of the denominator. Explain why this is true.

   b. Determine the roots for each of the following rational functions.
      i. $f(x) = \dfrac{4x}{x+5}$
      ii. $g(x) = \dfrac{2x^3 + 2}{(x-2)(x+3)}$
      iii. $h(x) = \dfrac{4x+4}{(x+1)(x+2)}$

4. Use the information from Exercises #1-3 to graph the functions. Check your answers by graphing the functions using a calculator. If you made any mistakes, explore the reason why the graph behaves differently than you expected.
   a. $f(x) = \dfrac{4x}{x+5}$

   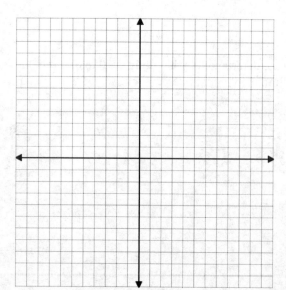

   b. $g(x) = \dfrac{2x^3 + 2}{(x-2)(x+3)}$

   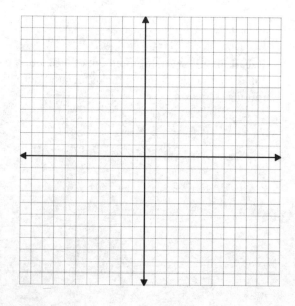

c. $h(x) = \dfrac{4x+4}{(x+1)(x+2)}$

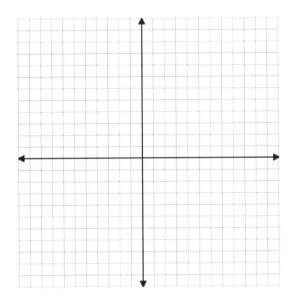

5. A rational function can change from positive to negative or negative to positive at its roots and its vertical asymptotes (but does not have to).

   a. Determine the vertical asymptote of $f(x) = \dfrac{(x-3)^2}{x}$.

   b. Determine the root(s) of $f(x) = \dfrac{(x-3)^2}{x}$.

   c. We observe that *f* may have different signs on either side of $x = 0$ and $x = 3$. Determine if *f* is positive or negative on each of the three intervals $x < 0$, $0 < x < 3$, and $x > 3$ by choosing an input value from each interval and determining the sign of the associated output.

d. Did the function change signs from one interval to the next? If not, explain why not.

6. For each of the given functions, determine any vertical asymptote(s), horizontal asymptote, vertical intercept, root(s), and/or holes. Use this information to sketch the graphs and then use the calculator to confirm your answers.

   a. $g(x) = \dfrac{2x^2}{(x+3)(x-4)}$

   b. $h(x) = \dfrac{x^2 - 3x - 4}{x^2 - x - 2}$

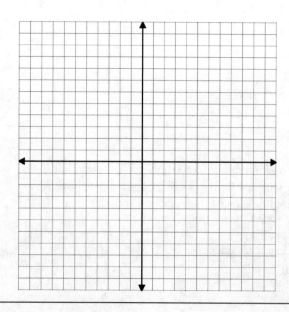
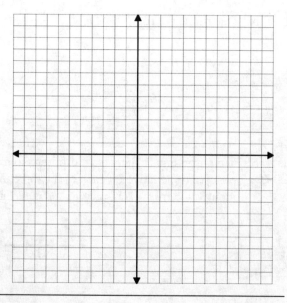

Algebra II
Module 6: Investigation 4
Solving Rational Equations #1: Using Common Denominators

A rational equation is a statement of equality involving one or more rational expressions. For example, $\frac{7}{x+1} = \frac{x}{10}$ is a rational equation. In this investigation we want to develop powerful techniques that will help us solve all rational equations including more complicated examples such as $\frac{x}{x-1} + \frac{5}{x^2-1} = \frac{x-7}{x+1}$.

1. Consider the equation $\frac{\square}{3} = \frac{2}{3}$ where the box represents an unknown number.
   a. What number must be inside the box for the equation to be true?

   b. Solve the following equations for x based on your answer to part (a).
   i. $\frac{x}{3} = \frac{2}{3}$
   ii. $\frac{2x}{3} = \frac{2}{3}$
   iii. $\frac{x-7}{3} = \frac{2}{3}$
   iv. $\frac{4(x+5)}{3} = \frac{2}{3}$

2. a. What number must be inside each box for the equations to be true?
   i. $\frac{\square}{7} = \frac{5}{7}$
   ii. $\frac{157}{212} = \frac{\square}{212}$
   iii. $\frac{\square}{13} = -\frac{4}{13}$
   iv. $\frac{0.9}{4.2} = \frac{\square}{4.2}$

   b. Solve the following equations for x using your answers to part (a).
   i. $\frac{x+1}{7} = \frac{5}{7}$
   ii. $\frac{157}{212} = \frac{3x-5}{212}$
   iii. $\frac{10x}{13} = -\frac{4}{13}$
   iv. $\frac{0.9}{4.2} = \frac{0.5x+0.3}{4.2}$

3. For each equation, do the following.
   i. Get a common denominator on both sides of the equation.
   ii. What number must be placed in the box to make the equation true?
   iii. If the box were replaced with the expression $2x+3$, what value(s) of x would make the equation true?
   a. $\frac{\square}{24} = \frac{7}{6}$
   b. $\frac{\square}{60} = -\frac{2}{5}$
   c. $\frac{9}{10} = \frac{\square}{5}$

Algebra II

Module 6: Investigation 4
Solving Rational Equations #1: Using Common Denominators

4. Solve each equation by first getting a common denominator on both sides of the equation.

   a. $\dfrac{x}{7} = \dfrac{117}{21}$

   b. $\dfrac{x+6}{15} = \dfrac{2}{3}$

   c. $\dfrac{20}{32} = \dfrac{2x+1}{8}$

   d. $\dfrac{6x-7.5}{3} = 1.5$

   e. $\dfrac{x}{5} = \dfrac{x+2}{3}$

   f. $\dfrac{x+8}{6} = \dfrac{x-3}{7}$

5. Solve each of the following equations.

   a. $\dfrac{x}{8} + \dfrac{3}{4} = \dfrac{13}{8}$

   b. $\dfrac{5}{4} = \dfrac{5}{12} + \dfrac{2x}{3}$

   c. $\dfrac{7}{2} + \dfrac{x-1}{7} = \dfrac{1}{14}$

   d. $\dfrac{17}{4} = \dfrac{x}{4} + \dfrac{2x+5}{4}$

   e. $\dfrac{x+1}{6} + \dfrac{x}{3} = \dfrac{x}{12}$

   f. $\dfrac{2x-3}{5} + \dfrac{4x+1}{6} = \dfrac{5x}{3}$

Algebra II

Module 6: Investigation 4
Solving Rational Equations #1: Using Common Denominators

6. Rewrite each equation so that both sides have a common denominator.

   a. $-\dfrac{1}{4} = \dfrac{6x}{x+8}$

   b. $\dfrac{x+1}{x} = 3.5$

   c. $\dfrac{1}{3x-2} = \dfrac{2}{x}$

   d. $\dfrac{x}{2} + \dfrac{1}{4x} = \dfrac{x+1}{x}$

   e. $\dfrac{3}{x} + \dfrac{x}{x+5} = \dfrac{7}{x}$

   f. $\dfrac{8}{x-1} + \dfrac{4}{x+2} = \dfrac{1}{x+3}$

7. Solve each of the following equations by first getting a common denominator on both sides of the equation.

   a. $\dfrac{x+1}{x} = \dfrac{7}{5}$

   b. $\dfrac{12}{x-6} = \dfrac{3}{x}$

   c. $\dfrac{x}{x+1} = -0.25$

Algebra II
Module 6: Investigation 4
Solving Rational Equations #1: Using Common Denominators

8. Let $f(x) = \dfrac{x+5}{x-3}$ and $g(x) = \dfrac{2}{x-1}$. Find the solution(s) to the system. Check your solutions by graphing the functions.

9. Let $f(x) = (x+1) + \dfrac{x+5}{x}$ and $g(x) = \dfrac{x+11}{2}$. Find the solution(s) to the system. Check your solutions by graphing the functions.

10. Let $f(x) = \dfrac{3x}{2(3x+14)} + \dfrac{1}{x+5}$ and $g(x) = \dfrac{x}{3x+14}$. Find the solution(s) to the system. Check your solutions by graphing the functions.

Algebra II
Module 6: Investigation 5
Solving Rational Equations #2: Dealing with Extraneous Solutions

For Exercises #1-2, do the following.
  a) Solve the rational equation.
  b) Check your answers by substituting your solutions from part (a) back into the original equation.

1. $\dfrac{x^2}{x-5} = \dfrac{25}{x-5}$

2. $\dfrac{1}{x+3} + \dfrac{2}{x} = -\dfrac{3}{x(x+3)}$

In Exercises #1-2 we found that the process of solving a rational equation sometimes gives us answers that aren't actually solutions to the original equation. These are called **extraneous solutions**.

> **Extraneous Solutions**
>
> An **extraneous solution** is a solution to a modified form of an equation that is not a solution to the original equation.

Let's explore why extraneous solutions can appear when solving rational equations.

3. Consider the rational equation $\dfrac{1}{x-6} + \dfrac{x}{x-2} = \dfrac{4}{(x-6)(x-2)}$.

  a. Before we attempt to solve the equation, take a minute to look at it closely. What values of x must immediately be restricted from the possible solutions? Why?

Algebra II

Module 6: Investigation 5
Solving Rational Equations #2: Dealing with Extraneous Solutions

b. The first steps in the solution process are to get common denominators on both sides of the equation and then combine any fractions on the same side of the equation. Then we set the numerators equal.

$$\frac{1}{x-6} + \frac{x}{x-2} = \frac{4}{(x-6)(x-2)}$$

$$\left(\frac{1}{x-6}\right)\left(\frac{x-2}{x-2}\right) + \left(\frac{x}{x-2}\right)\left(\frac{x-6}{x-6}\right) = \frac{4}{(x-6)(x-2)}$$

$$\frac{x-2}{(x-6)(x-2)} + \frac{x(x-6)}{(x-6)(x-2)} = \frac{4}{(x-6)(x-2)}$$

$$\frac{x-2+x(x-6)}{(x-6)(x-2)} = \frac{4}{(x-6)(x-2)}$$

$$x-2+x(x-6) = 4$$

Complete the solution process by solving for x.

c. Compare your answers in parts (a) and (b). What do you notice?

d. Why is it possible to get extraneous solutions when solving a rational equation?

Algebra II

Module 6: Investigation 5
Solving Rational Equations #2: Dealing with Extraneous Solutions

For Exercises #4-9, solve each rational equation. *Don't forget to check for extraneous solutions.*

4. $\dfrac{x^3}{x-3} = \dfrac{27}{x-3}$

5. $\dfrac{3}{x} = 1 + \dfrac{4}{x+1}$

6. $\dfrac{x-2}{x} - \dfrac{x-3}{x-6} = \dfrac{1}{x}$

7. $\dfrac{3x}{2x-2} + \dfrac{6x-9}{2x-2} = 3$

8. Let $f(x) = \dfrac{x-2}{x} - \dfrac{x-3}{x-6}$ and $g(x) = \dfrac{1}{x}$. Find the solution(s) to the system. Check for extraneous solutions and verify your answers by graphing the functions.

9. Let $f(x) = 1 + \dfrac{2x}{x-1}$ and $g(x) = \dfrac{x+3}{x^2-1}$. Find the solution(s) to the system. Check for extraneous solutions and verify your answers by graphing the functions.

Algebra II

Module 6: Investigation 6
Equivalent Forms of Rational Expressions

In Module 5 we used long division to determine whether a given value of x was a zero for a polynomial function. For example, we tested whether $x = -1$ was a zero of the polynomial function $g(x) = x^3 + 2x^2 - 5x - 6$ with long division using the following process:

I. If $x = -1$ is a zero of g, then $g(x)$ must have a factor $(x - (-1))$ [also written $(x + 1)$].

II. Complete the division of $x^3 + 2x^2 - 5x - 6$ by $(x + 1)$. If the remainder is 0, then $x = -1$ is a zero of the function.

$$\begin{array}{r} x^2 + x - 6 \\ x+1 \overline{\smash{)}\, x^3 + 2x^2 - 5x - 6} \\ \underline{-(x^3 + 1x^2)} \\ x^2 - 5x \\ \underline{-(x^2 + 1x)} \\ -6x - 6 \\ \underline{-(-6x - 6)} \\ 0 \end{array}$$

The remainder is 0, so $(x + 1)$ is a factor of g, and we can say $g(x) = (x+1)(x^2 + x + 6)$.

We could have written the quotient as $\dfrac{x^3 + 2x^2 - 5x - 6}{x+1}$ instead, which looks just like the rational expressions we've been working with in this module. Using the results from long division, the quotient can be rewritten, or simplified, to $\dfrac{x^3 + 2x^2 - 5x - 6}{x+1} = \dfrac{(x+1)(x^2 + x - 6)}{(x+1)} = x^2 + x - 6$. However, we must remember that $x \neq -1$ based on the original quotient.

1. Use long division to rewrite $\dfrac{x^3 + 7x^2 + 7x - 15}{x+5}$.

Module 6: Investigation 6
© 2014 Carlson, O'Bryan & Joyner

Algebra II

Module 6: Investigation 6
Equivalent Forms of Rational Expressions

In the previous examples we confirmed that a number or expression was a factor by completing the long division and showing that the remainder was zero. When the numerator and denominator of a rational expression *do not* share a common factor, the result is a non-zero remainder. The resulting expression will therefore look different from the previous examples.

Let's first explore this concept without variables. In Example 1, 7 is a factor of 196. In Example 2, 6 is not a factor of 233.

Ex. 1: $\dfrac{196}{7}$

$$\begin{array}{r} 28 \\ 7\overline{)196} \\ -14 \\ \hline 56 \\ -56 \\ \hline 0 \end{array}$$

$$\dfrac{196}{7} = 28 + \dfrac{0}{7}$$
$$= \boxed{28}$$

Ex 2: $\dfrac{233}{6}$

$$\begin{array}{r} 38 \\ 6\overline{)233} \\ -18 \\ \hline 53 \\ -48 \\ \hline 5 \end{array}$$

$$\dfrac{233}{6} = 38 + \dfrac{5}{6}$$
$$= \boxed{38\tfrac{5}{6}}$$

In the first case, we can say that 7 is a factor of 196, and that 196 is 28 times as large as 7 (or that $28(7) = 196$). In the second case, 6 is not a factor of 233, but we can still make a comparison of 6 to 233: 233 is $38\tfrac{5}{6}$ times as large as 6 (or $6(38\tfrac{5}{6}) = 233$).

We can use the same methods when applying long division to rational expressions with variables. For example, we begin by using long division with the quotient $\dfrac{x^2 + 2x - 6}{x + 3}$.

$$\begin{array}{r} x - 1 \\ x+3\overline{)x^2 + 2x - 6} \\ \underline{-(x^2 + 3x)} \\ -x - 6 \\ \underline{-(-x - 3)} \\ -3 \end{array}$$

We can then rewrite the quotient as follows.

$$\dfrac{x^2 + 2x - 6}{x + 3} = \boxed{(x-1) + \dfrac{-3}{x+3} \text{ or } (x-1) - \dfrac{3}{x+3}}$$

Module 6: Investigation 6
© 2014 Carlson, O'Bryan & Joyner

2. Use long division to re-write the rational expressions.

   a. $\dfrac{3x^2 + 9x - 13}{x + 6}$

   b. $\dfrac{2x^3 - 4x^2 - 5x + 4}{x - 3}$

c. $\dfrac{x^4 - 5x^3 + 2x^2 - 12x + 18}{x^2 - 4x + 5}$

Algebra II

Module 6 Homework
Rational Functions

I. VERTICAL ASYMPTOTES OF RATIONAL FUNCTIONS

1. A beverage company has just completed brewing a large batch of tea (1,500 gallons). They will add sweetener to the tea and then bottle it for distribution.
    a. Suppose 10 gallons of corn syrup is added to the tea as a sweetener. What is the ratio of tea to corn syrup in the mixture? What if 50 gallons of corn syrup is added instead? 200 gallons?
    b. Define a function f whose input x is the amount of corn syrup added (in gallons) and whose output r is the ratio of tea to corn syrup in the mixture. What is the practical domain for this function?
    c. Graph f.
    d. Complete the table of values for f.
    e. What does it mean to say "as $x \to 0^+$"?
    f. What happens to r as $x \to 0^+$? Why does this make sense?

x	$r = f(x)$
0.001	
0.01	
0.1	
1	

2. Orange juice concentrate is made by taking the juice from oranges and removing the water. Consumers can buy the concentrate and add water to produce normal orange juice. While each brand will suggest the amount of water to add to the concentrate, consumers can vary this amount to produce stronger or weaker orange juice depending on personal taste. Jermaine dumps one can of orange juice concentrate into a pitcher.
    a. Suppose Jermaine adds water to the concentrate until there are three cans worth of liquid. What is the ratio of concentrate to water? What if he adds water until there are four cans of liquid?
    b. Define a function f whose input x is the total volume of the mixture (in cans) and whose output r is the ratio of concentrate to water in the mixture. What is the practical domain for this function?
    c. Graph f.
    d. Complete the table of values for f.
    e. What does it mean to say "as $x \to 1^+$"?
    f. What happens to r as $x \to 1^+$? Why does this make sense?

x	$r = f(x)$
1.01	
1.1	
1.5	
2	

3. A weight loss center uses the following pricing plan. Joining the program costs a one-time fee of $50.00, and members don't have to pay anything more unless they lose weight. However, members pay the center $3.00 per pound they lose while in the program.
    a. Define a function f to express the cost C for someone joining the program and losing x pounds. What is the practical domain of this function?
    b. Describe how to determine the average cost (in dollars per pound lost) for a member who loses x pounds in the program.
    c. Complete the given table to determine the average cost (in dollars per pound lost) for a member who loses x pounds in the program.
    d. Write a formula for function g that determines the member's average cost (in dollars per pound lost) based on the number of pounds lost.

x	Total Cost $f(x)$	Average Cost $g(x)$
0.01		
0.1		
1		
10		
100		

    e. When x is positive and getting very close to 0 (which we write as $x \to 0^+$), what happens to the average cost per pound? Why does this happen?
    f. Does the average cost per pound have a maximum value? Explain.

Algebra II

Module 6 Homework
Rational Functions

4. A cylindrical container is being designed to carry liquid that must be insulated (see diagram below). The interior cylinder will hold the liquid, and the insulation must be 2 inches thick.

   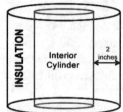

   a. If the radius of the entire container (interior cylinder and insulation) is 4.5 inches, what is the radius of the interior cylinder? What if the radius of the entire container is 5.9 inches? x inches?
   b. The formula for the volume of a cylinder is given by $V = \pi r^2 h$. Suppose the interior cylinder must hold exactly 100 in^3 of liquid. How tall must the container be if the radius of the entire container is 4.5 inches? What if the radius of the entire container is 5.9 inches?
   c. Write a formula for function f whose input x is the radius of the entire package (in inches) and whose output h is the height of the package (in inches) so that the interior cylinder holds 100 in^3.
   d. What happens to the value of h as $x \to 2^+$? What does this mean in the context of this problem?

5. Let $f(x) = \dfrac{1}{x-9}$.

   a. Complete the tables of values. Show the calculations that provided your answers.

x	f(x)
8	
8.9	
8.99	
8.999	

x	f(x)
9.001	
9.01	
9.1	
10	

   b. How do the output values of f change as $x \to 9^-$? Why does this happen?
   c. How do the output values of f change as $x \to 9^+$? Why does this happen?
   d. Use a calculator to graph f. Explain how the graph supports your conclusions in parts (b) and (c).
   e. What changes if the definition of f becomes $f(x) = \frac{1}{9-x}$? Why does this happen?
   f. What changes if the definition of f becomes $f(x) = \frac{1}{x+9}$? Why does this happen?

6. Let $f(x) = \dfrac{-1}{x+6}$.

   a. Complete the following tables of values. Show the calculations that provided your answers.

x	f(x)
−7	
−6.1	
−6.01	
−6.001	

x	f(x)
−5.999	
−5.99	
−5.9	
−5	

   b. How do the output values of f change as $x \to -6^-$? Why does this happen?
   c. How do the output values of f change as $x \to -6^+$? Why does this happen?
   d. Use a calculator to graph f. Explain how the graph supports your conclusions in parts (b) and (c).
   e. What changes if the definition of f becomes $f(x) = \frac{1}{x+6}$? Why does this happen?
   f. What changes if the definition of f becomes $f(x) = \frac{-1}{x-6}$? Why does this happen?

7. Let $f(x) = \dfrac{2}{x+7}$.

   a. How do the output values of f change as $x \to -7^-$?
   b. How do the output values of f change as $x \to -7^+$?
   c. Use a calculator to graph f. Explain how the graph supports your conclusions in parts (a) and (b).
   d. What changes if the definition of f becomes $f(x) = -\frac{2}{x+7}$? Why does this happen?

Algebra II

Module 6 Homework
Rational Functions

8. Let $g(x) = \dfrac{5x}{(x+1)(x-3)}$.

   a. How do the output values of g change as $x \to -1^-$? As $x \to -1^+$?
   b. How do the output values of g change as $x \to 3^-$? As $x \to 3^+$?
   c. Use a calculator to graph g. Explain how the graph supports your conclusions in parts (a) and (b).
   d. What changes if the definition of g becomes $g(x) = -\dfrac{5x}{(x+1)(x-3)}$? Why does this happen?

9. For each of the following functions, predict where a vertical asymptote will exist and then test your prediction using a graphing calculator.

   a. $f(x) = \dfrac{x-10}{x+10}$
   b. $g(x) = \dfrac{x^2}{5x-9}$
   c. $h(x) = \dfrac{x}{x^2+2}$
   d. $m(x) = \dfrac{6}{x^2-1}$
   e. $n(x) = \dfrac{5x+5}{x+1}$
   f. $p(x) = \dfrac{6-x}{(x-4)(x-5)}$

10. For each of the following functions, predict where a vertical asymptote will exist and then test your prediction using a graphing calculator.

    a. $f(x) = \dfrac{x+5}{x-4}$
    b. $g(x) = \dfrac{x^2}{x^2+9}$
    c. $h(x) = \dfrac{x}{x^2-4}$
    d. $m(x) = \dfrac{x-1}{x^2-1}$
    e. $n(x) = \dfrac{x^2}{(x-2)(x+1)}$
    f. $p(x) = \dfrac{2+2x}{x+1}$

II. END BEHAVIOR OF RATIONAL FUNCTIONS

11. Let $f(x) = \dfrac{3x}{x-2}$.

    a. Complete the table of values.
    b. How does $f(x)$ vary as x increases without bound? Why does this happen?
    c. How does $f(x)$ vary as x decreases without bound? Why does this happen?
    d. Using a calculator, graph f. Explain how the graph supports your conclusions in parts (b) and (c).

x	$3x$	$x-2$	$f(x) = \dfrac{3x}{x-2}$
−1,000,000			
−10,000			
−100			
100			
10,000			
1,000,000			

12. Let $f(x) = \dfrac{4x}{2x+5}$.

    a. Complete the following table of values.
    b. How does $f(x)$ vary as x increases without bound? Why does this happen?
    c. How does $f(x)$ vary as x decreases without bound? Why does this happen?
    d. Using a calculator, graph f. Explain how the graph supports your conclusions in parts (b) and (c).

x	$4x$	$2x+5$	$f(x) = \dfrac{4x}{2x+5}$
−1,000,000			
−10,000			
−100			
100			
10,000			
1,000,000			

13. A beverage company has just completed brewing a large batch of tea (1,500 gallons). They will add corn syrup to the tea and then bottle it for distribution. Function *f* relates the ratio *r* of tea to corn syrup in the mixture when *x* gallons of corn syrup are added. Thus $f(x) = \frac{1500}{x}$ where $r = f(x)$.

    a. Complete the table of values.
    b. What happens to *r* as *x* increases without bound? What does this mean in the context of this problem?
    c. Is it practical in this context to allow *x* to increase without bound? Explain.

x	r = f(x)
100	
10,000	
1,000,000	
10,000,000	

14. A cylindrical container is being designed to carry liquid that must be insulated (see diagram below). The interior cylinder will hold the liquid, and the insulation must be 2 inches thick. Function *f* is defined by $f(x) = \frac{200}{\pi(x-2)^2}$ where $h = f(x)$ and *x* is the radius of the entire package (in inches) and *h* is the height of the package (in inches) necessary so that the interior container holds 200 in³ of liquid.

x	h = f(x)
10	
100	
1,000	
10,000	

    a. Complete the table of values.
    b. What happens to the value of *h* as *x* increases without bound? What does this mean in the context of this problem?
    c. Does it make sense in this context to allow *x* to increase without bound? Explain.

15. The wildlife game commission brought 5 antelope into a 20-acre wildlife refuge. Function *h* defines the number of antelope, *n*, in the refuge as a function of time *t* (measured in years) since the antelope were introduced.

    $$h(t) = \frac{6t+5}{0.5t+1} \text{ where } n = h(t)$$

    a. Graph of *h* and label the axes.
    b. Describe how the number of antelope in the refuge changes based on the model as the number of years since they were introduced increases.
    c. According to the model, does the antelope population have a maximum value? If so, what is the largest population? If not, why not?

16. Clark's Soda Company incurred a start-up cost of $1,276 for equipment to produce a new soda flavor. The cost of producing the drink is $0.26 per can. The company generates revenue by selling the soda for $0.75 per can. (*Assume they sell every can they produce.*)
    a. Write a formula for function *f* that determines the cost (including the start-up cost) in dollars to produce *x* cans of soda.
    b. Write a formula for function *g* that determines the revenue (in dollars) generated from selling *x* cans of soda.
    c. Write a formula for function *h* that determines the profit (in dollars) from selling *x* cans of soda. (Recall that profit = revenue – cost).
    d. Write a formula for function *j* that determines the average cost per can of producing the soda.
    e. How does *j(x)* change as the number of cans produced gets larger and larger? Will the average cost function *j* ever reach a minimum value? Explain.

Algebra II
Module 6 Homework
Rational Functions

17. A salt cell (used to create salt chlorine for swimming pools) is cleaned using a mixture of water and acid. There are currently 10 liters of water in a large vat. A technician will then add acid to the vat.
Let x = the number of liters of acid added to the 10 liters of water
Let A = the percentage of the solution that is acid (called *acidity*)
a. Complete the following table of the acidity of the mixture (represented as a percentage) as acid is added to the 10 liters of water:

liters of acid, x	0	0.5	1.0	1.5	2.0	2.5	3.0	3.5
Acidity of mixture, A								

b. Write a formula for function f that calculates the acidity of the mixture (or percentage of the solution that is acid) based on the number of liters of acid x added to the vat.
c. Use a calculator to graph f.
d. Describe how the acidity of the mixture changes as the number of liters of acid added to the solution increases without bound.
e. Is it possible for the mixture to ever reach 100% acid? Explain.

18. A mosquito repellent is made by mixing a moisturizing lotion with a chemical called DEET. A tank currently contains 97 ounces of lotion and 4 ounces of DEET in the tank. As time elapses, more lotion flows into the tank at a constant rate of 12 ounces per second and more DEET flows into the tank at a constant rate of 6.4 ounces per second. The company needs to monitor the percentage of DEET in the mixture.
a. Write a formula for function f that calculates the number of ounces of DEET in the tank as a function of elapsed time (measured in seconds).
b. Write a formula for function g that calculates the total number of ounces of the mixture (both lotion and DEET) given the elapsed time (measured in seconds).
c. Write the formula for function h that determines the percentage of DEET in the mixture as a function of elapsed time (measured in seconds).
d. Complete the table. Round your answers to the nearest ten-thousandths.

Number of seconds elapsed, t	Percentage of DEET in the mixture, $h(t)$
0	
50	
100	
200	
500	

e. Describe how the percentage of DEET in the mixture changes as time increases. Will the percentage of DEET in the mixture ever reach a maximum? Explain.
f. Describe the end-behavior (as t increases without bound) of h.
g. Use a calculator to graph h and then explain how the graph supports your answer to part (f).

19. For each function, state the horizontal asymptote (if one exists).
a. $f(x) = \dfrac{x}{2x-3}$
b. $g(x) = \dfrac{x^2}{x+5}$
c. $h(x) = \dfrac{4x+1}{2x-10}$
d. $n(x) = \dfrac{17x+200}{10x^3 - 100x^2}$
e. $r(x) = \dfrac{4x^2 + x + 11}{(3x+1)(2x-3)}$
f. $q(x) = \dfrac{x^2(x+5)}{(x-8)(x-5)}$

Algebra II

Module 6 Homework
Rational Functions

20. For each function, state the horizontal asymptote (if one exists).

   a. $q(x) = \dfrac{2x^3 + 7}{5x^4 - 10}$
   b. $p(x) = \dfrac{5x^2}{x(x-4)}$
   c. $w(x) = \dfrac{(x+9)(x-5)x}{(x+6)(x+2)}$

   d. $s(x) = \dfrac{x^3(1+x)}{2x^3 - 4x}$
   e. $m(x) = \dfrac{x^2 - 2}{x^2 + 3x + 2}$
   f. $p(x) = \dfrac{5x^2 + x}{x^2}$

21. For each of the functions, state the horizontal asymptote, vertical asymptote(s), and hole(s) if any exist.

   a. $f(x) = \dfrac{(x-3)(x+1)}{(2x+10)(x-3)}$
   b. $g(x) = \dfrac{2x^2 + 2x}{3x^3 - 3}$
   c. $h(x) = \dfrac{x^2 + 6x - 7}{(x-1)(x+5)}$

22. For each of the functions, state the horizontal asymptote, vertical asymptote(s), and hole(s) if any exist.

   a. $m(x) = \dfrac{9x^3}{(x+7)x}$
   b. $n(x) = \dfrac{x^2 - 25}{(x-5)(3x-1)}$
   c. $p(x) = \dfrac{(x+7)(x-3)(2x)}{(x+1)(x-9)(x-2)}$

III. GRAPHING RATIONAL FUNCTIONS

23. Find the horizontal and vertical asymptotes and the roots for each of the following functions.

   a. $q(x) = \dfrac{3x^3 + 3}{5x^4 - 10x^2}$
   b. $r(x) = \dfrac{x^2 + 6x - 7}{(3x+1)(2x-3)}$
   c. $w(x) = \dfrac{x(x+9)(x-5)}{x(x-5)(x+2)}$

   d. $s(x) = \dfrac{x^3(1+x)}{2x^3 - 18x}$
   e. $n(x) = \dfrac{x^2 - 9}{10x^3 - 9x^2}$
   f. $p(x) = \dfrac{5x^2 + x}{x^2}$

24. Find the horizontal and vertical asymptotes and the roots for each of the following functions.

   a. $f(x) = \dfrac{x(2x-3)}{2x-3}$
   b. $g(x) = \dfrac{x^2 + 5x + 6}{x+5}$
   c. $h(x) = \dfrac{4x+1}{2x-10}$

   d. $m(x) = \dfrac{x^2 - 2}{x^2 + 3x + 2}$
   e. $p(x) = \dfrac{5x^2}{x(x-4)}$
   f. $q(x) = \dfrac{x^2(x+5)}{(x-8)(x-5)}$

For the functions in Exercises #25-28, identify the *x*-intercepts (roots), *y*-intercept, horizontal asymptote, vertical asymptotes, and the function's domain. (*State "DNE" if an intercept or asymptote does not exist.*) Use this information to sketch a graph of the function.

25. a. $a(x) = \dfrac{3}{x-7}$

   b. $b(x) = \dfrac{x}{2x+6}$

26. a. $d(x) = \dfrac{9x}{3x+3}$

   b. $f(x) = \dfrac{-x+3}{2x-1}$

27. a. $t(x) = \dfrac{-5}{x+2}$

   b. $h(x) = \dfrac{14x}{3x-4}$

28. a. $k(x) = \dfrac{x-11}{x^2 - 5x + 6}$

   b. $q(x) = \dfrac{(x^2 + 2x - 3)}{x^2 - 1}$

Algebra II

Module 6 Homework
Rational Functions

29. Use the graph of *f* to identify the vertical and horizontal intercepts, the vertical and horizontal asymptotes, and the domain and range of the function. Then write a possible formula for the function.

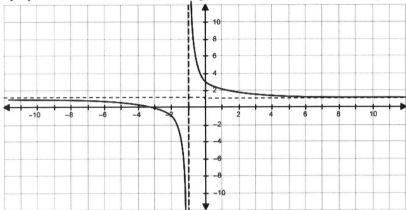

30. Use the graph of *g* to identify the vertical and horizontal intercepts, the vertical and horizontal asymptotes, and the domain and range of the function. Then write a possible formula for the function.

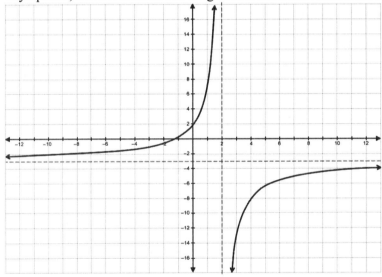

IV. SOLVING RATIONAL EQUATIONS #1: USING COMMON DENOMINATORS

31. Solve the following equations for *x*.

    a. $\dfrac{x}{5} = \dfrac{6}{5}$
    b. $\dfrac{5x}{3} = \dfrac{9}{3}$
    c. $\dfrac{x+6}{11} = \dfrac{19}{11}$
    d. $\dfrac{2(x-3)}{223} = \dfrac{24}{223}$

32. Solve the following equations for *x*.

    a. $\dfrac{x}{51} = \dfrac{-2}{51}$
    b. $\dfrac{12x}{78} = \dfrac{76.8}{78}$
    c. $\dfrac{0.4x-7}{121} = -\dfrac{19.1}{121}$
    d. $\dfrac{2(x+5)}{43} = \dfrac{-12.4}{43}$

Module 6: Homework

Algebra II — Module 6 Homework — Rational Functions

33. For each equation, do the following:
    i. Get common denominators on both sides of the equation.
    ii. Tell what the number inside the box must be for the equation to be true.
    iii. If the box were replaced with the expression $4x-1$, what value(s) of x would make the equation true?

    a. $\dfrac{\Box}{24} = \dfrac{5}{8}$
    b. $\dfrac{\Box}{20} = -\dfrac{1}{5}$
    c. $\dfrac{4}{40} = \dfrac{\Box}{8}$

34. For each equation, do the following:
    i. Get common denominators on both sides of the equation.
    ii. Tell what the number inside the box must be for the equation to be true.
    iii. If the box were replaced with the expression $2x+1$, what value(s) of x would make the equation true?

    a. $\dfrac{\Box}{12} = \dfrac{7}{4}$
    b. $\dfrac{\Box}{7} = -\dfrac{8}{28}$
    c. $-\dfrac{9}{32} = \dfrac{\Box}{8}$

35. Solve each of the equations by first getting a common denominator on both sides of the equation.

    a. $\dfrac{x}{7} = \dfrac{94}{28}$
    b. $\dfrac{x-3}{28} = \dfrac{5}{4}$
    c. $\dfrac{8}{56} = \dfrac{4x-7}{8}$
    d. $\dfrac{6}{2} = \dfrac{4x+3}{7}$
    e. $\dfrac{2x}{6} = \dfrac{-x-4}{4}$
    f. $\dfrac{2x-3}{5} = \dfrac{x+5}{6}$

36. Solve each of the equations by first getting a common denominator on both sides of the equation.

    a. $\dfrac{x}{2} = \dfrac{3}{12}$
    b. $\dfrac{2-x}{14} = \dfrac{4}{7}$
    c. $\dfrac{9}{27} = \dfrac{3x+12}{3}$
    d. $\dfrac{9}{3} = \dfrac{5x-1}{8}$
    e. $\dfrac{2x}{12} = \dfrac{-x-7}{2}$
    f. $\dfrac{2x+1}{5} = \dfrac{2x+1}{4}$

37. Solve each of the following equations.

    a. $\dfrac{x}{10} + \dfrac{4}{5} = \dfrac{17}{10}$
    b. $\dfrac{3}{4} = \dfrac{4}{20} + \dfrac{3x}{5}$
    c. $\dfrac{4}{3} + \dfrac{2x+1}{8} = \dfrac{2}{24}$
    d. $\dfrac{-4}{3} = \dfrac{2x}{3} + \dfrac{3x-4}{3}$
    e. $\dfrac{x-1}{7} - \dfrac{x}{14} = \dfrac{x}{28}$
    f. $\dfrac{x+5}{4} + \dfrac{3x-2}{5} = \dfrac{-6x}{10}$

38. Solve each of the following equations.

    a. $\dfrac{x}{8} + \dfrac{1}{2} = \dfrac{5}{2}$
    b. $\dfrac{1}{3} = \dfrac{5x}{12} + \dfrac{3}{4}$
    c. $\dfrac{3x+5}{2} + \dfrac{5}{8} = \dfrac{11}{16}$
    d. $\dfrac{-9}{7} = \dfrac{4x}{7} + \dfrac{2x-5}{7}$
    e. $\dfrac{x-5}{27} + \dfrac{2x}{9} = \dfrac{x}{3}$
    f. $\dfrac{2x-3}{3} - \dfrac{4x+1}{10} = \dfrac{4x}{5}$

Algebra II

Module 6 Homework
Rational Functions

39. Solve each of the following equations by first getting a common denominator on both sides of the equation.

   a. $\dfrac{x-1}{x} = \dfrac{9}{4}$

   b. $\dfrac{1}{3} = -\dfrac{5x}{x-4}$

   c. $\dfrac{3}{2x+5} = \dfrac{5}{3x}$

   d. $\dfrac{2x}{3} - \dfrac{2}{6x} = \dfrac{2x+5}{x}$

   e. $\dfrac{2}{2x} + \dfrac{x}{x-4} = -\dfrac{1}{2x}$

   f. $\dfrac{5}{x+1} + \dfrac{x}{x-2} = \dfrac{3}{x+1}$

40. Solve each of the following equations by first getting a common denominator on both sides of the equation.

   a. $\dfrac{x-1}{x} = \dfrac{5}{4}$

   b. $\dfrac{4}{x+3} = \dfrac{2}{x}$

   c. $\dfrac{x+4}{x+2} = \dfrac{3}{x-1}$

   d. $-\dfrac{4}{2x} + \dfrac{0.5x}{2} = \dfrac{4x-1}{4x}$

   e. $(x-1) + \dfrac{x-4}{2x} = \dfrac{4x-15}{2}$

   f. $\dfrac{2x}{3(2x-5)} + \dfrac{3}{x+1} = -\dfrac{5x}{2x-5}$

V. Solving Rational Equations #2: Dealing with Extraneous Solutions

41. Consider the rational equation $\dfrac{x}{x-2} + \dfrac{1}{x-4} = \dfrac{2}{(x-2)(x-4)}$.

   a. Before we attempt to solve the equation, take a minute to look at it closely. What values of x must immediately be restricted from the possible solutions? Why?
   b. Solve the rational equation for x by getting a common denominator on both sides.
   c. Compare your answers to parts (a) and (b). What do you notice?
   d. Why is it possible to get extraneous solutions when solving a rational equation?

42. Consider the rational equation $\dfrac{3}{x-3} + \dfrac{3}{2} = \dfrac{x}{x-3}$.

   a. Before we attempt to solve the equation, take a minute to look at it closely. What values of x must immediately be restricted from the possible solutions? Why?
   b. Solve the rational equation for x by first getting a common denominator on both sides.
   c. Compare your answers to parts (a) and (b). What do you notice?
   d. Why is it possible to get extraneous solutions when solving a rational equation?

For Exercises #43-48, solve each rational equation. *Don't forget to check for extraneous solutions.*

43. $\dfrac{x^2}{x-4} = \dfrac{16}{x-4}$

44. $\dfrac{5}{x-5} = 1 + \dfrac{9}{x-5}$

45. $\dfrac{x}{x-1} - 2 = \dfrac{1}{x-1}$

46. $\dfrac{1}{x-2} - \dfrac{3}{x+2} = \dfrac{6x}{(x-2)(x+2)}$

47. $\dfrac{x+2}{x-1} - \dfrac{2}{x(x-1)} = \dfrac{2x+1}{x}$

48. $\dfrac{10}{x^2-2x} = \dfrac{5}{x-2} - \dfrac{4}{x}$

Module 6: Homework

Algebra II

Module 6 Homework
Rational Functions

VI. EQUIVALENT FORMS OF RATIONAL EXPRESSIONS

For Exercises #49-54, use long division to rewrite the given rational functions. *Don't forget to pay attention to the domain of the function.*

49. $\dfrac{x^3 - 4x^2 + 5x - 6}{x - 3}$

50. $\dfrac{2x^3 - 5x^2 - 13x + 4}{x - 4}$

51. $\dfrac{x^4 - x^3 - 2x^2 - x - 1}{x + 1}$

52. $\dfrac{-x^4 + 2x^3 + 8x^2 + 3x + 6}{x + 2}$

53. $\dfrac{2x^4 + 6x^3 + 3x^2 + 6x - 9}{x + 3}$

54. $\dfrac{3x^4 - 5x^3 + 3x^2 - 4x + 3}{x - 1}$

For Exercises #55-60, use long division to rewrite the given rational functions.

55. $\dfrac{x^2 + 3x - 1}{x - 2}$

56. $\dfrac{x^3 + 3x - 1}{x + 1}$

57. $\dfrac{2x^3 + 11x^2 + 3x - 30}{x + 4}$

58. $\dfrac{-x^4 + 2x^3 + 3x^2 + x - 17}{x - 2}$

59. $\dfrac{2x^4 + 8x^3 + x^2 + 12x + 3}{x^2 + 4x - 1}$

60. $\dfrac{3x^4 + x^3 - 11x^2 + 11}{3x^2 + x - 5}$

Algebra II

Module 7: Investigation 1
Introduction to Radical Functions

1. If an object starts at rest and accelerates at a rate of *a* feet per second each second, then the distance *d* that the object travels (in feet) after *t* seconds is given by $d = \frac{1}{2}at^2$. Suppose a car accelerates at a rate of 14 feet per second each second over the course of 8 seconds. Then $f(t) = 7t^2$ where $d = f(t)$.

    a. Complete the table relating the two quantities.

Time elapsed, in seconds	Distance traveled, in feet
0	
2	
3.5	
4	
5	
8	

    b. Describe the behavior of the function. In other words, explain how the two quantities co-vary.

    c. What is the practical domain of *f*? What is the practical range of *f*?

    d. Determine the formula for f^{-1}. What quantity is the input? What quantity is the output?

    e. Complete the table relating the two quantities.

distance traveled, in feet	time elapsed, in seconds
0	
28	
85.75	
112	
175	
448	

    f. Explain how the table you completed in part (e) is related to the table you completed in part (a).

Module 7: Investigation 1

© 2014 Carlson, O'Bryan & Joyner

g. Sketch a graph of f^{-1} on the given axes.

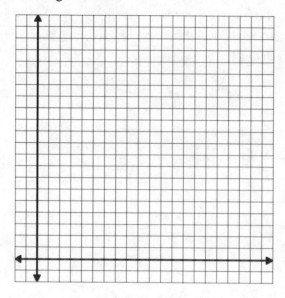

h. Describe the behavior of f^{-1}. In other words, explain how the two quantities change together.

The function you created in Exercise #1d is called a *radical function*. We will continue to explore these types of functions throughout this module.

2. When police are investigating traffic accidents, they can sometimes use skid marks to determine how fast a car was traveling. Suppose $g(d) = \sqrt{16.5d}$ where $S = g(d)$ represents the speed of a car (in miles per hour) before beginning its skid given the length of the skid marks d (in feet).
   a. Sketch a graph of g on the given axes.

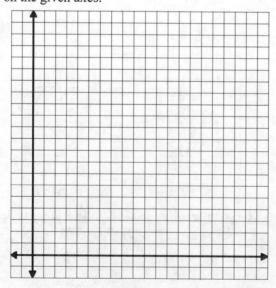

Algebra II
Module 7: Investigation 1
Introduction to Radical Functions

b. Determine the practical domain and range of the function. Explain your reasoning.

c. Describe how the quantities *length of the skid marks* and *speed of the car* change together.

d. Evaluate $g(28)$ and explain its meaning in the context of the problem.

3. $h(V) = \sqrt[3]{V}$ outputs the side length of a cube (in cm) given the cube's volume (in cm^3).
   a. Complete the following table relating the two quantities.

volume of cube, in cubic centimeters	side length of cube, in centimeters
0	
1	
8	
42.875	
216	

   b. Sketch a graph of h on the given axes.

   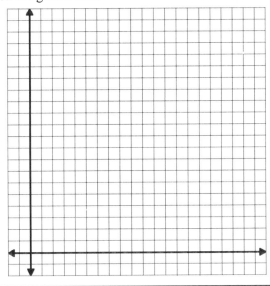

Algebra II

Module 7: Investigation 1
Introduction to Radical Functions

c. Explain how the quantities *volume of the cube* and *side length* change together.

d. Determine the practical domain and range of the function.

e. Evaluate $h(64)$ and explain its meaning in the context of the problem.

f. Explain how you can determine the value of $h^{-1}(9)$ and explain the meaning of the solution in the context of the problem.

Algebra II

Module 7: Investigation 2
Working with the Square Root Function

1. The function $f(d) = \frac{1}{4}\sqrt{d}$ where $t = f(d)$ is used to determine the amount of time t (in seconds) it will take an object to hit the ground when dropped from a height of d feet above the ground.

   a. Complete the table relating the two quantities.

drop height (in feet)	time elapsed (in seconds)
–1	
0	
1	
2	
4	
9	
10	

   b. Use your table to sketch a graph of f on given axes.

   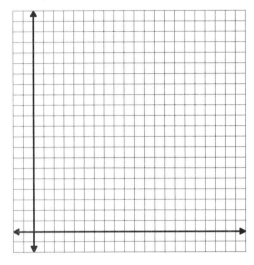

   c. State the function's practical domain and range.

   d. Evaluate $f(60)$ and explain its meaning in the context of the problem.

   e. Suppose it takes an object 6.7 seconds to hit the ground after being dropped. Explain how you can determine the distance of the drop.

Module 7: Investigation 2

2. The basic square root function is defined as $f(x)=\sqrt{x}$ and its graph is given.

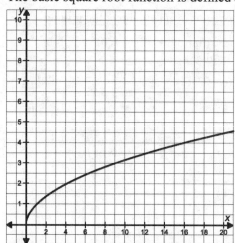

   a. What is the function's domain? Explain why this makes sense.

   b. What is the function's range?

   c. How is $f(x)=\sqrt{x}$ related to $g(x)=x^2$?

3. Given the functions $f(x)=\sqrt{x}$ and $g(x)=-\sqrt{x}$,
   a. Complete the following tables.

x	f(x)
−4	
−1	
1	
9	
12	

x	g(x)
−4	
−1	
1	
9	
12	

   b. Sketch a graph of g on the given axes.

   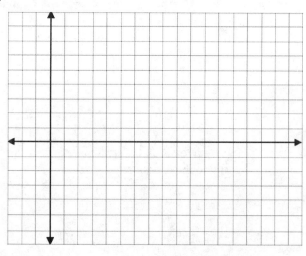

   c. Explain how g is related to f.

d. State the domain and range of g and explain your choices.

4. Given the functions $f(x)=\sqrt{x}$ and $h(x)=\sqrt{-x}$,
   a. Complete the following tables.

x	f(x)
−9	
−4	
−2	
2	
4	
9	

x	h(x)
−9	
−4	
−2	
2	
4	
9	

   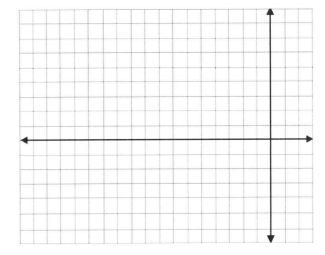

   b. Sketch a graph of h on the given axes.

   c. Explain how h is related to f.

   d. State the domain and range of h and explain your choices.

5. Given the functions $f(x)=\sqrt{x}$ and $k(x)=\sqrt{x+5}$,
   a. Complete the following tables.

x	f(x)
−4	
−1	
4	
9	
16	
25	

x	k(x)
−9	
−6	
−1	
4	
11	
20	

b. Sketch a graph of *k* on the given axes.

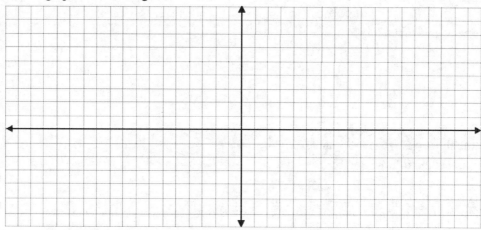

c. Explain how *k* is related to *f*.

d. State the domain and range of *k*.

6. Given the functions $f(x) = \sqrt{x}$ and $m(x) = \sqrt{x+4}$,
   a. Complete the following tables.

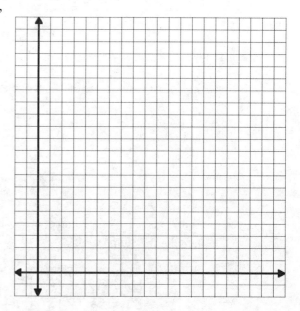

x	f(x)
−4	
0	
1	
5	
9	

x	m(x)
−4	
0	
1	
5	
9	

b. Sketch a graph of *m* on the given axes.

c. Explain how *m* is related to *f*.

d. State the domain and range of *m*.

Algebra II

Module 7: Investigation 2
Working with the Square Root Function

7. Let $f(x) = \sqrt{2x+3}$.

   a. What is the domain for this function? Pick a value outside of the domain and explain why it shouldn't be included.

   b. What is the range for this function? Pick a value outside of your range and explain why it is not possible for your function to take on this value.

   c. Sketch a graph of the function on the given axes.

   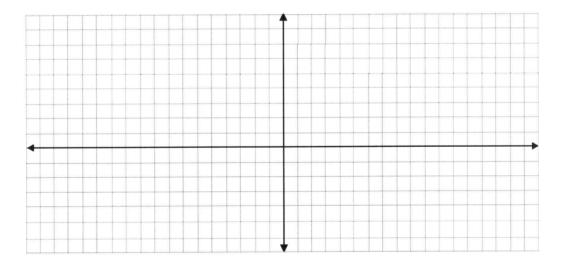

Algebra II

Module 7: Investigation 2
Working with the Square Root Function

8. Complete the following for each function:
   i. Without graphing, predict the function's domain.
   ii. Without graphing, predict the function's range.
   iii. Graph the function using a table of values or your calculator. If you were wrong in parts (i) or (ii), explain <u>why</u> the function behaves differently from what you expected (*you might have to plug in some numbers and perform calculations to understand*).

   a. $f(x) = 3\sqrt{x-4}$ b. $f(x) = \sqrt{-x+1} + 6$

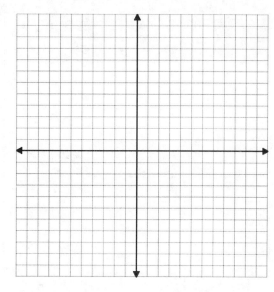
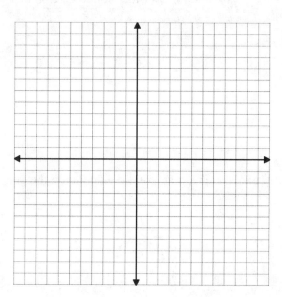

   c. $f(x) = -\sqrt{6-x} - 2$

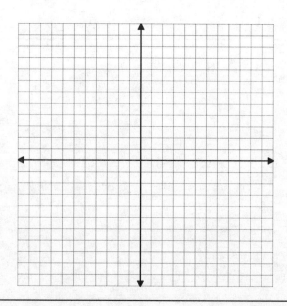

Algebra II

Module 7: Investigation 3
Working with the Cube Root Function

1. Complete the following for each function:
   i. Without graphing, predict the function's domain.
   ii. Without graphing, predict the function's range.
   iii. Graph the function using a table of values or your calculator. If you were wrong in parts (i) or (ii), explain <u>why</u> the function behaves differently from what you expected.

   a. $f(x) = 3\sqrt{x} - 2$

   b. $f(x) = -\sqrt{x+9} + 1$

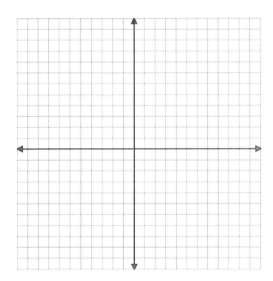

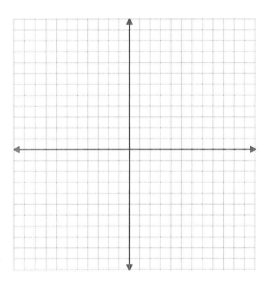

2. The radius length $r = f(v)$ of a sphere (in feet) is given by $f(v) = \sqrt[3]{\dfrac{3v}{4\pi}}$ where v is the volume of the sphere (in cubic feet). Also, $f^{-1}(r) = \dfrac{4}{3}\pi r^3$ where $v = f^{-1}(r)$.

   a. Explain what it means for these functions to be inverses.

   b. Complete the following tables.

v	$f(v)$
0	
4.189	
113.1	
523.6	
2000	

r	$f^{-1}(r)$
0	
1	
3	
5	
7.816	

Algebra II

Module 7: Investigation 3
Working with the Cube Root Function

c. Graph f on the left axes and f^{-1} on the right axes.

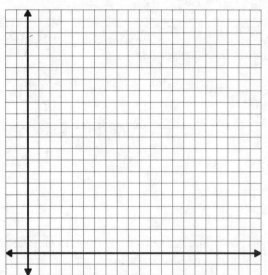 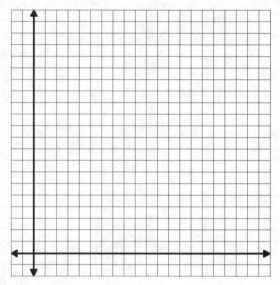

d. What is the practical domain of f? What is the practical range of f?

e. Suppose we examine f outside of the context of the sphere. Does this change the domain and range of f? If so, how?

f. Describe how v and r co-vary.

3. Function f in Exercise #2 is called a **cube root function** because the cube root appears in the function rule. The most basic cube root function is $f(x) = \sqrt[3]{x}$.
   a. What is the domain of f? Explain why this makes sense.

   b. What is the range of f? Explain why this makes sense.

c. Graph f on the given axes.

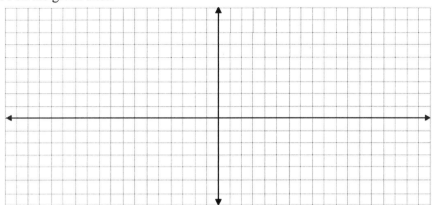

4. Transformations of f in Exercise #3 will have similar results to transformations made to the square root function we studied in Investigation 2. For example, consider the functions $f(x) = \sqrt[3]{x}$ and $g(x) = \sqrt[3]{x+2}$.

   a. Complete the following tables.

x	f(x)
−8	
−1	
0	
1	
8	

x	g(x)
−10	
−3	
−2	
−1	
6	

   b. Graph g on the given axes.

   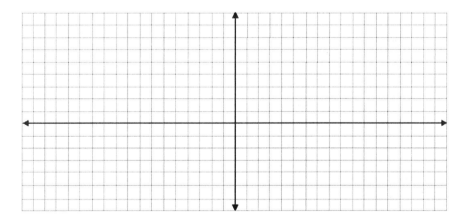

   c. Explain how g is related to f.

d. State the domain and range of g and explain your choices.

5. Complete the following for each function:
   i. Without graphing, predict the function's domain.
   ii. Without graphing, predict the function's range.
   iii. Graph the function using a table of values or your calculator. If you were wrong in parts (i) or (ii), explain <u>why</u> the function behaves differently from what you expected.

   a. $f(x) = \sqrt[3]{x-3} + 4$

   b. $f(x) = \sqrt[3]{6-x} - 2$

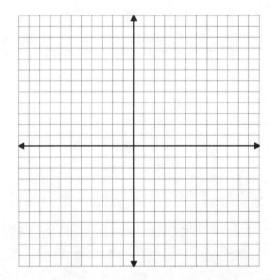
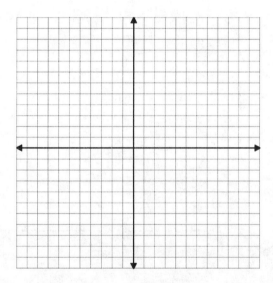

   c. What do you think is true about the domain and range for any cube root function? Explain.

Algebra II

Module 7: Investigation 4
Solving Radical Equations

1. Graph each function without using a calculator and state the domain and range. Then check your answers using a graphing calculator. *If you made a mistake, think about your answers and explain why the function behaves differently from what you expected.*

   a. $f(x) = 3\sqrt{x-6} + 1$

   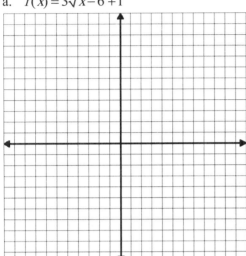

   b. $f(x) = \sqrt[3]{x-8} + 2$

   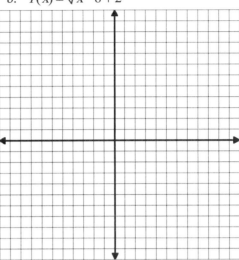

2. Let $f(x) = \sqrt{7x-12}$.

   a. Evaluate $f(3)$.

   b. Determine x if $f(x) = 10$.

   c. Determine the value of x such that $f(x) = x$. Check your answers by substituting them back into the original equation.

Algebra II

Module 7: Investigation 4
Solving Radical Equations

3. Solve each of the following equations algebraically and demonstrate what the solutions represent graphically.

   a. $\sqrt{4x-3}+1=5$

   b. $-2\sqrt{x+6}+7=3$

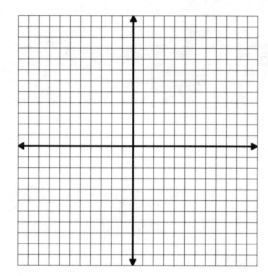
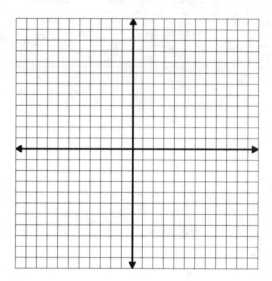

   c. $3\sqrt{4-x}=12$

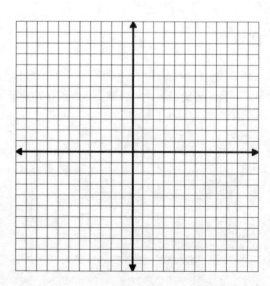

Algebra II

Module 7: Investigation 4
Solving Radical Equations

4. Let $f(x) = \sqrt{x}$ and $g(x) = x - 2$. Find the solution(s) to the system. Check your answers by substituting them back into the original functions and/or graphing the functions.

5. Let $f(x) = \sqrt{x+2} - 2$ and $g(x) = x$. Find the solution(s) to the system. Check your answers by substituting them back into the original functions and/or graphing the functions.

6. Let $f(x) = \sqrt{5x-6}$ and $g(x) = 2x$. Find the solution(s) to the system. Check your answers by substituting them back into the original functions and/or graphing the functions.

In Exercise #4 we saw that the process of solving a radical equation can generate ***extraneous solutions***.

> **Extraneous Solutions**
>
> An ***extraneous solution*** is a solution to a modified form of an equation that is not a solution to the original equation.

Let's explore why extraneous solutions can appear when solving radical equations.

7. Let $f(x) = \sqrt{x+3}$ and $g(x) = x+1$. If we want to solve the system, we begin by setting $g(x) = f(x)$, which yields the equation $\sqrt{x+3} = x+1$. The first step in the solution process is to square both sides, giving you $x+3 = (x+1)^2$. You've changed how your equation looks, but have you also changed the possible solutions? Will every value that makes $x+3 = (x+1)^2$ true also make $\sqrt{x+3} = x+1$ true? Let's explore.

   a. If $a = b$ can you always conclude that $a^2 = b^2$?

   b. If $c^2 = d^2$ does this automatically mean $c = d$? (for example, if $c^2 = 25$ and $d^2 = 25$, do c and d have to be the same number?) Why or why not?

   c. Solve the equation $x+3 = (x+1)^2$ and then substitute your answers into the equation $\sqrt{x+3} = x+1$ to determine which solution is extraneous. *Pay close attention to what happens when you substitute the extraneous solution.*

   d. Why do extraneous solutions sometimes appear when solving radical equations?

Algebra II

Module 7: Investigation 4
Solving Radical Equations

8. Solve each of the following equations algebraically and demonstrate what the solution represents graphically.

   a. $\sqrt[3]{2x-4} - 3 = -1$

   b. $-1 \cdot \sqrt[3]{x+3} = 1$

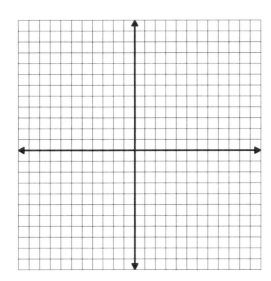
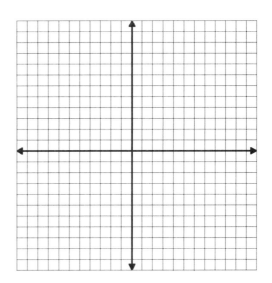

c. $6 \cdot \sqrt[3]{5-2x} - 1 = 4$

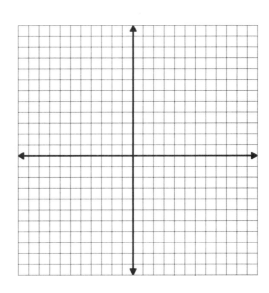

Algebra II

Module 7: Investigation 5
Inverses of Radical and Power Functions

We've looked at inverse functions several times in this course. Recall that two functions f and g are inverses of each other if they express the same relationship between two quantities but have their input and output quantities switched. When two functions are inverses we typically use the notation f^{-1} to communicate this relationship. Therefore, if f has an inverse function, we call it f^{-1}, and if $y = f(x)$ then $x = f^{-1}(y)$.

Inverses of Radical Functions

1. In a previous investigation we examined the function $f(d) = \frac{1}{4}\sqrt{d}$ where $t = f(d)$, which determines the amount of time (in seconds) it takes an object to hit the ground when dropped from a height of d feet above the ground.

    a. What does $f(3)$ represent?

    b. What does $f^{-1}(3)$ represent?

    c. Determine the formula for f^{-1} and then explain what it models.

    d. Suppose it takes an object 6.7 seconds to hit the ground. Explain how you can determine the drop height i) using f and ii) using f^{-1}.

Algebra II
Module 7: Investigation 5
Inverses of Radical and Power Functions

For Exercises #2-5, do the following.
a) State the domain and range of the function.
b) Find the inverse of the function.
c) State the domain and range of the inverse function.

2. $f(x) = 3\sqrt{x}$ where $y = f(x)$

3. $g(r) = -2\sqrt[3]{r} - 6$ where $w = g(r)$

4. $h(t) = \sqrt[4]{t+3} - 1$ where $d = h(t)$

5. $j(x) = -\sqrt[9]{2x-1}$ where $n = j(x)$

Algebra II

Module 7: Investigation 5
Inverses of Radical and Power Functions

Inverses of Power Functions

> **Power Function**
> A *power function* is a function of the form $f(x) = ax^n$ where a and n are non-zero real number constants.

6. In a previous worksheet we examined the function $g(r) = \frac{4}{3}\pi r^3$ where r is the radius of a sphere, v is the volume of the sphere, and $v = g(r)$.

   a. What does $g(8)$ represent?

   b. What does $g^{-1}(8)$ represent?

   c. Find the formula for g^{-1} and then explain what it models.

   d. Suppose you have a sphere with a volume of 15 cubic units. Explain how you can determine the radius of the sphere i) using g and ii) using g^{-1}.

Algebra II

Module 7: Investigation 5
Inverses of Radical and Power Functions

For Exercises #7-9, find the inverse of each power function. Restrict the domain of the original function if necessary so that the inverse is a function.

7. $f(x) = 4x^2$ where $y = f(x)$

8. $g(x) = \frac{5}{7}x^{11}$ where $z = g(x)$

9. $h(p) = -p^{-2}$ where $y = h(p)$

Algebra II

Module 7 Homework
Radical Functions

INVESTIGATION 1: INTRODUCTION TO RADICAL FUNCTIONS

1. Suppose a car accelerates at a rate of 9 feet per second each second over the course of 6 seconds. The function $f(t) = 4.5t^2$ where $d = f(t)$ then represents the distance in feet the car has traveled after t seconds during this time.

   Table 1 (Function f)

Time elapsed, in seconds	Distance traveled, in feet
0	
2	
2.4	
4	
4.7	
6	

   Table 2 (Function f^{-1})

Distance traveled, in feet	Time elapsed, in seconds
0	
18	
40.5	
72	
91.125	
162	

   a. Complete Table 1 relating the two quantities for f.
   b. Describe the behavior of the function. That is, explain how the two quantities co-vary.
   c. State the practical domain and range of f.
   d. Determine the formula for f^{-1}. What quantity is the input? What quantity is the output?
   e. Complete Table 2 relating the two quantities for f^{-1}.
   f. Explain how the table you completed in part (e) is related to the table you completed in part (a).
   g. Sketch a graph of the function f^{-1}.
   h. Describe the behavior of the function. That is, explain how the two quantities co-vary.

2. The formula $f(r) = 4\pi r^2$ gives the surface area of a spherical fitness stability ball in terms of its radius r, measured in some units. A particular fitness stability ball advertises that it can be aired up to have a maximum radius of 18 inches.

   Table 1 (Function f)

Radius, in inches	Surface Area, in inches squared
0	
5.5	
11	
14.25	
18	

   Table 2 (Function f^{-1})

Surface Area, in inches squared	Radius, in inches
0	
28.2743	
380.1327	
2,123.7166	
4071.5041	

   a. Complete Table 1 relating the two quantities for f.
   b. Describe the behavior of the function. That is, explain how the two quantities co-vary.
   c. State the practical domain and range of f for the specific stability ball described.
   d. Determine the formula for f^{-1}. What quantity is the input? What quantity is the output?
   e. Complete Table 2 relating the two quantities for f^{-1}.
   f. Explain how the table you completed in part (e) is related to the table you completed in part (a).
   g. Sketch a graph of the function f^{-1}.
   h. Describe the behavior of the function. That is, explain how the two quantities co-vary.

Algebra II

Module 7 Homework
Radical Functions

3. A commonly used formula to determine a person's Body Surface Area (BSA), measured in square meters is $BSA = \sqrt{\dfrac{w \cdot h}{3131}}$ where w is the person's weight in pounds and h is the person's height in inches. Suppose you wanted to examine how BSA co-varied with weight in people who were 69 inches tall. The function relating the quantities of BSA and weight is $f(w) = \sqrt{\dfrac{69w}{3131}}$.

   a. Sketch a graph of the function f on a set of axes.
   b. Describe how the quantities weight and BSA co-vary.
   c. Determine a practical domain and range of the function f. Explain your reasoning.
   d. Evaluate $f(150)$ and explain its meaning in the context of the problem.
   e. Explain how you can determine the value of $f^{-1}(2.099)$ and explain the meaning of the solution in the context of the problem.

4. The function $g(A) = \sqrt{\dfrac{A}{\pi}}$ where $r = g(A)$ determines the radius of a circle given its area A.

   a. Sketch a graph of the function.
   b. Describe how the quantities area and radius co-vary.
   c. Determine the domain and range of the function.
   d. Determine $g(120)$ and explain its meaning in the context of the problem.
   e. Explain how you can determine the value of $g^{-1}(14.1)$ and explain the meaning of the solution in the context of the problem.

INVESTIGATION 2: WORKING WITH THE SQUARE ROOT FUNCTION

5. The function $f(d) = \sqrt{\dfrac{d}{9.8}}$ where $t = f(d)$ can be used to determine the amount of time in seconds it will take an object to hit the ground when dropped a distance of d meters.

   a. Complete the table relating the two quantities.

Distance dropped, in meters	Time elapsed, in seconds
0	
4	
6.1	
9.3	
15	

   b. Use your table to sketch a graph of f.
   c. State the practical domain and range of the function.
   d. Evaluate $f(14)$ and explain its meaning in this context.
   e. Suppose it takes an object 3.7 seconds to hit the ground after being dropped. Determine how far the object fell.

Algebra II

Module 7 Homework
Radical Functions

6. The function $f(h) = \sqrt{1.5h}$ where $d = f(h)$ determines the horizontal distance that one can see (in miles) given a height of their eye h above sea level (in feet) *assuming flat terrain*. For example, if a person is standing on a boat and the height of their eye above sea level is 10 feet, then the horizontal distance they are able to see to is $\sqrt{1.5 \cdot 10} \approx 3.87$ miles.

   a. Complete the table relating the two quantities.

Height of eye above sea level (in feet)	Horizontal distance that one can see (in miles)
0	
1.5	
3	
4.5	
6	

   b. Sketch a graph of f.
   c. Explain how the quantities change together and what this means in the context of the problem.
   d. Determine $f(8.2)$ and explain its meaning in this context.
   e. Determine h when $f(h) = 5.6$ and explain its meaning in this context.

7. Given the functions $f(x) = \sqrt{x}$ and $k(x) = \sqrt{x+2}$,
   a. Complete the following tables.

x	f(x)
−2	
−1	
0	
1	
2	

x	k(x)
−3	
−2	
−1	
0	
1	

   b. Sketch a graph of k.
   c. Explain how function k is related to function f.
   d. State the domain and range of k.

8. Given the functions $f(x) = \sqrt{x}$ and $m(x) = 2\sqrt{x} - 1$,
   a. Complete the following tables.

x	f(x)
−2	
−1	
0	
1	
2	

x	m(x)
−2	
−1	
0	
1	
2	

   b. Sketch a graph of m.
   c. Explain how function m is related to function f.
   d. State the domain and range of m.

Algebra II

Module 7 Homework
Radical Functions

9. Given the function $g(x) = \sqrt{x+1} + 2$,
   a. What is the domain for this function? Pick a value outside of the domain and explain why it shouldn't be included.
   b. What is the range for this function? Pick a value outside of your range and explain why it is not possible for your function to take on this value.
   c. Sketch a graph of the function.

10. Given the function $f(x) = -2\sqrt{x-1}$,
    a. What is the domain for this function? Pick a value outside of the domain and explain why it shouldn't be included.
    b. What is the range for this function? Pick a value outside of your range and explain why it is not possible for your function to take on this value.
    c. Sketch a graph of the function.

11. Complete the following for each function:
    i. Without graphing, predict the domain of f.
    ii. Without graphing, predict the range of f.
    iii. Graph the function using a table of values or your calculator. If you were wrong in parts (a) or (b), explain <u>why</u> the functions behaves differently than what you expected (you might have to plug in some numbers and perform calculations to understand).

    a. $f(x) = \sqrt{-x+1} - 2$
    b. $f(x) = -\sqrt{3+x}$
    c. $f(x) = 4\sqrt{2x-1} - 4$

12. Complete the following for each function:
    i. Without graphing, predict the domain of f.
    ii. Without graphing, predict the range of f.
    iii. Graph the function using a table of values or your calculator. If you were wrong in parts (a) or (b), explain <u>why</u> the functions behaves differently than what you expected (you might have to plug in some numbers and perform calculations to understand).

    a. $f(x) = -2\sqrt{x-3}$
    b. $f(x) = \sqrt{7-x} - 1$
    c. $f(x) = \sqrt{2x+4} + 2$

Algebra II

Module 7 Homework
Radical Functions

INVESTIGATION 3: WORKING WITH THE CUBE ROOT FUNCTION

13. A company that manufactures boxes creates boxes with the width and height having equal measure and the length of the box being 3.5 times the as long as the width. Thus we can find the width of the box using the function $f(v) = \sqrt[3]{\dfrac{v}{3.5}}$ where v is the volume of the box in cubic inches and $x = f(v)$, where x is the width of the box in inches. This function is the inverse function of $g(x) = 3.5x^3$.

    a. Complete the following tables relating the quantities. What should you find given that the functions are inverses of one another?

x	v
0	
1	
2	
3	
4	

v	x
0	
3.5	
28	
94.5	
224	

    b. What is the practical domain of f? What is the practical range of f?
    c. Restricting the values to their practical domain and range, sketch a graph of f on one set of axes and g on a separate set of axes. Label each set of axes.
    d. Describe how v and x co-vary.
    e. Consider function f outside of the context of the box. How does this impact the domain and range?

14. A high-diver's trajectory from a 10-meter board was studied over hundreds of dives and a model of his average dive was created. The function $f(x) = \sqrt[3]{-x+10}$ where $t = f(x)$ gives the number of seconds since his feet left the board t in terms of his height above the water x in meters. This function is the inverse function of $g(t) = -t^3 + 10$.

    a. Complete the following tables relating the quantities. What should you find given that the functions are inverses of one another?

t	x
0	
0.5	
1	
1.5	
2	

x	t
10	
9.875	
9	
6.625	
2	

    b. Considering that the model serves to describe the diver's motion from the board to the surface of the water, sketch a graph of the function f on one axis and the function g on another axis while attending to their practical domains and ranges. Label each set of axes.
    c. Describe how t and x co-vary.
    d. Suppose we were to examine the function f outside of the context of the diver. How does this change the domain and range of f?

Algebra II

Module 7 Homework
Radical Functions

15. Given the functions $f(x) = \sqrt[3]{x}$ and $g(x) = \sqrt[3]{x-4}$,
    a. Complete the following tables.

x	f(x)
−27	
−1	
0	
1	
8	

x	g(x)
	−3
	−1
	0
	1
	2

    b. Sketch a graph of g.
    c. Explain how function g is related to function f.
    d. State the domain and range of g.

16. Given the functions $f(x) = \sqrt[3]{x}$ and $g(x) = -2 \cdot \sqrt[3]{x}$,
    a. Complete the following tables.

x	f(x)
−8	
−1	
0	
1	
27	

x	g(x)
−8	
−1	
0	
1	
27	

    b. Sketch a graph of g on the axes below.
    c. Explain how function g is related to function f.
    d. State the domain and range of g.

17. Complete the following for each function.
    i. Without graphing, predict the domain of f.
    ii. Without graphing, predict the range of f.
    iii. Describe how you think the graph of this function will compare to the function $g(x) = \sqrt[3]{x}$
    iv. Graph the function using a table of values or your calculator. If you were wrong in parts (i), (ii), or (iii) explain <u>why</u> the functions behaves differently than what you expected.
    a. $f(x) = 3 \cdot \sqrt[3]{x}$
    b. $f(x) = \sqrt[3]{-x+1}$
    c. $f(x) = -3 \cdot \sqrt[3]{x} + 3$

18. Complete the following for each function.
    i. Without graphing, predict the domain of f.
    ii. Without graphing, predict the range of f.
    iii. Describe how you think the graph of this function will compare to the function $g(x) = \sqrt[3]{x}$
    iv. Graph the function using a table of values or your calculator. If you were wrong in parts (i), (ii), or (iii) explain <u>why</u> the functions behaves differently than what you expected.
    a. $f(x) = \sqrt[3]{x} + 2$
    b. $f(x) = \sqrt[3]{x-3}$
    c. $f(x) = -5 \cdot \sqrt[3]{x-2}$

INVESTIGATION 4: SOLVING RADICAL EQUATIONS

19. Without using a calculator, sketch each of the following graphs and state the domain and range. Then check your answers using a graphing calculator or table of values.
    a. $f(x) = -\sqrt{x+2}$
    b. $f(x) = \sqrt{3-x} - 3$
    c. $f(x) = 2 \cdot \sqrt[3]{x-2}$

Algebra II

Module 7 Homework
Radical Functions

20. Without using a calculator, sketch each of the following graphs and state the domain and range. Then check your answers using a graphing calculator or table of values.
    a. $f(x) = \sqrt{-x-1}$
    b. $f(x) = -4 \cdot \sqrt[3]{x+7} - 2$
    c. $f(x) = -\sqrt{x-3} + 1$

21. Given the following function: $f(x) = \sqrt{3x+5}$.
    a. Evaluate $f(2)$.
    b. Determine x if $f(x) = 7$.
    c. Determine the value of x such that $f(x) = x$. Don't forget to check for extraneous solutions.!

22. Given the following function: $f(x) = 2 - \sqrt{5-x}$.
    a. Evaluate $f(-3)$.
    b. Determine x if $f(x) = -6$.
    c. Determine the value of x such that $f(x) = x$. Don't forget to check for extraneous solutions.

23. Solve each of the following equations algebraically and graphically for x.
    a. $\sqrt{2x+4} - 5 = 2$
    b. $-\sqrt{-x-4} - 1 = -7$
    c. $4\sqrt{1+2x} = 20$

24. Solve each of the following equations algebraically and graphically for x.
    a. $\sqrt{3x-2} + 4 = 6$
    b. $-\sqrt{-x+1} + 8 = 5$
    c. $-3\sqrt{5x+5} + 10 = 1$

25. Let $f(x) = \sqrt{x}$ and $g(x) = x - 3$. Find the solution(s) to the system. Check your answers by substituting them back into the original functions and/or graphing the functions.

26. Let $f(x) = \sqrt{-x+1} - 3$ and $g(x) = x$. Find the solution(s) to the system. Check your answers by substituting them back into the original functions and/or graphing the functions.

27. Solve each of the following equations algebraically and graphically for x.
    a. $\sqrt[3]{x+3} + 1 = 3$
    b. $2 \cdot \sqrt[3]{x-7} + 2 = -4$
    c. $-1 \cdot \sqrt[3]{2-x} - 3 = -1$

28. Solve each of the following equations algebraically and graphically for x.
    a. $\sqrt[3]{x+2} - 3 = -4$
    b. $-1 \cdot \sqrt[3]{x+1} = 1$
    c. $\sqrt[3]{2-3x} + 2 = 1$

INVESTIGATION 5: INVERSES OF RADICAL AND POWER FUNCTIONS

29. When investigating traffic accidents, it is useful to be able to determine the speed a car was traveling. When skid marks are apparent, their length can be used to help determine the speed of a car when it began to skid to a stop. Suppose the function $f(d) = \sqrt{16.5d}$ where $S = f(d)$ represents the speed of a car in miles per hour on a specific road given the length of the skid marks d in feet.
    a. What does $f(30)$ represent?
    b. What does $f^{-1}(30)$ represent?
    c. Find the formula for f^{-1} then explain what it models.
    d. d. Suppose a set of skid mark is 47 feet long. Explain how you can estimate the speed of the car i) using f and ii) using f^{-1}.

Algebra II

Module 7 Homework
Radical Functions

30. The surface area s of a cone whose base has a radius of 3 feet is modeled by the function $g(h) = 3\pi\sqrt{9+h^2}$ where h is the height of the cone in feet and $s = g(h)$.
    a. What does $g(2.3)$ represent?
    b. What does $g^{-1}(2.3)$ represent?
    c. Find the formula for g^{-1} then explain what it models.
    d. Suppose a cone has a surface area of 51 ft². Explain how you can estimate the height of the cone i) using g and ii) using g^{-1}.

For Exercises #31-36, do the following.
   a) State the domain and range of the function.
   b) Find the inverse of the function.
   c) State the domain and range of the inverse function.

31. $f(x) = -14\sqrt{x}$

32. $g(y) = 4\sqrt[3]{y} + 2$

33. $j(x) = 5 - \sqrt[7]{10-2x}$

34. $h(t) = -2\sqrt[6]{t-2} + 5$

35. $k(x) = \sqrt[4]{2x+6} - 4$

36. $m(y) = 3\sqrt[5]{-2x+1} - 5$

37. The function $f(r) = \pi r^2$ determines the area A of a circle with respect to the radius r of the circle where $A = f(r)$.
    a. What does $f(12)$ represent?
    b. What does $f^{-1}(12)$ represent?
    c. Find the formula for f^{-1} then explain what it models.
    d. Suppose a circle has an area of 22 units². Explain how you can determine the radius of the circle i) using f and ii) using f^{-1}.

38. The function $g(s) = s^3$ determines the volume V of a cube with respect to the side length s of the cube where $V = g(s)$.
    a. What does $g(8.2)$ represent?
    b. What does $g^{-1}(8.2)$ represent?
    c. Find the formula for g^{-1} then explain what it models.
    d. Suppose a cube has a volume of 58.2 units³. Explain how you can determine the length of the side of the cube i) using g and ii) using g^{-1}.

For Exercises #39-44, find the inverse of each power function. Restrict the domain of the original function if necessary so that the inverse is a function.

39. $f(x) = -12x^2$

40. $g(x) = 12x^7$

41. $h(x) = 6x^{-5}$

42. $j(x) = \frac{8}{15}x^{-8}$

43. $k(x) = -0.672x^4$

44. $m(x) = -\frac{14}{5}x^{-4}$

Algebra II — **Module 8: Trigonometric Functions**
Online Supplemental Module

The investigations and homework for Module 8: Trigonometric Functions is available online at www.rationalreasoning.net.

Investigation 1: Introduction to Angle Measure

Investigation 2: Angle Measure in Radians

Investigation 3: Representing Circular Motion

Investigation 4: Evaluating the Sine and Cosine Functions and Circular Motion

Investigation 5: Using the Sine and Cosine Functions in Applied Settings

Investigation 6: Transformations of the Sine Function

Investigation 7: Periodicity in Non-Circle Contexts

Investigation 8: Introducing the Inverse Sine and Inverse Cosine Functions

Investigation 9: The Tangent Function

Investigation 10: The Pythagorean Identity

Algebra II

Module 9: Reference Page
Coins, Dice, and Cards

In this module we will be working with probabilities involving coins, cards, and dice. Use this page as a reference if you are unfamiliar with these objects.

Coins

Every coin has two sides. One side is typically referred to as *heads* and the other side is referred to as *tails*. This is true even if there is no "head" shown on the coin – one side is selected to represent *heads* and the other side represents *tails*.

Heads    Tails

In this module we will assume that all coins are *fair coins*, meaning that each side is equally likely to land facing up when you flip it.

Dice

A standard six-sided die has six sides that represent the numbers 1 through 6.

    one   two   three   four   five   six

In this module we will assume that all dice are *fair dice*, meaning that each side is equally likely to land facing up when you roll it.

Note that it's also possible to have dice with different numbers of sides.

You will often be asked to consider the possible outcomes when rolling two dice together, usually two standard six-sided dice. The 36 possible results are shown here, along with the resulting sums. We also show the sums in a histogram.

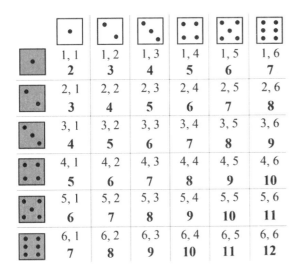
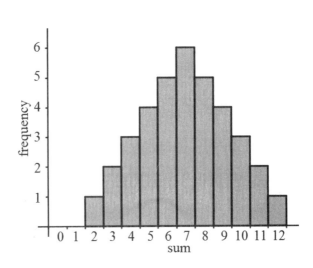

Module 9: Reference Page
© 2014 Carlson, O'Bryan & Joyner

Algebra II

Module 9: Reference Page
Coins, Dice, and Cards

Cards

A standard deck of cards contains 52 cards.
- The cards are separated into four **suits** (spades♠, clubs♣, hearts♥, and diamonds♦) and each suit contains 13 cards with different **ranks** (*ace, 2, 3, 4, 5, 6, 7, 8, 9, 10, jack, queen, king*).
- ***Face card*** is a common term used to refer to jacks, queens, and kings. (The name comes from the fact that these cards usually show drawings of people).
- Two suits (26 cards) are colored black (spades♠ and clubs♣).
- Two suits (26 cards) are colored red (hearts♥ and diamonds♦).

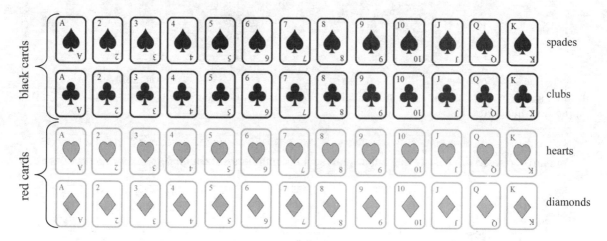

Some common examples of playing cards.

8 of hearts　　　　　　Queen of Clubs　　　　　　King of Diamonds

Algebra II

Module 9: Investigation 1
Sample Spaces

Probability is the study of how often we expect an event to occur if we repeat some activity a large number of times. We can use the probability of an event occurring to help us make informed decisions. Before we study probability, however, we need to understand how to think about and organize the possible outcomes of an activity.

Stochastic Process

A ***stochastic process*** is an activity that
1. we can repeat (or imagine repeating) as many times as we want and
2. has a range of possible outcomes that can happen and these outcomes occur at random (meaning we can't predict the exact outcome ahead of time).

Ex 1: Flipping a coin is a stochastic process because we could flip the coin as many times as we want and because the multiple outcomes (getting heads or tails) occur at random.

Ex 2: Shuffling a deck of cards and then turning over the top card is a stochastic process.

---

1. Explain why shuffling a deck of cards and then turning over the top card is a *stochastic process*.

2. Come up with a third example of a *stochastic process* and justify your choice.

Sample Space for a Stochastic Process

The ***sample space*** for a stochastic process is a list of all possible outcomes for the activity. By convention we write this list inside of a set of braces as follows.

{list of possible outcomes}

The ***size of the sample space*** is the number of possible outcomes.

Ex: Suppose we are flipping a coin (a repeatable activity with multiple possible outcomes that occur at random). Then the coin can land either ***heads up*** or ***tails up***. The sample space is

{heads, tails}   or   {H, T}

The size of the sample space is **2** because there are 2 possible outcomes.

3. Provide the sample space and its size for each of the following activities.
   a. Flipping a coin two times in a row.

   b. Rolling a six-sided die once.

   c. Flipping a coin three times in a row.

   d. Choosing a first letter from the choices A, B, and C and choosing a second letter from the choices D and E.

Sometimes it can be difficult to keep track of the possible outcomes to make sure we count them all, especially when we are performing more than one action (like flipping a coin twice, rolling a die three times, etc.). One helpful method is to draw a *tree diagram* showing the possible results at each step.

For example, the following tree diagram represents the possible outcomes (and thus the sample space) for flipping a coin twice.

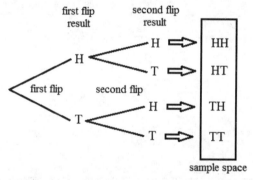

4. Draw tree diagrams to show the sample spaces for each of the following activities and then state the size of the sample spaces.
   a. Choosing a first letter from the choices A, B, and C and a second letter from the choices D and E.
   b. Flipping a coin three times.

Algebra II

Module 9: Investigation 1
Sample Spaces

c. Rolling a six-sided die and then flipping a coin.

Another benefit of drawing a tree diagram is that it helps us visualize an important property of sample spaces. In the diagram for the activity "flipping a coin twice" we can clearly see that there are two possible results for the first flip, and then for each of these results there are two possible results for the second flip. If we multiply 2 by 2 we get 4, which is the size of the sample space. This is an example of the *multiplication counting principle*.

5. Show or explain how we can use multiplication to determine the size of the sample space for each activity in Exercise #4.
    a.

    b.

    c.

Multiplication Counting Principle

When performing an activity with multiple steps, the total number of possible outcomes can be determined by multiplying the number of possible outcomes for each step.

Ex: Suppose you want to order a sandwich meal at a restaurant. There are six different sandwiches available (turkey, ham, vegetarian, roast beef, club, and pastrami) and five different sides to choose from (chips, fries, side salad, pasta salad, fresh fruit). Then there are (6)(5) = 30 total possible sandwich meals you could order.

Algebra II

Module 9: Investigation 1
Sample Spaces

6. For each of the given activities, do the following.
   i. Draw a tree diagram and use it to determine the sample space of the possible outcomes.
   ii. Use the multiplication counting principle to verify the size of the sample space you found in part (i).

   a. Choosing an outfit to wear given that you have a green shirt, a blue shirt, and a white shirt and green pants, blue pants, and white paints *assuming that you don't want to wear the same color shirt and pants.*

   b. We want to choose a first letter and a second letter from A, B, C, and D *assuming that we don't want the same letter twice.*

Algebra II **Module 9: Investigation 1**
Sample Spaces

Sometimes it's better to *imagine* the tree diagram rather than actually drawing it because the sample space is so large.

7. For each of the following activities, imagine the associated tree diagram and use the multiplication counting principle to determine the size of the sample space.
   a. Rolling two six-sided dice (one is white and one is red).

   b. Rolling three six-sided dice (white, red, and blue).

   c. Writing a random "word" formed by any two letters (the "word" does not have to be a word from the dictionary). *Assume that you can use the same letter more than once.*

   d. Writing a random "word" formed by any three letters (the "word" does not have to be a word from the dictionary). *Assume that you **cannot** use the same letter more than once.*

8. Suppose a certain state issues license plates that contain three digits (from 1 to 9) followed by three letters. Examples of license plates formed this way include 368EKB and 117WCT.
   a. Assuming that we are allowed to repeat digits and letters on a plate (such as 244TUQ or 139**AA**H), how many unique license plates could the state issue?

   b. How many unique license plates could the state issue if they allow repeated digits but ***not*** repeated letters on a plate?

c. How many unique license plates could the state issue if they did *not* repeat digits or letters on a plate?

d. Suppose the state expects to issue about 10,000,000 unique license plates over time. Review your answers to parts (a) through (c) and write a recommendation to help the state decide how to make sure there are enough unique options available.

Module 9: Investigation 2
Events, Venn Diagrams, and Complements

In Investigation #1 we thought about the sample space for an activity – a list of all of the possible outcomes for the activity. Sometimes, however, we are interested in a subset of those outcomes. For example, when flipping a coin three times we might want to focus only on the outcomes where all three flips land heads-up or tails-up.

> ### Event
> An *event* is a subset of the sample space for an activity. In other words, an event describes only the outcomes that meet a narrower set of requirements.
>
> ---
>
> *Ex:* Suppose we are rolling a standard six-sided die with a sample space $\{1, 2, 3, 4, 5, 6\}$. "**Rolling an even number**" is an example of an event with outcomes $\{2, 4, 6\}$. This is a subset of the sample space because all of the outcomes are taken from the sample space.

1. List and count the outcomes for each event.
   a. Activity: flipping a coin twice
      Event: getting exactly two tails

   b. Activity: flipping a coin three times
      Event: getting heads exactly twice

   c. Getting at least two tails when flipping a coin three times.

   d. Getting a sum of 7 on a roll of two six-sided dice.

2. Come up with your own stochastic process and event. Then list and count the outcomes from the sample space for your chosen event. *Do not repeat one of the activities/events from Exercise #1.*

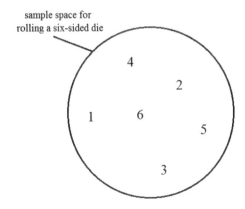

We often use a **Venn diagram** to visually represent the sample space for an activity and/or specific events within the sample space. *Note that we have written the specific outcomes in the diagram. This is not required.*

We then represent **subsets** of the sample space by highlighting regions in the diagrams. These subsets represent the outcomes for specific events within the sample space.

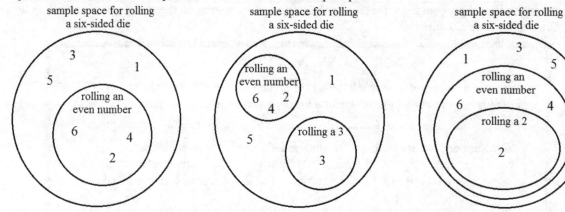

3. Show how you can use Venn diagrams to visualize the events in Exercise #1.

4. In Exercise #1b you listed the outcomes for the event "getting exactly two heads when you flip a coin three times."
   a. What are the outcomes from the sample space that *aren't* part of this event?

   b. How many outcomes does this include?

   c. Where are the outcomes from part (a) represented in the Venn diagram for the activity?

Algebra II

Module 9: Investigation 2
Events, Venn Diagrams, and Complements

5. In Exercise #4d you listed the outcomes for the event "getting a sum of 7 on a roll of two six-sided dice."
   a. What are the outcomes from the sample space that *aren't* part of this event?

   b. How many outcomes does this include?

   c. Where are the outcomes from part (a) represented in the Venn diagram for the event?

In Exercises #4 and #5 you considered an event and its **complement**.

Complement of an Event

The *complement* of an event is…

_____

Ex: Suppose we are rolling a six-sided die with the sample space {1, 2, 3, 4, 5, 6}. We are interested in the event "rolling an even number" with {2, 4, 6} as its list of possible outcomes.

The complement of this event is…

---

6. For each given event, describe the complement of the event and list the outcomes for the complement.
   a. getting at least one head when flipping a coin four times

   b. reaching into a bag full of blue, red, and green marbles and drawing a blue marble

   c. getting a sum greater than 5 when rolling two six-sided dice

Algebra II
Module 9: Investigation 2
Events, Venn Diagrams, and Complements

7. What are some characteristics of an event and its complement? List as many as you can think of, then use a Venn diagram to represent the relationship between an event and its complement.

Sometimes an event doesn't have any possible outcomes in the sample space for an activity. For example, when rolling one standard six-sided die the event "rolling a 7" is impossible. The possible outcomes for this event are represented by ***the empty set***, written as { } or ∅, indicating that there are no outcomes from the sample space where the event occurs.

8. Provide an example of a stochastic process and event where no outcomes from the sample space satisfy the event.

9. Provide an example of a stochastic process and event where no outcomes from the sample space satisfy the complement of the event.

Algebra II

Module 9: Investigation 3
Introduction to Probability

The study of probability involves thinking about what we ***expect*** to happen before we actually perform some repeatable activity (stochastic process). But at the core of probability is ***randomness***. When we perform the stochastic process "flipping a coin" several times we expect that it should land heads-up half of the time and tails-up half of the time. However, we don't know ahead of time exactly what will happen because the outcome is random. It's possible for the actual outcome to be different from what we expect.

*Note: When we perform a stochastic process one time we refer to this instance as a **trial**, and every trial produces an **outcome**.*

1. Suppose you have a standard six-sided die.
   a. If you roll the die once, is any particular outcome more likely than the other outcomes?

   b. If you perform 1000 trials what do you expect to happen? Explain your thinking.

   c. Suppose you roll the die 1000 times. Answer each of the following questions and justify your response.
      i. Do you expect to roll "6" more often than "2"?

      ii. Do you expect to roll "odd" more often than "even"?

      iii. Do you expect to roll "odd" more often than "multiple of 3"?

2. Suppose you have a coin and that you flip it <u>twice</u> and record your results (this is one trial). You repeat this process for a total of 100 trials.
   a. How many times do you expect to get heads on both flips? Why?

   b. How many times do you expect to get at least one heads? Why?

Algebra II

Module 9: Investigation 3
Introduction to Probability

   c. Do you expect to get exactly two heads more often than exactly one heads on each pair of flips? Why?

3. a. If you flip a coin, what is the probability of getting heads? What does this number mean to you?

    b. Perform 10 trials of flipping a coin and record your results. Then compare your results to the probability you found in part (a).

    c. Combine all of your results together as a class to determine the portion of all flips that landed heads up.

4. a. What is the probability of rolling a number evenly divisible by 3 on a standard six-sided die?

    b. Perform 12 trials of rolling a six-sided die and record your results. Then compare your results to the probability you found in part (a).

    c. Combine all of your results together as a class to determine the portion of all rolls with a number evenly divisible by 3.

Algebra II

Module 9: Investigation 3
Introduction to Probability

5. a. If you flip a coin twice, what is the probability of getting two heads?

    b. Perform 10 trials of flipping a coin twice and record your results. Then compare your results to the probability you found in part (a).

    c. Combine all of your results together as a class to determine the portion of all trials where both flips landed heads up.

6. a. If you roll two six-sided dice, what is the probability of rolling a total of 7?

    b. Perform 12 trials of rolling two six-sided dice and record your results. Then compare your results to the probability you found in part (a).

    c. Combine all of your results together as a class to determine the portion of all trials where the sum of the dice was 7.

In Exercises #3-6 we noticed that reality didn't necessarily match our expectations. For example, when flipping a coin 10 times we expect 5 heads and 5 tails, but you might have flipped 3 heads and 7 tails or 6 heads and 4 tails. However, when we pooled the class results the outcomes were likely to be close to our expectations.

Algebra II

Module 9: Investigation 3
Introduction to Probability

Probability as Long-Term Relative Frequency

As the number of trials gets very large, the actual results of a stochastic process should match our expectations. Therefore, we can define *the probability that an event occurs* as its *long-term relative frequency*.

The *long-term relative frequency* is defined as the portion of the total trials where the event occurs as the number of trials grows infinitely large.

---

7. Suppose one of your friends asks you "How do I calculate a probability?" Based on your work in Exercises #3-6, write an explanation for how to determine the probability of an event. (*Assume all individual outcomes are equally likely.*)

Calculating a Probability (*for equally likely outcomes*)

(*assuming that all outcomes are equally likely*)

We can calculate the *probability of an event occurring* by finding the ratio

_____

Ex: When flipping a coin three times, the probability of getting all tails is

because…

---

8. Explain what each of the following statements means based on the definition of probability.
   a. The probability of getting exactly 5 heads when flipping a coin 6 times is 3/32 (or 0.09375).

   b. The probability of winning the Powerball lottery jackpot when buying a single ticket is 1/175,223,510.

Algebra II

Module 9: Investigation 3
Introduction to Probability

In Exercises #9-13, determine the probability that each of the following events occur.

9. rolling a number greater than 4 on a standard six-sided die

10. rolling doubles with two standard six-sided dice

11. rolling a standard six-sided die twice and getting a larger number on the second roll

12. correctly answering four true-false questions on a test by random guessing

13. correctly answering two multiple choice questions (A-B-C-D options) by random guessing

So far we've focused on contexts where each outcome in the sample space is **equally likely**, meaning that we don't expect any outcome to happen more often than any other outcome. (For example, when rolling a six-sided die, we don't *expect* a 6 to come up more often than a 1.)

Suppose a spinner for a board game is designed as follows. Use this spinner in Exercises #14 and #15.

14. a. What is the sample space for the activity *spinning the dial once*?

b. Explain why the probability of spinning "1" on a single spin is not $\frac{1}{4}$ even though the sample space for *spinning the dial once* contains four outcomes.

Algebra II

Module 9: Investigation 3
Introduction to Probability

15. What complications happen when outcomes in a sample space are not equally-likely? How you might adjust for this when calculating the probability for an event?

16. Suppose you place 40 white Ping-Pong balls in a box along with 10 orange Ping-Pong balls and then shake the box.
    a. If you reach into the box without looking and pull out a ball at random, what do you expect to happen? Explain.

    b. Suppose you repeat the activity in part (a) 250 times (replacing the Ping-Pong ball you drew each time and shaking the box before pulling a new one). What do you expect to happen in these 250 trials?

Algebra II
Module 9: Investigation 4
Probability Spaces and the Probability of Complements

In Investigation 3 we learned what it means to talk about *the probability of an event occurring*. In this investigation we will discuss how to represent probability using mathematical notation and examine the range of values for a probability.

Probability Notation

The *probability of an event A occurring* can be represented using the notation $P(A)$.
Note: We refer to the probability that A's complement occurs as $P(\text{not } A)$.

Ex. 1: When rolling a standard six-sided die, $P(2) = 1/6$ and $P(\text{not } 2) = 5/6$.

Ex. 2: When flipping a coin, $P(\text{heads}) = 1/2$ and $P(\text{not heads}) = 1/2$.

---

1. Write each of the following probabilities using probability notation and determine their values.
   a. the probability of getting two tails when flipping a coin tw0 times

   b. the probability of getting three heads when flipping a coin three times

   c. the probability of getting exactly one heads when flipping a coin three times

   d. the probability of getting a total of three when rolling two standard six-sided dice

2. Is it possible for the probability of an event occuring to have each of the following values? If not, explain your thinking.
   a. $P(A) = 0.712$
   b. $P(A) = 0$

   c. $P(A) = 2$
   d. $P(A) = -1$

Probability Values

The *probability of an event A occurring* will always have a value between _____ and _____.

If $P(A) =$ _____, then the outcomes for event A...

If $P(A) =$ _____, then the outcomes for event A...

Module 9: Investigation 4

© 2014 Carlson, O'Bryan & Joyner

Algebra II
Module 9: Investigation 4
Probability Spaces and the Probability of Complements

3. Suppose we roll a standard six-sided die.
   a. Find $P(5)$.

   b. What is the complement of the event in part (a) and what is the probability that the complement occurs?

4. Suppose we roll a standard six-sided die.
   a. Find $P(\text{odd})$.

   b. What is the complement of the event in part (a) and what is the probability that the complement occurs?

5. Suppose there are 25 pieces of chocolate candy in a bag and 7 of them are blue. You draw one of the pieces of candy at random.
   a. Find $P(\text{blue})$.

   b. What is the complement of the event in part (a) and what is the probability that the complement occurs?

6. Look back at your work in Exercises #3-5.
   a. What is the relationship between the probability of an event and the probability of its complement?

   b. Why does this relationship make sense?

Probability of Complementary Events

If A is some event, then
$$P(A) + P(\text{not } A) =$$

Therefore, we can also conclude

$P(A) =$ $\qquad$ $P(\text{not } A) =$

Algebra II

Module 9: Investigation 4
Probability Spaces and the Probability of Complements

7. If you roll two standard six-sided dice, $P(\text{at least one odd number}) = \frac{27}{36} = \frac{3}{4}$.

   a. What is the complement of the event "rolling at least one odd number"?

   b. What is the probability of the complement? Express your answer as a fraction, decimal, and percent.

8. Suppose you flip a coin three times.
   a. Find $P(\text{exactly two heads})$.

   b. What is the complement of the event in part (a)?

   c. What is the probability of the complement? Express your answer as a fraction, decimal, and percent.

9. The probability of winning the PowerBall jackpot if you buy 70 tickets is $\frac{1}{2,503,193}$. What is the probability of not winning the PowerBall lottery if you buy 70 tickets? Express your answer as a fraction, decimal, and percent.

Probability Space

A ***probability space*** is a combination of a sample space for an activity along with an assignment of probability for each outcome.

Ex: For the activity "flipping a coin" the *probability space* is as follows.

{heads, tails}
$P(\text{heads}) = 1/2$
$P(\text{tails}) = 1/2$

10. Suppose a spinner for a board game is designed as follows. Give the probability space for a single spin. (*Estimate the probabilities if necessary.*)

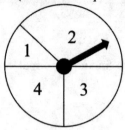

11. In a bag of marbles, 5 are clear, 6 are blue, 12 are red, and 7 are green. Give the probability space for randomly drawing a marble from the bag.

12. Consider your answers to Exercises #10-11. What is the sum of the probabilities for a probability space? Why does this make sense?

Algebra II
Module 9: Investigation 5
Unions and Intersections

Earlier in this module we represented the sample space for a stochastic process using Venn diagrams and learned that events represent subsets of the outcomes in the sample space. Sometimes we might be interested in more than one event, and sometimes those events share common outcomes.

For example, suppose we roll a standard six-sided die and are interested in the two events "odd number" ({1, 3, 5}) and "multiple of 3" ({3, 6}). Let's represent these events using a Venn diagram.

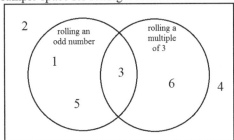

Notice how the two events share "3" as an outcome? This is called the ***intersection*** of the two events, and describes the outcome of the event "rolling an odd number ***and*** a multiple of 3." In other words, the intersection represents the outcome(s) where *both* events occur. We can write the intersection using set notation as {3}.

Intersection

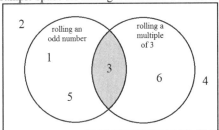

Union

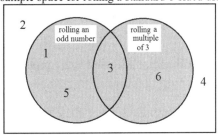

However, we also might be interested in *either* of the two events occurring – we don't really care which one. The outcomes for "rolling an odd number ***or*** a multiple of 3" are {1, 3, 5, 6}. This is called the ***union*** of the two events. In other words, the union represents the outcome(s) where one or more of the events occur.

For Exercises #1-3, do the following.
   a) Draw a Venn diagram to represent the sample space of the activity and the outcomes for events A and B.
   b) Write the set of outcomes for the *intersection* of the events and count them.
   c) Write the set of outcomes for the *union* of the events and count them.

1. Activity: rolling two standard six-sided dice
   Event A: rolling doubles
   Event B: rolling a sum of 10 or greater

Algebra II

Module 9: Investigation 5
Unions and Intersections

2. Activity: flipping a coin three times
   Event A: getting at least one heads
   Event B: getting at least one tails

3. Activity: rolling two standard six-sided dice
   Event A: rolling at least one "2"
   Event B: rolling a sum of 10 or greater

4. a. In Exercise #1, what is the probability that you roll doubles and get a sum of 10 or greater?

   b. In Exercise #2, what is the probability that you get at least one heads and at least one tails?

   c. In Exercise #3, what is the probability that you roll at least one "2" and get a sum of 10 or greater?

5. If A and B are two events for some activity, what is the general process we can follow to find the probability that both A and B occur?

Working with the union of two events involves adding probabilities. For example, when rolling a standard six-sided die, $P(1) = 1/6$ and $P(2) = 1/6$. But $P(1 \text{ or } 2) = 2/6$ or $1/3$. We may also notice that

$$P(1 \text{ or } 2) = P(1) + P(2)$$
$$= \tfrac{1}{6} + \tfrac{1}{6}$$
$$= \tfrac{2}{6}$$

Algebra II

Module 9: Investigation 5
Unions and Intersections

6. Find the probability that each of the following events occur.
   a. $P(3$ or even$)$ when rolling a standard six-sided die.

   b. $P($two heads or two tails$)$ when flipping a coin twice.

   c. $P($heads or tails$)$ when flipping a coin.

7. What did each of the pairs of events in Exercise #6 have in common?

8. Does the process we just explored work to determine $P($at least one heads or at least one tails$)$ when flipping a coin three times? Explore this question and explain your findings.

Probability of Unions

If A and B are two events for an activity, then

$P(A$ or $B) =$

We _____ because...

In Exercises #9-11, find each probability.

9. Imagine rolling a standard six-sided die.
   a. $P(3$ and odd$)$　　　　　　　　b. $P(4$ and $6)$

   c. $P(3$ or odd$)$　　　　　　　　d. $P($odd or even$)$

10. Imagine flipping a coin and rolling a standard six-sided die.
    a. $P($heads and $4)$　　　　　　b. $P($tails and odd$)$

    c. $P($heads or $4)$　　　　　　　d. $P($tails or odd$)$

11. Imagine rolling two standard six-sided dice.
    a. P(3 on a die and 4 on a die)
    b. P(both dice roll even)
    c. P(two even numbers or doubles)
    d. P(at least one die rolls even)

12. The following diagrams provide some information about students at a high school. (*The numbers represent counts – how many students at the school have the given characteristics.*)

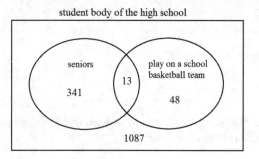
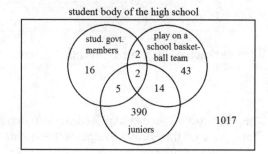

   a. How many students play on a school basketball team?

   b. How many students attend the school?

   c. What is the probability that a randomly chosen student from the school is a senior?

   d. What is the probability that a randomly chosen student from the school is a senior and plays on a school basketball team?

   e. What is the probability that a randomly chosen student from the school is a senior or plays on a school basketball team?

   f. What is the probability that a randomly chosen student from the school is a junior or a member of student government?

   g. What is the probability that a randomly chosen student from the school is a junior who plays on a school basketball team but is not in student government?

   h. What is the probability that a randomly chosen student from the school is not a junior?

   i. Suppose we randomly select a student from the high school's junior class. What is the probability that the student is a member of student government?

Algebra II   **Module 9: Investigation 6**
                                                                           Conditional Probability

1. Answer the following survey questions about yourself. (*Please answer only "yes" or "no".*)
   a) Do you enjoy reading?   b) Do you enjoy watching sports on TV?

We can combine the information for the entire class into a ***two-way frequency table*** that summarizes the survey results.

	Enjoy watching sports on TV	Don't enjoy watching sports on TV	Total
Enjoy reading			
Don't enjoy reading			
Total			

2. Suppose your teacher randomly selects one student from the class by pulling a name out of a hat. Calculate each of the following probabilities.
   a. the chosen student doesn't enjoy reading

   b. the chosen student enjoys watching sports on TV and enjoys reading

   c. the student enjoys watching sports on TV

3. Now suppose that your teacher separates the class and groups all of the students who enjoy reading in one area and then selects one of those students at random.
   a. What's the probability that the chosen student enjoys watching sports on TV?

   b. What's the probability that the chosen student doesn't enjoy watching sports on TV?

4. Now suppose that your teacher separates the class and groups all of the students who don't enjoy watching sports on TV in one area and then selects one of those students at random.
   a. What's the probability that the chosen student enjoys reading?

   b. What's the probability that the chosen student doesn't enjoy reading?

5. Compare Exercise #2 with Exercises #3 and #4. What was the important difference between these exercises?

Algebra II

Module 9: Investigation 6
Conditional Probability

In Exercises #3 and #4 you dealt with a *conditional probability* because you wanted to know the probability of an event (such as "the chosen student enjoys watching sports on TV") given that another event has occurred ("the chosen student enjoys reading").

Conditional Probability

Conditional probability refers to the probability of an event occurring given that another event has occurred.

Let A and B be two events. Then $A \mid B$ represents "event A occurs given that event B has occurred" and $P(A \mid B)$ represents "the probability that A occurs given that B has occurred."

Ex: Suppose a box contains colored slips of paper with numbers on them. There are 5 blue slips of paper with the numbers 1 through 5 and 7 red slips of paper with the numbers 6 through 12.
- $P(2 \mid \text{blue})$ means "the probability that we draw the number 2 given that we drew a blue slip of paper" and this probability is 1/5.
- $P(\text{red} \mid 5)$ means "the probability that we draw a red slip of paper given that we drew the number 5" and this probability is 0 because we know that 5 only appears on a blue slip of paper.

---

6. Suppose you shuffle a standard deck of 52 cards, draw a card at random and set it aside, then reshuffle the deck and draw a second card at random. Rewrite each of the following statements in words, and then determine the probability.

   a. $P(\text{second card is black} \mid \text{first card is black})$

   b. $P(\text{second card is black} \mid \text{first card is red})$

   c. $P(\text{second card is a heart} \mid \text{first card is a heart})$

   d. $P(\text{second card is red} \mid \text{first card is a diamond})$

Algebra II

Module 9: Investigation 6
Conditional Probability

7. Suppose you roll a pair of standard six-sided dice. One die lands on the table showing "6" and the other die falls on the floor and rolls away so that you can't see it. Determine each of the following probabilities.
   a. The probability that the second die shows "5".

   b. The probability that the total of the dice is 11 or more.

   c. The probability that the total of the dice is 7 or more.

8. How does the probability in Exercise #7c compare to the probability of getting a total of 7 or more when rolling two dice? Explain.

Use the following information for Exercises #9-11.
In a paper published in *The Canadian Journal of Public Health*, PJ Villeneuve and Y. Mao estimated that 172 of every 1000 men who smoke will develop lung cancer during their lifetimes, and 116 of every 1000 women who smoke will develop lung cancer during their lifetimes. In comparison, 13 of every 1000 men and 14 of every 1000 women who are non-smokers will develop lung cancer during their lifetimes.[1]

In a separate study published in *Clinical Cancer Research*, J. Samet and his author team estimate that about 10% to 15% of people who develop lung cancer were non-smokers.[2] Assume that the conclusions of these studies are accurate for the entire population.

9. Determine the following probabilities.
   a. P(will develop lung cancer | male smoker)

   b. P(will develop lung cancer | female non-smoker)

   c. P(will not develop lung cancer | male smoker)

   d. P(will not develop lung cancer | female non-smoker)

10. Estimate the following probabilities.
    a. $P(\text{non-smoker} \mid \text{person develops lung cancer})$

    b. $P(\text{smoker} \mid \text{person develops lung cancer})$

11. a. Explain in your own words the difference in meaning between
    $P(\text{will develop lung cancer} \mid \text{smoker})$ and $P(\text{smoker} \mid \text{person develops lung cancer})$.

    b. Based on the information we have, which probability in part (a) is larger?

Algebra II

Module 9: Investigation 6
Conditional Probability

Use the following information for Exercises #12-16.
Medical tests for a disease are not always perfect, and companies try to create new tests that are more accurate or cost less than existing tests. Suppose that a company is developing a new test for a disease. The company recruited 1,718 people (731 people who they know already have the disease and 987 people who they know do not have the disease). The company then used the new test on each patient and recorded the results.

	patient tested positive	patient tested negative	Total
patient has the disease	714	17	
patient does **not** have the disease		934	987
Total			

12. Complete the table by filling in the missing entries.

13. Based on the results of this test, rewrite the following statements in the form $P(A|B)$ and use the given information to estimate the probability.
   a. The probability that a person who has the disease will test positive.

   b. The probability that a person who does not have the disease will test negative.

14. A *false positive* occurs when someone who does not have the disease tests positive.
   a. Write the statement "the probability that a false positive occurs" in the form $P(A|B)$.

   b. Based on the given results, what is the probability that a false positive occurs?

Algebra II

Module 9: Investigation 6
Conditional Probability

15. A ***false negative*** occurs when someone who has the disease tests negative.
    a. Write the statement "the probability that a false negative occurs" in the form $P(A|B)$.

    b. Based on the given results, what is the probability that a false negative occurs?

16. The ***accuracy*** of a test is defined as the probability that the test provides the correct diagnosis for a patient. How can we calculate the accuracy of this test?

17. Two events A and B are subsets of the sample space of some stochastic activity. There are 280 total outcomes in the sample space, 126 outcomes for event A and 84 outcomes for event B. In addition, $P(A|B) = 2/3$ and $P(B \text{ doesn't occur} | A \text{ doesn't occur}) = 63/77$.

    Using this information, complete the two-way frequency table.

	event B occurs	event B does not occur	*total*
event A occurs			
event A does not occur			
total			

---

[1] Villeneuve, PJ & Mao, Y. "Lifetime probability of developing lung cancer, by smoking status, Canada." *Canadian Journal of Public Health*, Nov-Dec 1994, 85(6): 385-8.
[2] Samet, J., et. al. "Lung Cancer in Never Smokers: Clinical Epidemiology and Environmental Risk Factors." *Clinical Cancer Research*, September 15, 2009, 15(18): 5626-45.

Algebra II   **Module 9: Investigation 7**
                                                              Independence

In Exercises #1-3, you are given two events for a compound stochastic activity. Think about each activity and then answer the given questions.

1. Activity: Rolling two standard six-sided dice (one blue and one red).
   Event A: The red die rolls an odd number.
   Event B: The blue die rolls a "6".
   a) Find $P(A)$.   b) Find $P(B)$.   c) Find $P(A|B)$.   d) Find $P(B|A)$.

2. Activity: Shuffling a deck of cards, drawing a card and putting it back in the deck, reshuffling, and drawing a second card.
   Event A: The first card drawn is a red card.
   Event B: The second card drawn is a black card.
   a) Find $P(A)$.   b) Find $P(B)$.   c) Find $P(B|A)$.

3. Activity: Randomly pulling a marble from a bag containing 3 blue and 2 red marbles, setting it aside, and then randomly pulling a second marble.
   Event A: The first marble drawn is blue.
   Event B: The second marble drawn is blue.
   a) Find $P(A)$.   b) Find $P(B|A)$.   c) True or False: $P(B|A) = P(B)$ in this context

4. Reflect on your answers to Exercises #1-3. Discuss anything important that you note.

Algebra II

Module 9: Investigation 7
Independence

5. a. Come up with a stochastic activity and two events *A* and *B* such that $P(A|B) = P(A)$ and $P(B|A) = P(B)$.

   b. Come up with a stochastic activity and two events *A* and *B* such that $P(B|A) \neq P(B)$.

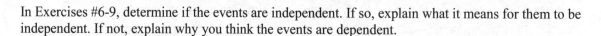

Independent Events

Two events are said to be **independent** if…

This means that if *A* and *B* are two independent events,

$P(A|B) =$ and/or $P(B|A) =$

Note: If two events are not independent, we say they are **dependent**.

Ex: When rolling two standard six-sided dice, the events "rolling a 6 on the first die" and "rolling a 6 on the second die" are independent.

In Exercises #6-9, determine if the events are independent. If so, explain what it means for them to be independent. If not, explain why you think the events are dependent.

6. A student from your math class is selected at random. You are interested in the probability of the events "the student has an A in the class" and "the student earned an A on the last test."

7. The following spinner is used in a certain board game. On a player's turn, he/she spins twice. You are interested in the probability of the events "the first spin lands on 2" and "the second spin lands on 2."

8. A professional basketball player is fouled and earns two free throws. You are interested in the probability of the events "the player makes the first free throw" and "the player makes the second free throw."

9. A student is selected at random from your school. You are interested in the probability of the events "the student has brown hair" and "the student's parents have brown hair."

Algebra II

Module 9: Investigation 7
Independence

10. In Investigation 6 we considered the following context: *Medical tests for a disease are not always perfect, and companies try to create new tests that are more accurate or cost less than existing tests. Suppose that a company is developing a new test for a disease. The company recruited 1,718 people (731 people who they know already have the disease and 987 people who they know do not have the disease). The company then used the new test on each patient and recorded the results.*

	patient tested positive	patient tested negative	Total
patient has the disease	714	17	731
patient does **not** have the disease	53	934	987
Total	767	951	1718

There are four events in this context (as listed). Are the events independent? Justify your answer.
  Event A: patient tested positive   Event C: patient has the disease
  Event B: patient tested negative   Event D: patient does not have the disease

11. In Investigation 6 we considered the relationship between smoking and lung cancer. Suppose a 60-year old male is selected at random. Do you think the events "the man has been diagnosed with lung cancer" and "the man is a smoker" are independent? Justify your answer.

Algebra II

Module 9: Investigation 8
Multiplication Rule

1. Suppose you flip a coin once, record the outcome, and then flip the coin again.
   a. Provide the probability space for the activity.

   b. How is P(heads on the first flip and heads on the second flip) related to P(heads on the first flip) and P(heads on the second flip)?

2. Suppose you roll a standard six-sided die once, record the outcome, and then roll the die again. The sample space for the activity is given.

$$\begin{Bmatrix} (1,1) & (1,2) & (1,3) & (1,4) & (1,5) & (1,6) \\ (2,1) & (2,2) & (2,3) & (2,4) & (2,5) & (2,6) \\ (3,1) & (3,2) & (3,3) & (3,4) & (3,5) & (3,6) \\ (4,1) & (4,2) & (4,3) & (4,4) & (4,5) & (4,6) \\ (5,1) & (5,2) & (5,3) & (5,4) & (5,5) & (5,6) \\ (6,1) & (6,2) & (6,3) & (6,4) & (6,5) & (6,6) \end{Bmatrix}$$

   a. Are events such as "the first roll is 2" and "the second roll is 2" independent or dependent?

   b. Determine P(first roll is 2 and second roll is 2).

   c. Determine P(first roll is even and second roll is even).

   d. Determine P(first roll is 6 and second roll is odd).

Algebra II

Module 9: Investigation 8
Multiplication Rule

3. Suppose you have a bag of marbles with 3 red marbles and 2 blue marbles. You randomly select one of the marbles, set it aside, and then randomly select a second marble. The sample space for the activity is given. *Note that we have numbered the marbles to help us keep track of the possible outcomes.*

$$\begin{Bmatrix} (R_1, R_2) & (R_1, R_3) & (R_1, B_1) & (R_1, B_2) \\ (R_2, R_1) & (R_2, R_3) & (R_2, B_1) & (R_2, B_2) \\ (R_3, R_1) & (R_3, R_2) & (R_3, B_1) & (R_3, B_2) \\ (B_1, R_1) & (B_1, R_2) & (B_1, R_3) & (B_1, B_2) \\ (B_2, R_1) & (B_2, R_2) & (B_2, R_3) & (B_2, B_1) \end{Bmatrix}$$

a. Are the events "first marble drawn is blue" and "second marble drawn is blue" independent? What about the events "first marble drawn is blue" and "second marble drawn is red"? Justify your answer.

b. Determine each probability using the given sample space.
   i) P(first draw red and second draw red) = P(R, R)

   ii) P(R, B)   iii) P(B, R)   iv) P(B, B)

c. After completing Exercises #1 and #2, your classmate said "To find the probability of two events, we multiply the probability that the first happens and the probability that the second happens. Therefore, I get the following answers for part (b)."

   $P(R, R) = \frac{3}{5} \cdot \frac{3}{5} = \frac{9}{25} = 0.36$ $P(R, B) = \frac{3}{5} \cdot \frac{2}{5} = \frac{6}{25} = 0.24$

   $P(B, R) = \frac{2}{5} \cdot \frac{3}{5} = \frac{6}{25} = 0.24$ $P(B, B) = \frac{2}{5} \cdot \frac{2}{5} = \frac{4}{25} = 0.16$

   Discuss what you think of this answer.

Multiplication Rule for Compound Events

If *A* and *B* are two events for an activity, then

$$P(A \text{ and } B) =$$

Note: If *A* and *B* are <u>independent</u> events, then

$$P(A \text{ and } B) =$$

because…

Ex 1: When rolling two standard six-sided dice, the events "rolling a 6 on the first die" and "rolling a 6 on the second die" are independent, so

$P(\text{first die is 6 and second die is 6}) =$

Ex 2: If you randomly draw a card from a standard 52-card deck, set it aside, reshuffle, and then randomly draw a second card, then

$P(\text{first card is red and second card is red}) =$

---

4. Suppose you randomly select a card from a standard 52-card deck, set it aside, reshuffle, and then randomly draw a second card. Determine each of the following probabilities.

   a. *P*(first card black and second card black)   b. *P*(first card queen and second card king)

   c. *P*(first card ace and second card ace)   d. *P*(first card heart and second card heart)

Algebra II
Module 9: Investigation 8
Multiplication Rule

5. Suppose you have a bag with green and clear marbles and you repeat the activity in Exercise #3. Assume the following information.

sample space: $\{(G, G), (G, C), (C, G), (C, C)\}$

$P(\text{green on first draw}) = \frac{7}{10}$ $P(\text{clear on first draw}) = \frac{3}{10}$

$P(\text{green on second draw} | \text{green on first draw}) = \frac{2}{3}$ $P(\text{clear on second draw} | \text{clear on first draw}) = \frac{2}{9}$

$P(\text{clear on second draw} | \text{green on first draw}) = \frac{1}{3}$ $P(\text{green on second draw} | \text{clear on first draw}) = \frac{7}{9}$

  a. Find each of the following probabilities.
     i) $P(C, G)$ ii) $P(C, C)$

     iii) $P(G, C)$ iv) $P(G, G)$

  b. What is the sum of the four probabilities in part (a)? What is the significance of this?

  c. If there are 150 marbles in the bag, how many are green and how many are clear?

6. The following spinner is used in a certain board game. On a player's turn, he or she spins twice. Determine each of the following probabilities. *Assume that the events in the sample space are independent.*

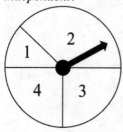

  a. $P(\text{first spin 4 and second spin 4})$

  b. $P(\text{first spin 1 and second spin 2})$

  c. $P(\text{first spin is not 1 and second spin is not 3})$

Algebra II **Module 9: Investigation 8**
Multiplication Rule

7. In previous investigations we considered the following context: *Medical tests for a disease are not always perfect, and companies try to create new tests that are more accurate or cost less than existing tests. Suppose that a company is developing a new test for a disease. The company recruited 1,718 people (731 people who they know already have the disease and 987 people who they know do not have the disease). The company then used the new test on each patient and recorded the results.*

	patient tested positive	patient tested negative	Total
patient has the disease	714	17	731
patient does **not** have the disease	53	934	987
Total	767	951	1718

Event A: patient tested positive   Event C: patient has the disease
Event B: patient tested negative   Event D: patient does not have the disease

Suppose you randomly select one of the patients. Confirm each of the following equalities.

a) $P(A \text{ and } C) = P(A)P(C|A)$ b) $P(B \text{ and } D) = P(B)P(D|B)$

c) $P(C \text{ and } B) = P(C)P(B|C)$ d) $P(B \text{ and } A) = P(B)P(A|B)$

8. A professional basketball player has a career free-throw percentage of 90.4%. *Assume that making free throws are independent events and that this is an accurate predictor of the player's ability to make any given free throw.*
   If the player shoots two free throws, determine each of the following probabilities.
   a. The probability that the player makes both free throws.

   b. The probability that the player makes the first free throw and misses the second free throw.

c. The probability that the player misses both free throws.

d. The probability that the player makes exactly one of the two free throws.

9. Suppose you randomly select a card from a standard 52-card deck, set it aside, reshuffle, randomly draw a second card, reshuffle, etc.
   a. What is the probability that the first five cards are all hearts?

   b. What is the probability that the first five cards are all the same suit?

   c. What is the probability that the first four cards are all aces?

   d. What is the probability that the first four cards all have the same rank?

Algebra II

Module 9: Investigation 9
Permutations and Combinations

1. Suppose you have 10 different books and are trying to decide how to order them on a shelf.
   a. How many unique arrangements can you create if you only put 5 books on the shelf?

   b. How many unique arrangements can you create if you put all 10 books on the shelf?

Factorial
A *factorial* is…
We use an exclamation mark to help us represent factorials, such as…

2. How many different ways can you arrange 8 books on a shelf? 12 books?

3. Suppose you want to randomly select letters from the word BANQUET.
   a. Suppose you choose 4 letters at random and write them down in the order you select them. What is the probability that you choose the letters B, A, N, and Q and that they appear in alphabetical order?

   b. Suppose you choose 5 letters at random and write them down in the order you select them. What is the probability that you choose the letters B, A, N, Q, and U and that they appear in alphabetical order?

4. Suppose you have a lock that requires a specific three-number pattern to unlock. Suppose there are 40 numbers shown on the lock, and you know that the numbers can't be repeated.
   a. How many unique combinations are possible?

   b. What is the probability that you can randomly guess the combination on 5 attempts?

When we have a compound event where our choices are "used up" and where order is important in determining whether or not the outcome is unique, we call each outcome a ***permutation*** of the choices.

Permutations

A ***permutation*** is an outcome of a compound event where order is important in determining if the outcome is unique and no selection can be used more than once. The total number of permutations is the size of the sample space for the activity.

Each of the following scenarios represents a permutation of a compound event.
   i) One way of lining up all of the students in a class of 25.
   ii) One selection of the starting line-up for a basketball team by assigning 5 players from a team of 12 to each of the 5 positions.
   iii) One specific 3-letter "word" formed from the letters in the word NEUTRAL.
   iv) One selection and ordering of 5 books from a pile of 20 books.

5. Consider the examples given. Can you come up with a general process for computing the number of unique permutations of r items when we have n items to choose from? (*For example, choosing how many ways to order r books on a shelf if we have a total of n books to choose from.*)

Algebra II

Module 9: Investigation 9
Permutations and Combinations

> **Permutation Notation**
>
> The notation $_nP_r$ is used to represent the total number of permutations of size r that can be formed from n total options.

6. Represent the total number of permutations possible in the examples listed before Exercise #5 using permutation notation.

7. Determine the number of permutations indicated.
   a. $_5P_3$
   b. $_6P_2$
   c. $_{10}P_7$

   d. $_{12}P_{12}$
   e. $_3P_5$

8. Suppose you are playing a card game with a standard 52-card deck. You are curious how many unique hands are possible if you shuffle the deck and draw 5 cards at random. Your friend says the answer is $52 \cdot 51 \cdot 50 \cdot 49 \cdot 48 = 311,875,200$ hands. Do you agree? Explain your thinking.

Algebra II

Module 9: Investigation 9
Permutations and Combinations

When we have a compound event where our choices are "used up" and where changing the order of the choices does not create a unique outcome, we call each outcome a *combination* of the choices.

Combinations

A *combination* is an outcome of a compound event where changing the order of selection does not create a unique outcome and no selection can be used more than once. The total number of combinations is the size of the sample space for the activity.

The notation $_nC_r$ is used to represent the total number of combinations of size r that can be formed from n total options.

Each of the following scenarios represents a combination of a compound event.
  i) A specific group of 4 students chosen from a class to help a teacher carry books to the office.
  ii) A selection of 6 books chosen from the library to check out and take home.
  iii) A selection of 3 people from your group of friends to share tickets to a football game.

9. An ice cream stand offers bowls of ice cream with two scoops for $3.49. They have 12 different flavors.
   a. How many unique flavor choices could you make assuming that you choose a different flavor for each scoop?

   b. You can add an additional scoop for $0.99. How many unique flavor choices could you make assuming that you choose three different flavors for a 3-scoop bowl?

   c. How many unique flavor choices could you make assuming that you choose four different flavors for a 4-scoop bowl?

Module 9: Investigation 9
© 2014 Carlson, O'Bryan & Joyner

Algebra II

Module 9: Investigation 9
Permutations and Combinations

10. You want to plant a flower garden at your house. The nursery has 18 different flower varieties available for purchase.
    a. If you want to include 4 different types of flowers in your garden, how many unique combinations of flowers could you choose?

    b. If you want to include 9 different types of flowers in your garden, how many unique combinations of flowers could you choose?

    c. If you want to include 12 different types of flowers in your garden, how many unique combinations of flowers could you choose?

11. Use what you've learned in this investigation to answer the following questions.
    a. In your class of 32 students, 4 students will be chosen to represent your class in a competition at the next pep assembly. How many unique groups of students can be created to represent your class?

    b. In your class of 32 students, 4 students will be chosen to read aloud for 4 specific parts in a play. How many unique ways can students be chosen to read the parts of the play?

Algebra II

Module 9: Investigation 9
Permutations and Combinations

   c. A home alarm system has a 5-digit code to turn the alarm on or off. If the panel contains the digits 1-9, how many unique codes can be created?

   d. A school locker has a 4-number combination chosen from the numbers 0 to 59 (no repeats allowed). How many unique combinations can be chosen for the lock?

   e. A state lottery randomly chooses 5 numbers from 1 to 45 to determine the jackpot winner. How many unique sets of numbers can be drawn? (*Note: The winning numbers are always reported in ascending order, not in the order they are chosen.*)

Investigation 10: *Extensions using Permutations and Combinations* and associated homework is available online at www.rationalreasoning.net.

Algebra II

Module 9: Homework
Probability

INVESTIGATION 1: SAMPLE SPACES

In Exercises #1-4, provide the sample space and its size for each activity.

1. The following spinner is used in a board game. A player spins the spinner twice in a turn.

2. Randomly guessing answers on a three-question True/False quiz.

3. Two players are playing rock-paper-scissors. Suppose each player randomly selects either rock, paper, or scissors.

4. Randomly guessing answers on a two-question multiple choice quiz (answer options A, B, C, D available).

In Exercises #5-8, do the following for each activity.
   a) Draw a tree diagram and use it to determine the sample space of possible outcomes.
   b) Use the multiplication counting principle to verify the size of the sample space you found in part (a).

5. The following spinner is used in a board game. A player spins the spinner twice in a turn.

6. Rolling a standard four-sided die twice. *A four-sided die can roll a 1, 2, 3, or 4.*

7. A bag contains a green marble, a blue marble, and a red marble. You reach into the bag and randomly draw a marble, set it aside, and then repeat the process.

8. The letters A, E, N, and Q are written on cards and then shuffled. You draw one letter, set it aside, reshuffle the cards, and then draw a second letter.

9. In the Women's National Basketball Association (WNBA) Finals series, two teams play until one of them has won two games. Use a tree diagram to model all of the possible ways that the series can end. (*Call the two teams Team A and Team B*).

10. When two basketball teams play in a "best of five" series, they compete until one team has won three games. Use a tree diagram to model all of the possible ways that the series can end. (*Call the two teams Team A and Team B*).

11. Do you think the activities described in Exercises #9 and #10 qualify as stochastic activities? Explain.

Algebra II

Module 9: Homework
Probability

In Exercises #12-17, do the following.
    a) Provide two possible outcomes for the activity.
    b) Imagine the tree diagram representing the sample space for the activity and use the multiplication counting principle to determine the size of the sample space.

12. Rolling 4 six-sided dice (imagine each die is a different color).

13. Rolling 5 six-sided dice (imagine each die is a different color).

14. Choosing 3 letters at random to form a "word". (The "word" does not have to be a word from the dictionary). *Assume that you can use the same letter more than once.*

15. Choosing 5 letters at random to form a "word". (The "word" does not have to be a word from the dictionary). *Assume that you can use the same letter more than once.*

16. Repeat Exercise #14 if you **don't** use the same letter more than once.

17. Repeat Exercise #15 if you **don't** use the same letter more than once.

18. Suppose license plates in a certain state are formed by using 3 letters followed by 4 numbers (the digits 0 through 9).
    a. How many unique license plates are possible? (Assume you can use letters and numbers more than once).

    b. To avoid confusion between the number "1" and the letter "I" and between the number "0" and the letter "O" the state decides not to include the letters I and O on its license plates. Repeat part (a) given this new information.

19. Modern computers can operate so quickly that they can perform billions of processes every second. Computer hackers make use of this power to try to steal information by correctly guessing user passwords. To combat this, security experts suggest using complicated passwords that combine uppercase and lowercase letters, numbers, and symbols.
    a. How many unique passwords of length 8 can be created using only lowercase letters?
    b. How many unique passwords of length 8 can be created using only letters (uppercase and lowercase)?
    c. How many unique passwords of length 8 can be created using any combination of uppercase letters, lowercase letters, numbers from 0 to 9, and symbols from the set {!, #, $, %, &, *}?

Algebra II   **Module 9: Homework**
Probability

INVESTIGATION 2: EVENTS, VENN DIAGRAMS, AND COMPLEMENTS

In Exercises #20-29, do the following.
    a) List and count the outcomes for the event.
    b) Draw a Venn diagram to represent the activity's sample space and the event's outcomes.
    c) Describe the outcomes for the event's complement.
    d) List and count the outcomes for the event's complement.

20. Activity: rolling a standard six-sided die
    Event: rolling a number greater than 4

21. Activity: spinning the given spinner once
    Event: getting an even number

22. Activity: randomly drawing a marble from a bag containing a blue, a clear, a red, and a yellow marble
    Event: drawing a yellow marble

23. Activity: flipping a coin three times
    Event: getting exactly three tails

24. Activity: rolling two six-sided dice
    Event: getting a sum greater than 14

25. Activity: spinning the spinner shown in Exercise #21 two times
    Event: spinning the same number twice

26. Activity: flipping a coin four times
    Event: getting at least two tails

27. Activity: randomly drawing two marbles together from a bag containing one blue, one clear, one red, and one yellow marble
    Event: drawing two yellow marbles

28. Activity: rolling two six-sided dice
    Event: rolling at least one even number

29. Activity: randomly drawing two marbles together from a bag containing one blue, one clear, one red, and one yellow marble
    Event: drawing a yellow marble

30. What happens when you add the number of outcomes for an event to the number of outcomes for the event's complement? Why?

INVESTIGATION 3: INTRODUCTION TO PROBABILITY

31. Based on your work in this investigation, explain the difference between the probability of an event happening and the actual percentage of trials where the event occurs when you perform the activity.

32. Explain the meaning of *long-term relative frequency* in your own words.

In Exercises #33-36, explain what each statement means using the definition of probability as the long-term relative frequency of the event.

33. The probability of getting a sum of 12 when rolling 4 standard six-sided dice is 125/1296 (or about 0.096 or 9.6%).

34. The probability of getting a sum of 20 or more when rolling 4 standard six-sided dice is 35/648 (or about 0.054 or 5.4%).

35. The probability of randomly drawing two queens in a row from a deck of cards is 1/221 (or about 0.0045 or 0.45%).

36. The probability of randomly drawing three queens in a row from a deck of cards is 1/5525 (or about 0.00018 or 0.018%).

In Exercises #37-44, determine the probability that each event occurs and then explain how we should interpret this probability.

37. getting the same result (either all heads or all tails) on three flips of a coin

38. flipping a coin three times and getting exactly two heads

39. getting "2" on both spins of the given spinner

40. getting two odd numbers on two spins of the spinner in Exercise #39

41. rolling two standard six-sided dice and getting a sum of 8

42. rolling a standard six-sided die twice and getting a number equal or smaller on the second roll

43. correctly answering all 5 true-false questions on a quiz by random guessing

44. correctly answering 3 multiple choice questions (A-B-C-D options) by random guessing

45. Modern computers can operate so quickly that they can perform billions of processes every second. Computer hackers make use of this power to try to steal information by correctly guessing user passwords. To combat this, security experts suggest using complicated passwords that combine uppercase and lowercase letters, numbers, and symbols.
    a. How many unique passwords of length 5 can be created using only letters (uppercase and lowercase)?
    b. How many unique passwords of length 5 can be created using any combination of uppercase letters, lowercase letters, numbers from 0 to 9, and symbols from the set {!, #, $, %, &, *}?
    c. Assume a password of length 5 is randomly generated (so that it doesn't use real words or names for example). What is the probability that a hacking program could correctly guess a password from parts (a) and (b) with 100,000,000 random attempts using a password of length 5?

Algebra II

Module 9: Homework
Probability

   d. Discuss the benefits of using longer passwords and avoiding common words or names when generating passwords.

INVESTIGATION 4: PROBABILITY SPACES AND THE PROBABILITY OF COMPLEMENTS

In Exercises #46-49, write each probability using probability notation and then determine the value.

46. the probability of getting one heads and one tails when flipping a coin twice

47. the probability of getting four heads when flipping a coin four times

48. the probability of rolling a sum of 2 or 3 when rolling two standard six-sided dice

49. the probability of rolling a sum of 9 or more when rolling two standard six-sided dice

In Exercises #50-55, do the following.
   a) Find the probability of the event.
   b) State the event's complement.
   c) Find the probability of the complement.

50. Activity: rolling a six-sided die
    Event: rolling a number greater than 4

51. Activity: flipping a coin three times
    Event: getting exactly three tails

52. Activity: rolling two six-sided dice
    Event: getting a sum greater than 14

53. Activity: flipping a coin four times
    Event: getting at least two tails

54. Activity: randomly drawing two marbles together from a bag containing one blue, one clear, one red, and one yellow marble
    Event: drawing a yellow marble

55. Activity: randomly drawing two marbles together from a bag containing one blue, one clear, one red, one green, and one yellow marble
    Event: drawing a yellow marble

56. Suppose you have a bag of blue and red marbles and that the probability of drawing two blue marbles in a row is 13/24 (or about 0.542). What is the probability of not drawing two blue marbles in a row?

57. If you roll 4 standard six-sided dice, the probability of rolling a sum that is a prime number is 1/3. What is the probability of not rolling a prime sum?

58. If you roll 3 standard six-sided dice, the probability of rolling a sum of 7 or more is 49/54 (or about 0.907). What is the probability of rolling a 6 or less? Express your answer as a fraction, decimal, and percent.

59. The probability of winning the PowerBall jackpot if you buy 770 tickets is $\frac{1}{227,563}$. What is the probability of not winning the PowerBall lottery if you buy 770 tickets? Express your answer as a fraction, decimal, and percent.

Algebra II   Module 9: Homework
Probability

60. Suppose a spinner for a board game is designed as follows. Give the probability space for a single spin. (*Estimate the probabilities if necessary.*)

61. Suppose a spinner for a board game is designed as follows. Give the probability space for a single spin. (*Estimate the probabilities if necessary.*)

62. In a bag of marbles, 14 are purple, 13 are white, 7 are red, and 16 are yellow. Give the probability space for randomly drawing a marble from the bag.

63. In a high school there are 348 freshmen, 367 sophomores, 321 juniors, and 295 seniors. Give the probability space for randomly selecting a student from the school.

INVESTIGATION 5: UNIONS AND INTERSECTIONS

For Exercises #64-69, do the following.
  a) Draw a Venn diagram to represent the sample space of the activity and the outcomes for the given events.
  b) Write the set of outcomes for the *intersection* of the events and count them.
  c) Write the set of outcomes for the *union* of the events and count them.

64. Activity: rolling two standard six-sided dice
    Event A: rolling at least one "6"
    Event B: rolling a total of 9 or greater

65. Activity: flipping a coin twice
    Event A: getting heads on both flips
    Event B: getting tails on both flips

66. Activity: rolling two standard six-sided dice
    Event A: rolling at least one "5"
    Event B: rolling a total of 4 or less

67. Activity: choosing two different letters from the set {A, B, C, D, and E} assuming that order is not important (so for example AB and BA count as the same choice)
    Event A: letter A is chosen
    Event B: letter E is chosen

68. Activity: choosing two different letters from the set {A, B, C, D, and E} assuming that order is important (so for example AB and BA count as different choices)
    Event A: letter A is chosen
    Event B: letter E is chosen

Algebra II   **Module 9: Homework**
                                                                                                                Probability

69. Activity: choosing three different letters from the set {A, B, C, D, and E} assuming that order is not important (so for example ABC and BAC count as the same choice)
    Event A: letter A is chosen
    Event B: letter B is chosen

70. For Exercise #64, calculate $P(A \text{ and } B)$.   71. For Exercise #64, calculate $P(A \text{ or } B)$.

72. For Exercise #66, calculate $P(A \text{ or } B)$.   73. For Exercise #68, calculate $P(A \text{ or } B)$.

In Exercises #74-79, find each probability.

74. Imagine rolling a standard six-sided die.
    a. $P(2 \text{ and even})$ b. $P(4 \text{ or odd})$

75. Imagine rolling a standard six-sided die.
    a. $P(\text{even and less than 5})$ b. $P(\text{even or less than 4})$

76. Suppose you roll two standard six-sided dice.
    a. $P(\text{even sum and at least one 5})$ b. $P(\text{sum of 3 or 10})$ c. $P(\text{sum of 8 or doubles})$

77. Suppose you roll two standard six-sided dice.
    a. $P(\text{odd sum and at least one 2})$ b. $P(\text{sum of 6 or 7 or 8})$ c. $P(\text{sum less than 6 or doubles})$

78. Suppose you draw a card at random from a standard 52-card deck.
    a. $P(\text{club and even number})$ b. $P(\text{diamond and face card})$ c. $P(\text{spade or king})$

79. Suppose you draw a card at random from a standard 52-card deck.
    a. $P(\text{heart and ace})$ b. $P(\text{red or face card})$ c. $P(\text{diamond or face card})$

80. The following diagram provides some information about students at a high school. (*The numbers represent counts – how many students at the school have the given characteristics.*)

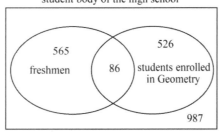

    a. How many students are enrolled in Geometry?
    b. What is the probability that a randomly chosen student at the school is a freshman?
    c. What is the probability that a randomly chosen student at the school is a freshman enrolled in Geometry?
    d. What is the probability that a randomly chosen student at the school is a freshman or is enrolled in Geometry?

81. The following diagram provides some information about teachers at a high school. (*The numbers represent counts – how many teachers at the school have the given characteristics.*)

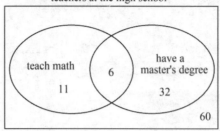

a. How many math teachers are at the school?
b. What is the probability that a randomly chosen teacher at the school has a master's degree?
c. What is the probability that a randomly chosen teacher at the school is a math teacher with a master's degree?
d. What is the probability that a randomly chosen teacher at the school teaches math or has a master's degree?

82. The following diagram provides some information about students at a high school. (*The numbers represent counts – how many students at the school have the given characteristics.*)

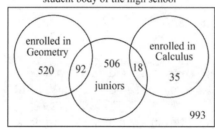

a. What is the probability that a randomly chosen student at the school is a junior enrolled in Geometry?
b. What is the probability that a randomly chosen student at the school is enrolled in Geometry and Calculus?
c. What is the probability that a randomly chosen student at the school is enrolled in Geometry or Calculus?
d. What is the probability that a randomly chosen student at the school is a junior or enrolled in Calculus?
e. Suppose we randomly select a junior from the high school. What is the probability that the student is enrolled in Geometry?

83. The following diagram provides some information about teachers at a high school. (*The numbers represent counts – how many teachers at the school have the given characteristics.*)

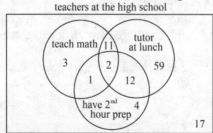

a. What is the probability that a randomly chosen teacher at the school teaches math and has 2nd hour prep?

Algebra II

Module 9: Homework
Probability

b. What is the probability that a randomly chosen teacher at the school tutors at lunch, teaches math, and has 2nd hour prep?
c. What is the probability that a randomly chosen teacher at the school teaches math or tutors at lunch?
d. What is the probability that a randomly chosen teacher at the school teaches math or doesn't have 2nd hour prep?
e. Suppose we randomly select a math teacher from the high school. What is the probability that the teacher tutors at lunch?

INVESTIGATION 6: CONDITIONAL PROBABILITY

84. The following table shows the results of a survey of 38 students at a school.
    Q1: Do you plan to buy a school yearbook?
    Q2: Do you plan to attend the yearbook signing party?

	plans to buy a yearbook	does not plan to buy a yearbook	Total
plans to attend the signing party	12	6	18
does not plan to attend the signing party	12	8	20
Total	24	14	38

  a. Suppose we randomly select one of the students who took the survey. Calculate and interpret the probability that the student plans to buy a yearbook.
  b. Suppose we randomly select one of the students who took the survey. Calculate and interpret the probability that the student does not plan to attend the signing party.
  c. Suppose we group all of the students from this survey who plan to buy a yearbook and choose one of them at random. Calculate and interpret the probability that the student plans to attend the yearbook signing party.
  d. Suppose we group all of the students from this survey who do not plan to attend the signing party and choose one of them at random. What is the probability that the student does not plan to buy a yearbook?
  e. Suppose the school has 1820 students. How many students do you estimate will buy a yearbook and attend the signing party?

85. The following table shows the results of a survey of 24 seniors at a high school.
    Q1: Do you play on a school sports team?
    Q2: Do you have an after-school job?

	plays on a school sports team	does not play on a school sports team	Total
has an after-school job	2	9	11
does not have an after-school job	5	8	13
Total	7	17	24

  a. Suppose we randomly select one of the students who took the survey. Calculate and interpret the probability that the student does not play on a sports team.

Algebra II **Module 9: Homework**
Probability

    b. Suppose we randomly select one of the students who took the survey. Calculate and interpret the probability that the student has an after-school job.

    c. Suppose we group all of the students from this survey who play on a school sports team and choose one of them at random. Calculate and interpret the probability that the student has an after-school job.

    d. Suppose we group all of the students from this survey who do not have an after-school job and choose one of them at random. What is the probability that the student does not play on a school sports team?

    e. Suppose the school has 312 seniors. How many seniors do you estimate play on a school sports team?

For Exercises #86-91, a box contains 15 blue and 12 yellow Ping-Pong balls. Suppose you shake the box, draw one Ping-Pong ball at random, set it to the side, shake the box, and draw a second Ping-Pong ball at random. Rewrite each of the following statements in words, and then determine the probability.

86. $P(\text{first draw blue})$

87. $P(\text{first draw yellow})$

88. $P(\text{second draw blue} \mid \text{first draw blue})$

89. $P(\text{second draw blue} \mid \text{first draw yellow})$

90. $P(\text{second draw yellow} \mid \text{first draw blue})$

91. $P(\text{second draw yellow} \mid \text{first draw yellow})$

For Exercises #92-97, suppose you shuffle a standard deck of 52 cards, draw a card at random and set it aside, then reshuffle the deck and draw a second card at random. Rewrite each of the following statements in words, and then determine the probability.

92. $P(\text{first card red})$

93. $P(\text{first card club})$

94. $P(\text{second card red} \mid \text{first card black})$

95. $P(\text{second card heart} \mid \text{first card club})$

96. $P(\text{second card spade} \mid \text{first card spade})$

97. $P(\text{second card king} \mid \text{first card king})$

98. Suppose you select a professional basketball team at random. Explain the difference between $P(\text{team had a winning record last year})$ and $P(\text{team had a winning record last year} \mid \text{team made the playoffs last year})$.

99. Suppose you select a city in the United States at random. Explain the difference between $P(\text{city has high pollution})$ and $P(\text{city has high pollution} \mid \text{city's population is at least 2 million})$.

100. Suppose you select a student at random from your school. Explain the difference between $P(\text{student has a 4.0 GPA} \mid \text{student is in honors math})$ and $P(\text{student is in honors math} \mid \text{student has a 4.0 GPA})$.

101. Suppose you select a person living in your state at random. Explain the difference between $P(\text{person has high cholesterol} \mid \text{person does not exercise regularly})$ and $P(\text{person does not exercise regularly} \mid \text{person has high cholesterol})$.

102. Suppose that a company is developing a new test for a disease. The company recruited 914 people (319 people who they know already have the disease and 595 people who they know do not have the disease). The company then used the new test on each patient and recorded the results.

	patient tested positive	patient tested negative	Total
patient has the disease	571		595
patient does **not** have the disease		293	319
Total			914

   a. Complete the table by filling in the missing entries.
   b. Rewrite the following statements in the form $P(A \mid B)$ and use the given information to estimate the probability.
      i) The probability that a person who has the disease will test positive.
      ii) The probability that a person who does not have the disease will test negative.
      iii) The probability that a patient tests negative if he or she has the disease (*called a false negative*).
      iv) The probability that a patient tests positive if he or she does not have the disease (*called a false positive*).

103. Suppose that a company is developing a new test for a disease. The company recruited 1,123 people (294 people who they know already have the disease and 829 people who they know do not have the disease). The company then used the new test on each patient and recorded the results.

	patient tested positive	patient tested negative	test results were inconclusive	Total
patient has the disease		28	4	294
patient does **not** have the disease	108	713		829
Total				1123

   a. Complete the table by filling in the missing entries.
   b. Rewrite the following statements in the form $P(A \mid B)$ and use the given information to estimate the probability.
      i) The probability that a person who has the disease will test positive.
      ii) The probability that a person who does not have the disease will test negative.
      iii) The probability that a patient tests negative if he or she has the disease (*called a false negative*).
      iv) The probability that a patient tests positive if he or she does not have the disease (*called a false positive*).

Algebra II

Module 9: Homework
Probability

104. Two events A and B are subsets of the sample space of some stochastic activity. There are 120 total outcomes in the sample space including 20 outcomes for event A and 22 outcomes for event B. In addition, $P(B \text{ doesn't occur} \mid A \text{ occurs}) = 9/10$ and $P(B \text{ doesn't occur} \mid A \text{ doesn't occur}) = 4/5$.

    Using this information, complete the two-way frequency table.

	event B occurs	event B does not occur	total
event A occurs			
event A does not occur			
total			

105. Three events A and B are subsets of the sample space of some stochastic activity. There are 360 total outcomes in the sample space including 72 outcomes where event A doesn't occur and 208 outcomes where event B doesn't occur. In addition, $P(A \text{ occurs} \mid B \text{ occurs}) = 16/19$ and $P(B \text{ doesn't occur} \mid A \text{ doesn't occur}) = 2/3$.

    Using this information, complete the two-way frequency table.

	event B occurs	event B does not occur	total
event A occurs			
event A does not occur			
total			

INVESTIGATION 7: INDEPENDENCE

In Exercises #106-111, you are given two events for a compound stochastic activity. Think about each activity and then answer the given questions.

106. Activity: Flipping two coins (one quarter and one nickel).
    Event A: The quarter lands heads up.
    Event B: The nickel lands heads up.
    a) Find $P(A)$.   b) Find $P(B)$.   c) Find $P(A|B)$.   d) Find $P(B|A)$.

107. Activity: Rolling two standard six-sided dice (one blue and one red).
    Event A: The red die rolls a 2 or 4.
    Event B: The blue die rolls a 1 or 3 or 5.
    a) Find $P(A)$.   b) Find $P(B)$.   c) Find $P(A|B)$.   d) Find $P(B|A)$.

108. Activity: Shuffling a deck of cards, drawing a card and putting it back in the deck, reshuffling, and drawing a second card.
    Event A: The first card drawn is a heart.
    Event B: The second card drawn is a club.
    a) Find $P(A)$.   b) Find $P(B)$.   c) Find $P(B|A)$.

Algebra II

Module 9: Homework
Probability

109. Activity: Randomly pulling a marble from a bag containing 4 blue and 4 red marbles, replacing it, and then randomly pulling a second marble.
    Event A: The first marble drawn is red.
    Event B: The second marble drawn is blue.
    a) Find $P(A)$.   b) Find $P(B)$.   c) Find $P(B|A)$.

110. Activity: Shuffling a deck of cards, drawing a card and setting it aside, reshuffling, and drawing a second card.
    Event A: The first card drawn is a heart.
    Event B: The second card drawn is a club.
    a) Find $P(A)$.   b) Find $P(B|A)$.   c) True or False: $P(B|A) = P(B)$ in this context

111. Activity: Randomly pulling a marble from a bag containing 4 blue and 4 red marbles, setting it aside, and then randomly pulling a second marble.
    Event A: The first marble drawn is red.
    Event B: The second marble drawn is blue.
    a) Find $P(A)$.   b) Find $P(B|A)$.   c) True or False: $P(B|A) = P(B)$ in this context

112. Explain in your own words what it means to say that two events are independent.

113. Explain in your own words what it means to say that two events are dependent.

In Exercises #114-119, determine if the events are independent. If so, explain what it means for them to be independent. If not, explain why you think the events are dependent.

114. Suppose you roll two dice, one of which is a standard six-sided die and the other is a standard twenty-sided die (with the results 1 through 20 possible). You are interested in the probability that the first die rolls "1" and the probability that the second die rolls "1".

115. The following spinner is used in a certain board game. On a player's turn, he or she spins twice. You are interested in the probability of the events "the first spin lands on 3" and "the second spin lands on 3."

116. Suppose your teacher needs two students to help carry textbooks to another classroom. He or she randomly chooses a student by drawing a name from a hat, and then randomly selects a second student by following the same process. You are interested in the probability that the first student chosen is a female and the probability that the second student chosen is female.

117. Suppose you randomly draw a card from a standard 52-card deck, replace it, shuffle the deck, and draw a second card. You are interested in the probability that the first card was a club and the probability that the second card is a diamond.

Algebra II — Module 9: Homework — *Probability*

118. Each day your math teacher randomly selects a student to present the solution to a homework problem. Once a student is selected, he or she is not chosen again until everyone has presented. You are interested in the probability that you are chosen today and the probability that you are chosen tomorrow.

119. Suppose you flip a coin and then roll a die according to the following rules. If the coin lands heads up you roll a standard six-sided die (with the results 1 to 6 possible). If the coin lands tails up you roll a standard eight-sided die (with the results 1 to 8 possible). You are interested in the events "the coin lands heads up" and "the die rolls 1".

120. Seniors at a certain high school completed a survey with the following results.

	student is on a school sports team	student is not on a school sports team	Total
student is taking a math class this year	56	224	280
student is not taking a math class this year	24	96	120
Total	80	320	400

    If we choose a senior at random, there is a chance that he or she is on a school sports team and a chance that he or she is taking a math class this year. Are these events dependent or independent? Justify your response.

121. Seniors at a certain high school completed a survey with the following results.

	student has a 4.0 GPA	student does not have a 4.0 GPA	Total
student is in the National Honor Society	4	11	15
student is not in the National Honor Society	1	384	385
Total	5	395	400

    If we choose a senior at random, there is a chance that he or she has a 4.0 GPA and a chance that he or she is in the National Honor Society. Are these events dependent or independent? Justify your response.

122. You and your friends are taking turns flipping a coin. The last four flips have all landed heads up. One friend thinks it's more likely that the next flip will land heads up because the last four flips have all landed heads up. Another friend thinks tails up is more likely because it's rare for five flips in a row to all land heads up. What is your reaction to this debate?

123. The "Gambler's Fallacy" is a common mistake people make when performing activities with **independent**, random outcomes. Many people wrongly believe that outcomes get "used up" and that if an event happens more often during one period it means that the event will happen less often later.
    a. A standard six-sided die is rolled several times, resulting in the outcomes 4, 3, 4, 4, 6, and 4. If someone believes the Gambler's Fallacy, what is he or she likely to say will happen on the next six rolls?
    b. Explain how the Gambler's Fallacy is related to the idea of independent vs. dependent events.
    c. While watching a baseball game on TV, you hear the announcers say that a player hasn't hit a homerun in two months and is therefore "due" (meaning they expect him to hit a homerun soon). What do you think about this conclusion?

Algebra II

Module 9: Homework
Probability

INVESTIGATION 8: MULTIPLICATION RULE

For the activities described in Exercises #124-135, determine the indicated probability. *Hint: It might be helpful to first decide if the given events are independent.*

124. What is the probability that you get a "5" followed by an even number on two rolls of a standard six-sided die?

125. What is the probability that you get an odd number followed by another odd number on two rolls of a standard six-sided die?

126. Suppose you have a bag of marbles containing 12 green marbles and 18 blue marbles. Now suppose that you randomly draw a marble, replace it, and then draw a second marble.
    a. What is the probability that the first marble drawn is green and the second marble drawn is blue?
    b. What is the probability that both marbles drawn are blue?

127. Suppose you have a standard deck of 52 cards and that you shuffle the deck, draw a card and replace it, reshuffle, and then draw a second card.
    a. What is the probability that the first card is a "10" and the second card is a jack?
    b. What is the probability that the first two cards are jacks?

128. The following spinner is used in a certain board game. On a player's turn, he or she spins twice. Determine each of the following probabilities.

    a. *P*(first spin 4 and second spin 6)
    b. *P*(first spin 3 and second spin not 3)

129. The following spinner is used in a certain board game. On a player's turn, he/she spins twice. Determine each of the following probabilities.

    a. *P*(first spin 2 and second spin 4)
    b. *P*(first spin is not 1 and second spin is not 4)

130. Suppose you have a bag of marbles containing 12 green marbles and 18 blue marbles. Now suppose that you randomly draw a marble, set it aside, and then draw a second marble.
    a. What is the probability that the first marble drawn is green and the second marble drawn is blue?
    b. What is the probability that both marbles drawn are blue?

131. Suppose you have a bag of marbles containing 10 purple marbles and 32 clear marbles. Now suppose that you randomly draw a marble, set it aside, and then draw a second marble.
    a. What is the probability that both marbles drawn are purple?
    b. What is the probability that both marbles drawn are clear?

Algebra II

Module 9: Homework
Probability

132. Suppose you have a standard deck of 52 cards and that you shuffle the deck, draw a card and set it aside, reshuffle, and then draw a second card.
   a. What is the probability that the first card is a "10" and the second card is a jack?
   b. What is the probability that the first two cards are jacks?

133. Suppose you have a standard deck of 52 cards and remove all of the spades, then you shuffle the deck, draw a card and set it aside, reshuffle, and then draw a second card.
   a. What is the probability that both cards are kings?
   b. What is the probability that both cards are black?

134. Suppose your math class has 18 girls and 16 boys. Each of the students writes his or her name on a slip of paper. Your teacher shuffles the slips, draws a name and sets it aside, shuffles the slips again, and draws a second name. Calculate and compare P(first student drawn is female and second student drawn is male) and P(one female and one male student drawn).

135. Suppose you have a standard deck of 52 cards and that you shuffle the deck, draw a card and set it aside, reshuffle, and then draw a second card. Calculate and compare P(first card is black and second card is red) and P(one black and one red).

For Exercises #136-139, suppose that a company is developing a new test for a disease. The company recruited 540 people (102 people who they know already have the disease and 438 people who they know do not have the disease). The company then used the new test on each patient and recorded the results.

	patient tested positive	patient tested negative	Total
patient has the disease	100	2	102
patient does **not** have the disease	25	413	438
Total	125	415	540

Event A: patient tested positive   Event C: patient has the disease
Event B: patient tested negative   Event D: patient does not have the disease

Suppose you randomly select one of the patients. Explain what each of the following equalities means in words and then confirm that each equation is true.

136. $P(A \text{ and } D) = P(A)P(D|A)$

137. $P(A \text{ and } C) = P(A)P(C|A)$

138. $P(B \text{ and } D) = P(B)P(D|B)$

139. $P(B \text{ and } C) = P(B)P(C|B)$

For Exercises #140-145, suppose you have a bag of red and yellow marbles and that you randomly draw a marble, set it aside, and then randomly draw a second marble.

sample space: $\{(R, R), (R, Y), (Y, R), (Y, Y)\}$

$P(\text{red on first draw}) = \frac{4}{9}$ $P(\text{yellow on first draw}) = \frac{5}{9}$

$P(\text{red on second draw}|\text{red on first draw}) = \frac{3}{7}$ $P(\text{red on second draw}|\text{yellow on first draw}) = \frac{16}{35}$

$P(\text{yellow on second draw}|\text{red on first draw}) = \frac{4}{7}$ $P(\text{yellow on second draw}|\text{yellow on first draw}) = \frac{19}{35}$

140. Find $P(R, R)$.   141. Find $P(Y, Y)$.   142. Find $P(R, Y)$.   143. Find $P(Y, R)$.

144. If there are 144 marbles in the bag, how many are red and how many are yellow?

145. If there are 720 marbles in the bag, how many are red and how many are yellow?

146. Suppose you randomly select a card from a standard 52-card deck, set it aside, reshuffle, randomly draw a second card, reshuffle, etc.
    a. What is the probability that the first five cards are all red?
    b. What is the probability that the first five cards are all the same color?
    c. What is the probability that the first four cards are all kings?
    d. What is the probability that the first four cards are *not* all the same rank?

147. Suppose you remove all of the hearts from a standard 52-card deck, shuffle the remaining cards and randomly draw one card, set it aside, reshuffle, randomly draw a second card, reshuffle, etc.
    a. What is the probability that the first five cards are all black?
    b. What is the probability that the first three cards are kings?
    c. What is the probability that you *don't* get three kings for your first three cards drawn?
    d. What is the probability that the first two cards drawn are diamonds and the next two cards drawn are clubs?

INVESTIGATION 9: PERMUTATIONS AND COMBINATIONS

In Exercises #148-153, do the following.
   a) Represent the total number of permutations using the notation $_nP_r$.
   b) Determine the total number of permutations.

148. How many unique ways can you arrange 6 books on a shelf?

149. How many unique ways can you arrange 15 people in a line?

150. Suppose you have 10 books. How many unique ways can you arrange 6 of the 10 books on a shelf?

151. How many unique 4-letter "words" can you form from the letters in MACHINE?

152. There are 32 people in your English class. How many unique ways can your teacher choose 4 people to read different parts in a play?

153. There are 15 people at a party and 7 different raffle prizes will be given out. How many unique ways can we distribute the raffle prizes to people at the party?

In Exercises #154-159, explain what the notation represents in words and then calculate the number of permutations.

154. $_5P_2$ 155. $_{10}P_5$ 156. $_{17}P_6$ 157. $_{100}P_4$ 158. $_8P_8$ 159. $_7P_8$

In Exercises #160-163, determine the indicated probability.

160. Suppose you want 5 people to form a line. Assuming that their positions are determined randomly, what's the probability that they are lined up in alphabetical order?

161. Suppose you randomly rearrange the letters in the word NORMAL. What is the probability that the letters end up arranged as ALMNOR or RONMLA?

Algebra II

Module 9: Homework
Probability

162. Suppose there are 8 people of different heights, and you want 5 of them to form a line. Assuming the people and positions are chosen randomly, what's the probability that the five tallest people are chosen and that they are lined up in order from taller to shorter?

163. Suppose you randomly select 3 letters from the word BOWLING and write them in the order they are selected. What is the probability that you get NOW?

In Exercises #164-169, do the following.
    a) Represent the total number of combinations using the notation $_nC_r$.
    b) Determine the total number of combinations.

164. Suppose there are 8 books you want to read at the library, but you are only allowed to check out 3 at a time. How many unique groups of 3 books could you select from the 8 you want to read?

165. Suppose you want to add 4 ingredients to a cookie recipe from the following list: peanuts, walnuts, raisins, milk chocolate chips, white chocolate chips, and rainbow sprinkles. How many unique combinations could you create?

166. Suppose there are 6 people in a family and they are given 3 tickets to a baseball game. How many unique groups of people could use the tickets?

167. There are 28 people in your math class. Your teacher needs 5 people to go to the office to pick up textbooks. How many unique groups of 5 people can be chosen?

168. There are 15 people at a party and 7 identical raffle prizes. How many ways can we distribute the raffle prizes to people at the party?

169. How many groups of 2 vowels (A, E, I, O, or U) can you select from the letters in the word BACKFIRE?

In Exercises #170-175, explain what the notation represents in words and then calculate the number of combinations.
170. $_5C_2$ 171. $_9C_4$ 172. $_{11}C_3$ 173. $_{100}C_5$ 174. $_6C_6$ 175. $_9C_{15}$

In Exercises #176-179, determine the probability that the given event occurs.

176. Suppose you randomly select 3 letters from the word SARDINE. What is the probability that all three letters are vowels (A, E, I, O, or U)?

177. Your teacher is randomly creating a group of 6 students from your class of 24 students. What is the probability that the group includes the 6 students with the highest grades in the class?

178. Suppose you randomly select 3 letters from the word SARDINE. What is the probability that all three letters are consonants (anything other than A, E, I, O, or U)?

179. Your teacher is randomly creating a group of 6 students from your class of 24 students. What is the probability that you are included in the group?

Algebra II   **Module 9: Homework**
                                                                                                Probability

In Exercises #180-188, do the following.
   a) Represent the total number of unique possibilities using the notation $_nP_r$ or $_nC_r$ (if possible).
   b) Determine the total number of unique possibilities.

180. Suppose you randomly guess the answers to 10 true-false questions on a test. How many different sequences of Ts and Fs could you choose for your final set of answers?

181. At a candy store you can create your own mix of jelly beans by adding scoops from 20 different flavors to a bag. If you select 3 different flavors (1 scoop each), how many different unique mixtures can you create?

182. A combination lock has the numbers 0 through 39 and requires the correct sequence of 4 different numbers to open. How many unique 4-number sequences are possible for the lock?

183. From 21 students in a PE class, a teacher must assign 9 students to the 9 different starting positions on a baseball team. How many different ways could he or she do this?

184. From the 15 members of student council a 4-person committee needs to be formed. How many 4-person committees are possible?

185. A car lot sells 4 styles of vehicles (compacts, trucks, sedans, and SUVs). If the sales manager records the style of the next 6 cars sold, how many different sequences of vehicle types are possible?

186. Your PE class has 30 people in it (you and 29 other students). You are asked to pick 4 other students to form a team with you. How many different teams could you create?

187. From 15 members of student council 3 people must be selected to take the roles of President, Vice President, and Treasurer. How many different ways can these roles be filled?

188. A lottery game randomly selects 4 numbers between 1 and 25. The numbers are always reported in ascending order regardless of the order in which they are drawn. A player wins $5,000 if he or she has a ticket matching all 4 numbers. How many different sets of 4 numbers can be chosen?

INVESTIGATION 10: EXTENSIONS USING PERMUTATIONS AND COMBINATIONS

> Investigation 10: *Extensions using Permutations and Combinations* and associated homework is available online at www.rationalreasoning.net.

Algebra II
Module 10: Investigation 1
Mean and Standard Deviation

When you're looking for a movie to watch, you might read the short plot synopsis for several movies to see if they sound interesting. However, there's no way to fully summarize a two-hour movie in a sentence or two. A lot of details have to be left out!

In mathematics, the **mean** (average) for a set of numbers is kind of like a plot summary. It gives you a snapshot of an entire set of numbers but you lose a lot of the details. When you have a set of numbers you can find the mean by summing all of the numbers and dividing by how many numbers are in the set.

1. Find the mean of each set of numbers.
   a. 8, 12, 6, 4, 6, 4, 9, 7, 2, 12

   b. –2, 1.2, 0.7, –1, 0.8, 0.3

   c. 14.75, 16.25, 19.5, 13, 15.25, 18.5, 20, 17.75, 15, 19.5

   d. –6.2, –0.2, –5.7, 1.3, –4, –8, 1.4, –1.1

A good way to interpret the mean is as a constant value that, if it replaced all of the actual numbers in a set, results in the same total sum. For example, In Exercise #1(a) you found the mean is 7 for the set of numbers 8, 12, 6, 4, 6, 4, 9, 7, 2, 12, and

$$\cancel{8} + \cancel{12} + \cancel{6} + \cancel{4} + \cancel{6} + \cancel{4} + \cancel{9} + \cancel{7} + \cancel{2} + \cancel{12} = 70$$
$$\phantom{\cancel{0}}7\,7\,7\,7\,7\,7\,7\,7\,7\,7$$
$$7 + 7 + 7 + 7 + 7 + 7 + 7 + 7 + 7 + 7 = 70$$

We can also visualize this by imagining a kind of modified histogram where each number is represented by a "chip." We can imagine moving the outermost pairs of values inward one step at a time. This simulates increasing the smallest value in the data set by 1 and decreasing the largest value in the data set by 1, which maintains the overall sum.

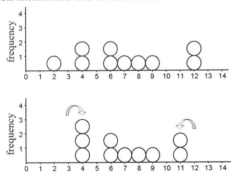
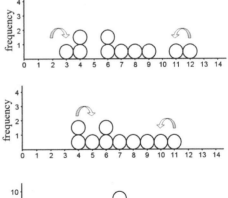

When all of the "chips" rest on the same value, we have found the mean. If every number in the set is replaced by 7 then we will maintain the same overall sum using a single constant in place of each number.

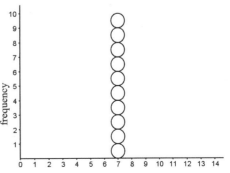

Module 10: Investigation 1

Algebra II

Module 10: Investigation 1
Mean and Standard Deviation

2. Explain how we can interpret each mean you calculated in Exercise #1.

Mean

The **mean** (average) of a set of numbers is the result of finding the overall sum and dividing by how many numbers were summed.

One way to interpret the mean is as the constant number that when used to replace each number in the set maintains the overall sum.

---

3. Sara scored 77, 89, 85, 83, and 90 on her last five chemistry lab reports. What was her mean score and how might we interpret this value?

4. In the 2012-2013 season, professional basketball player Kevin Durant played in 81 games. During those games, he played a total of 3,119 minutes and scored 2,280 points. Find the mean points scored and minutes played per game and explain how to interpret these values.

5. The average height of the 34 students in a math class is 64.2 inches. How might we interpret this value?

Algebra II

Module 10: Investigation 1
Mean and Standard Deviation

6. The mean is often used to help us understand very large numbers. For example, in 2014 the U.S. national debt was estimated to be about $17,500,000,000,000 (that's 17.5 trillion dollars). This number is so large that we can't really make sense of its size. So we might decide to divide the national debt by the U.S. population to obtain a *per capita* (meaning "per person") average.
   a. If the U.S. population in 2014 was about 315,000,000, what is the approximate national debt *per capita*?

   b. How might we interpret this value?

7. Find the mean for each of the following sets of numbers.
   a. 8, 10, 13, 10, 9, 12, 11, 7
   b. 10, 10, 10, 11, 10, 10, 9, 10

   c. 2, 9, 20, 8, 7, 14, 1, 19
   d. −30, 0, 100, 15, −50, −5, 60, −10

The largest drawback in using the mean to summarize a data set is that you lose a sense for how much variation existed in the original set. For example, in Exercise #7 you encountered several different sets of numbers with the same mean.

To deal with this issue we typically report a measure of variability along with a mean when summarizing a set of numbers.

8. Use the set of numbers in Exercise #7(a) and do the following.
   i. Take each number in the set and subtract the mean. (These values are called the ***deviations*** or ***deviations from the mean***).
   ii. Square each of the deviations. (These are called the ***squared deviations***.)
   iii. Find the mean of the squared deviations. (This value is called the ***variance*** for the data set.)
   iv. Take the square root of the variance. (This value is called the ***standard deviation***.)

original numbers	(i) deviations from the mean	(ii) squared deviations
8		
10		
13		
10		
9		
12		
11		
7		

(iii) variance:

(iv) standard deviation:

Algebra II

Module 10: Investigation 1
Mean and Standard Deviation

9. Repeat the process in Exercise #8 to find the standard deviation for each of the data sets in parts (b) – (d) of Exercise #7. (*You may use an Excel spreadsheet or the statistics functions on a graphing calculator if it helps.*)

Standard Deviation

The ***standard deviation*** is a measure of the "spread" for a set of numbers.

Small standard deviations indicate that the numbers are clustered close to the mean. Large standard deviations indicate that the numbers are spread out away from the mean.

Note: We use σ *(lowercase sigma) to represent the standard deviation for a set of numbers (and similarly* σ^2 *is used to represent the variance).*

---

10. Use the results of Exercises #7-9 to verify the idea that smaller standard deviations indicate that the numbers in a set are clustered more closely around the mean. (*No shown work is required.*)

11. What does it mean for a set of numbers if the standard deviation is 0?

12. You are given four histograms representing data sets with 20 numbers each and the same mean (9) but with different standard deviations. Match each standard deviation to the correct histogram.

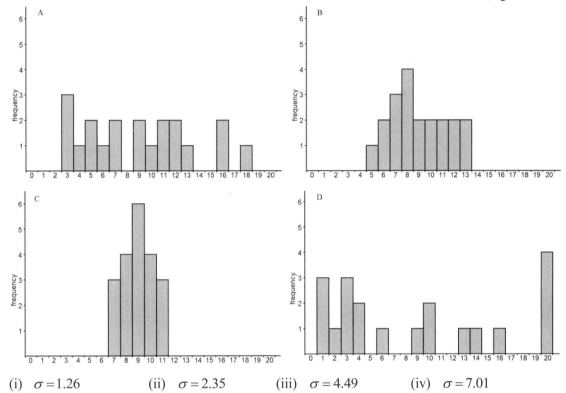

(i) $\sigma = 1.26$ (ii) $\sigma = 2.35$ (iii) $\sigma = 4.49$ (iv) $\sigma = 7.01$

13. In Exercise #3 we considered the following scenario: *Sara scored 77, 89, 85, 83, and 90 on her last five chemistry lab reports.*
    a. Calculate the standard deviation for Sara's scores.

    b. Suppose John and Charlene both have the same mean score on their five lab reports as Sara, but the standard deviations of their scores are about 1.8 and 6.3 respectively. What does this tell you about their lab report grades?

14. Tara and Keosha both play basketball. Over the course of the season, Tara average 18.4 points per game with a standard deviation of 2.3 points and Keosha averaged 20.7 points per game with a standard deviation of 5.9 points. What does this information tell us about their performances during the season? (*Assume they both played the same number of games during the season.*)

When studying series in Module 2 we learned how to represent the sum of a set of numbers using **sigma notation**. Take a moment to go back and review sigma notation now.

15. a. Suppose there are *n* numbers in a set and we denote the numbers by $x_1, x_2, ..., x_n$. Show how to represent the sum of these numbers *and* the mean using sigma notation. (*Note: We use $\bar{x}$ to stand for the mean.*)

   b. How can we represent the variance of the data set using sigma notation? (Look at the steps in Exercise #8 for how we calculate the variance of a data set.)

   c. How can we represent the standard deviation of the data set using sigma notation?

Sigma Notation for Representing the Mean and Standard Deviation

If we have a data set with *n* numbers ($x_1, x_2, ..., x_n$), $\bar{x}$ represents the mean of the data set, and σ represents the standard deviation, then

$$\bar{x} = \qquad \text{and} \qquad \sigma =$$

Algebra II

Module 10: Investigation 2
Mean and Expected Value

Lotteries are very popular in the United States. When most people think about lotteries they focus on the amount of money a jackpot winner can claim. However, lotteries are also huge money-makers for state governments. Let's explore why.

Wisconsin used to have a lottery scratcher game called *Moose on the Loose*. For $1 you could buy a ticket where you scratched off the surface to reveal three dollar amounts. If the dollar amounts all matched, you could trade in your ticket for the dollar value shown. The probability of winning different amounts is shown in the following table.

Moose on the Loose Payoffs and Probabilities

payoff	probability	payoff	probability
$0 (not a winner)	about 4/5	$20	1/241
$1	1/11	$50	1/931
$2	1/12	$100	1/2,401
$5	1/111	$500	1/28,801
$10	1/151		

http://www.wilottery.com/scratchgames/historical.aspx

1. Pick a payoff and probability and explain how to interpret the probability using the idea of long-term relative frequency.

2. a. Approximately how many $5 payoff tickets do you expect if 40,000 tickets are sold? What would be the total payoff on those tickets?

   b. Approximately how many $20 payoff tickets do you expect if 40,000 tickets are sold? What would be the total payoff on those tickets?

The *expected payoff for a single ticket* is an important value because it gives us information about the long-term expected payoffs from the game (*and it tells us whether or not playing the game is a wise investment*).

On a single ticket, there is a $\frac{1}{11}$ probability of winning $1, a $\frac{1}{12}$ probability of winning $2, and so on. If we multiply the probability of getting each payoff by the payoff amount and add all of the results, we get a value called the **expected value** (or **expected payoff**).

3. Use a calculator or spreadsheet software to approximate the expected payoff on a single ticket.
$$\tfrac{4}{5}(\$0)+\tfrac{1}{11}(\$1)+\tfrac{1}{12}(\$2)+\tfrac{1}{111}(\$5)+\tfrac{1}{151}(\$10)+\tfrac{1}{241}(\$20)+\tfrac{1}{931}(\$50)+\tfrac{1}{2,401}(\$100)+\tfrac{1}{28,801}(\$500)$$

Algebra II

Module 10: Investigation 2
Mean and Expected Value

The result of your calculation is related to the long-term relative frequency of payoff amounts. If you play the game many, many times then in the long run your total winnings will be about X per $1 spent, where X is the value from Exercise #3.

4. Consider your answer to Exercise #3. What is the significance of your result?

5. Let's consider the wider implications of your answer in Exercise #3.
   a. What is the total amount of money collected for 40,000 *Moose on the Loose* tickets?

   b. What is the expected amount of money the lottery commission will have to pay out to winners on these 40,000 tickets?

   c. The money collected but not paid out to winners is kept by the state government and used to pay for projects in its annual budget. What is the expected amount of money kept by the state from these 40,000 tickets?

Expected Value

The ***expected value*** is the weighted average of the outcomes from a stochastic activity and is calculated by multiplying each possible outcome value by the probability that the outcome occurs, and then summing the results.

Algebra II

Module 10: Investigation 2
Mean and Expected Value

6. Another scratcher game from Wisconsin was called *Pirate's Gold* with tickets costing $3 each. The probability of winning different amounts in the game is shown in the following table.

Pirate's Gold Payoffs and Probabilities

payoff	probability	payoff	probability
$0 (not a winner)	about 6/25	$50	1/1,176
$3	1/7	$100	1/3,201
$5	1/21	$250	1/64,001
$8	1/51	$500	1/96,001
$10	1/62	$5,000	1/480,001
$20	1/62	$35,000	1/480,001

http://www.wilottery.com/scratchgames/historical.aspx

a. Determine the expected payoff for a single ticket and explain how to interpret this value.

b. Estimate the total payouts on 100,000 tickets.

c. Estimate the amount of money kept by the state from 100,000 tickets.

7. Insurance companies frequently make use of expected value calculations when deciding how much to charge a customer for insurance. Suppose that, based on past experience and industry guidelines, they believe that they will have to pay out insurance claims with the following frequencies for customers each year.

approximate payout amount for insurance claims in one year for a customer	probability
$500	0.051
$1,000	0.041
$5,000	0.04
$10,000	0.038
$25,000	0.023
$50,000	0.011
$100,000	0.0013
$1,000,000	0.0002

Note: In this exercise we are ignoring the idea that different customers might have different risk levels. In real life insurance companies adapt the calculations to better model their risks, but the results are used in similar ways to what we demonstrate here.

a. What is the expected cost for insuring each customer?

Algebra II

Module 10: Investigation 2
Mean and Expected Value

   b. Suppose that the company has 1,750,000 customers, and they want to price their policies so that they earn $400,000,000 per year after payouts. How much should they charge a person for insurance for one year? (*Assume every person will pay the same rate.*)

   c. Why is it so important that the original probability estimates are accurate for the company?

8. A standard six-sided die has faces with values of 1, 2, 3, 4, 5, and 6. Each face has a $\frac{1}{6}$ probability of appearing when the die is rolled.
   a. What is the mean of the face values?

   b. What is the expected value on a single roll of the die?

   c. Did your answers to parts (a) and (b) match? If so, provide a mathematical justification for why the two calculations produce the same value. If not, explain why the difference occurs.

Algebra II

Module 10: Investigation 2
Mean and Expected Value

9. The following spinner is used in a board game.

   a. What is the mean of the possible values we could get?

   b. What is the expected value on a single spin?

   c. Did your answers to parts (a) and (b) match? If so, provide a mathematical justification for why the two calculations produce the same value. If not, explain why the difference occurs.

10. a. Complete the following statement. When each of the outcomes in the sample space has an equal chance of being chosen, then the expected value _____ the mean of the possible outcomes.

    b. Justify your answer in part (a) by using the general context of a stochastic process with n possible outcomes that have values of $a_1, a_2, ..., a_n$ to show that the calculations for mean and expected value will always be equal.

Algebra II

Module 10: Investigation 3
Sampling from a Population

Your teacher has a container with many similar objects inside. The objects have two different colors.
    Color A: _____   Color B: _____

You and your classmates have a job. Without opening the container and looking inside, and without dumping out the contents and counting, you have to predict what proportion (or percentage) of the objects in the bag have Color A.

1. One student in your class will draw 10 objects from the bag. Record the number of objects of Color A here: _____.

Sample and Population

A ***population*** in statistics describes all of the objects or people for whom you want to draw conclusions.

A ***sample*** is a subset of the entire population of interest.

> ***Ex:*** *You might be interested in the number of 2014 Ford F-150 trucks that report problems with their transmissions within the first year. Then <u>all</u> 2014 Ford F-150 trucks are the population. Instead of gathering data on the entire population, you might tag 300 new trucks and track their repair records to determine how many of them reported transmission problems. These 300 trucks form your sample.*

2. a. In Exercise #1, what is the population of interest? What are we doing to create samples of this population?

   b. In Exercise #1, what proportion of the sample are Color A?

   c. How confident are you that the same proportion of the total number of objects are Color A? Explain.

Algebra II

Module 10: Investigation 3
Sampling from a Population

3. As a class, repeat the experiment several times by using the following process.
   (i) Draw 10 objects. This constitutes a sample of the population.
   (ii) Count the number of objects with Color A in the sample and record this in the table.
   (iii) Determine the proportion of the sample with Color A and record it in the table.

	Number of Color A	Proportion of Sample		Number of Color A	Proportion of Sample
Sample 1			Sample 16		
Sample 2			Sample 17		
Sample 3			Sample 18		
Sample 4			Sample 19		
Sample 5			Sample 20		
Sample 6			Sample 21		
Sample 7			Sample 22		
Sample 8			Sample 23		
Sample 9			Sample 24		
Sample 10			Sample 25		
Sample 11			Sample 26		
Sample 12			Sample 27		
Sample 13			Sample 28		
Sample 14			Sample 29		
Sample 15			Sample 30		

4. Draw histograms showing the frequency and ***relative frequency*** for each proportion of the samples representing Color A.

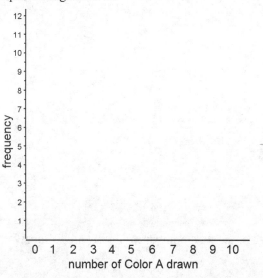
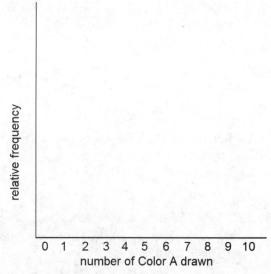

5. a. Determine the mean number of objects of Color A for all samples.

   b. How is this mean related to the histogram you drew?

Algebra II

Module 10: Investigation 3
Sampling from a Population

6. The previous exercises gave us evidence we can use to complete our original task.
   a. What conclusion(s) are you confident making based on the histogram?

   b. Suppose there are 100 total objects in the container.
      (i) Provide a range of possible values that you are confident contains the true number of Color A objects.

      (ii) Repeat if there are 55 total objects or 82 total objects.

7. Suppose you sample **20** objects from the container and that you get each of the following results. Give a ranking from 1 (very normal) to 4 (very unusual) for how unusual you would find the result.

Result	(very common)	(somewhat common)	(a little unusual)	(very unusual)
a. 2 objects are Color A	1	2	3	4
b. 11 objects are Color A	1	2	3	4
c. 13 objects are Color A	1	2	3	4
d. 9 objects are Color A	1	2	3	4
e. 4 objects are Color A	1	2	3	4
f. 19 objects are Color A	1	2	3	4

8. Let's call the container from the beginning of this investigation **Container 1**. Imagine that your teacher has another container with objects of Color A and Color B.
   a. Suppose you sample 20 objects at random and 16 of them are Color A. Which of the following conclusions do you think is more likely? Defend your answer.
      (i) The new container has a similar proportion of Color A in the population compared to Container 1.
      (ii) The new container has a very different proportion of Color A in the population compared to the original container.

b. Repeat part (a), but this time suppose that 8 of the 20 objects are Color A.

One important assumption in statistics is that samples drawn from a population are proportionally similar to the population (with only small variations). Suppose we randomly select samples from a population of people who like ice cream, and that 50% of the population of ice cream eaters prefer vanilla ice cream and 50% prefer chocolate. Then we expect that about 50% of each sample should prefer vanilla, but we can get some variation in these results.

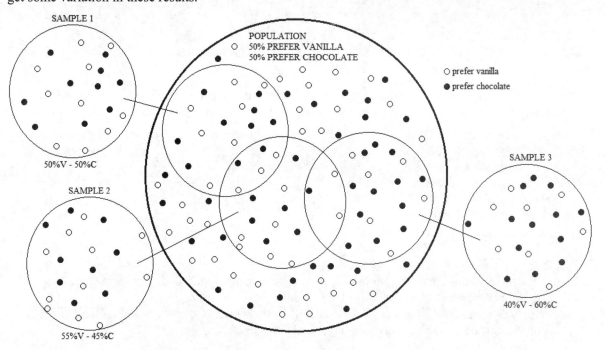

In order to make sure that our samples are good representations of the population, it's very important that the samples be selected *randomly*.

Random Sample

A sample is a ***random sample*** if
    a) all members of the population have an equal chance of being included and
    b) the selections are independent of one another.

If a sample is not generated randomly then there is a good chance that it could be ***biased***, meaning that not all of the members in the population are equally-likely to be included. Biased samples are often poor representations of the population.

Algebra II

Module 10: Investigation 3
Sampling from a Population

In the example of tracking the number of transmission problems in Ford F-150 trucks, a *random sample* could be chosen by listing all of the vehicle identification numbers (VIN) for all 2014 Ford F-150s manufactured in numerical order, and then using a random number generator to select 300 trucks from the list.

A non-random sample (or *biased sample*) might involve selecting 300 trucks that were all manufactured in the same week. The possibility exists that the condition of factory equipment, new employees, faulty parts, etc. could impact the results, and those conditions might not have existed during the entire year. Any conclusions we draw would only apply to trucks produced in a narrow time frame and not to all of the 2014 Ford F-150 trucks manufactured throughout the year.

9. Consider the following scenarios. For each,
    (i) describe how a random sample could be generated,
    (ii) describe a way to generate a biased sample, and
    (iii) explain a possible problem if the biased sample is used instead of the random sample.

   a. Your math teacher is interested in predicting how many of his/her 150 students would pass a college entrance exam.

   b. Your school's principal is interested in getting feedback from students about how many of them want a lunchtime tutoring program.

c. A factory manager is concerned about the quality of batteries produced at one of his factories, specifically the percentage of batteries that won't hold a charge.

10. Do you think sample size has an impact on how closely a sample might represent the population from which it's taken? Explain.

Algebra II

Module 10: Investigation 4
Performing Simulations

When elections are held, people often try to predict winners before all of the votes are counted by surveying voters. Their conclusions are based on probability. In real-life situations, polling and surveying is a lot like drawing objects from a container as we did in Investigation 3.

1. Explain how a poll asking voters which candidate they will vote for is similar to the activity from Investigation 3.

Now consider the following scenario. Your school is holding an election for student body president. Two students (Selene and Tyrese) are running. Furthermore, suppose that we can do the impossible – we know for a fact that ***exactly*** 50% of student voters support Selene and 50% support Tyrese.

2. Under this assumption, imagine conducting a random sample of 40 student voters where you ask "Who will you vote for?" What results should we expect?

Next, imagine selecting a single student voter at random. There is a 50% chance that the person will select either one of the candidates. In this way, randomly choosing a student and recording their response is a lot like flipping a fair coin and designating heads for Selene and tails for Tyrese.

In fact, we can ***simulate*** the process of polling student voters by flipping a coin repeatedly.

> **Simulation**
>
> A ***simulation*** duplicates a real-world event using less expensive and easily accessible tools. The results of the simulation can give researchers information about what to expect when the real event occurs.
>
> > ***Ex:*** Traffic engineers are interested in the impact of widening a certain freeway. They use sophisticated computer programs to simulate the effect of adding more lanes. The data they gather can help determine if the benefits are great enough to justify the millions of dollars in construction costs.

Simulations allow us to explore what might happen in the real world based on a set of assumptions. In this case, we are assuming a 50%-50% split in student voters. To simulate asking 40 people who they will vote for, we can flip a coin 40 times. Or, better yet, we can let a computer do the work for us!

3. a. As a class, simulate asking 40 student voters who they will vote for and tallying the number of Selene supporters.

Algebra II

Module 10: Investigation 4
Performing Simulations

b. Simulate polling 40 student voters three more times, and then discuss how your results compare to the 50%-50% split in the population of student voters.

4. Now, let's see what happens if we were to repeatedly poll 40 random student voters and ask them which candidate they support.
   a. Simulate the following process.
      (i) Choose a random sample of 40 student voters.
      (ii) Ask each of them who they will vote for and tally up the number of students who will vote for Selene.
      (iii) Repeat parts (i) and (ii) *500 times*.
   b. Draw a histogram to represent the results of your simulation.

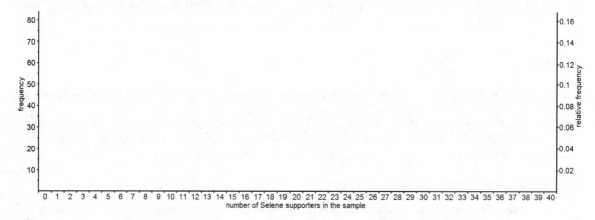

   c. Record the average number of Selene supporters per sample for 500 runs of the simulation and represent this value on your histogram. In addition, record the standard deviation.

5. With the simulation results from Exercise #4 in mind, suppose you randomly sample 40 real student voters (*still under the assumption that 50% of the population support each candidate*). Give a ranking of 1 (very common), 2 (somewhat common), 3 (a little unusual), or 4 (very unusual) for how unusual you would find each of the following results.

	Result	(very common)	(somewhat common)	(a little unusual)	(very unusual)
i.	18 voters support Selene	1	2	3	4
ii.	28 voters support Selene	1	2	3	4
iii.	7 voters support Selene	1	2	3	4
iv.	30 voters support Selene	1	2	3	4
v.	21 voters support Selene	1	2	3	4
vi.	16 voters support Selene	1	2	3	4

Algebra II

Module 10: Investigation 4
Performing Simulations

So far we have assumed that the students are split 50%-50% in their support of Selene and Tyrese. What if the split is different?

6. a. Simulate the process in Exercise #4 again using the assumption that 40% of students support Selene and 60% support Tyrese. Draw a histogram to represent the results of your simulation.

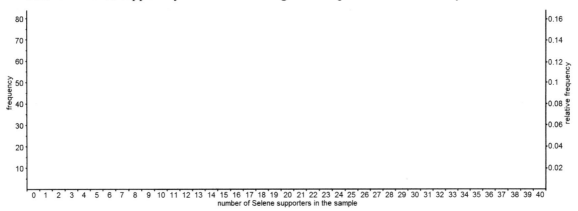

  b. Record the average number of Selene supporters per sample for 500 runs of the simulation and represent this value on your histogram. In addition, record the standard deviation.

  c. Assume that 40% of student voters support Selene and suppose you randomly sample 40 real student voters. Give a ranking of 1 (very common), 2 (somewhat common), 3 (a little unusual), or 4 (very unusual) for how unusual you would find each of the following results.

	Result	(very common)	(somewhat common)	(a little unusual)	(very unusual)
i.	18 voters support Selene	1	2	3	4
ii.	28 voters support Selene	1	2	3	4
iii.	7 voters support Selene	1	2	3	4
iv.	30 voters support Selene	1	2	3	4
v.	21 voters support Selene	1	2	3	4
vi.	16 voters support Selene	1	2	3	4

7. a. Simulate the process in Exercise #4 again using the assumption that 65% of students support Selene and 35% support Tyrese. Draw a histogram to represent the results of your simulation.

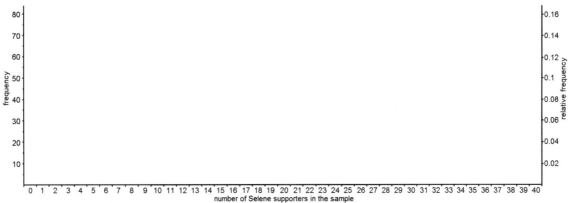

Module 10: Investigation 4
© 2014 Carlson, O'Bryan & Joyner

Algebra II   **Module 10: Investigation 4**
                                                                                                Performing Simulations

b. Record the average number of Selene supporters per sample for 500 runs of the simulation and represent this value on your histogram. In addition, record the standard deviation.

c. Assume that 65% of student voters support Selene and suppose you randomly sample 40 real student voters. Give a ranking of 1 (very common), 2 (somewhat common), 3 (a little unusual), or 4 (very unusual) for how unusual you would find each of the following results.

	Result	(very common)	(somewhat common)	(a little unusual)	(very unusual)
i.	18 voters support Selene	1	2	3	4
ii.	28 voters support Selene	1	2	3	4
iii.	7 voters support Selene	1	2	3	4
iv.	30 voters support Selene	1	2	3	4
v.	21 voters support Selene	1	2	3	4
vi.	16 voters support Selene	1	2	3	4

8. As a class, simulate the results of repeatedly sampling student voters under different assumptions about the population (the percentages of student voters who support Selene). Then answer the following questions.
    a. Discuss any similarities you find in the distribution of sample results.

    b. How do the different assumptions impact whether or not it would be unusual to get a sample containing 20 student voters supporting each candidate?

9. In most real-life applications of statistics, researchers don't have the money or resources to conduct repeated samples. Usually they have to carefully plan how they will collect one really good sample. Based on your work in this investigation, explain why one well-chosen sample can still help us draw conclusions about the population.

Algebra II

Module 10: Investigation 5
Hypothesis Testing

In the previous investigation we explored what might happen when samples are taken from populations with different characteristics. For example, if we surveyed 40 student voters from a population where 50% support Selene and 50% support Tyrese, and if we repeat the process 500 times, we might get the following result.

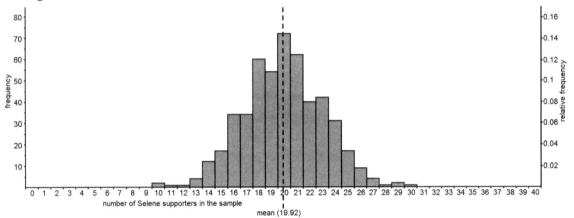

This histogram was generated using the assumption that there is a 50%-50% split in the population's support of the two candidates. In a real situation we wouldn't know if this assumption is true or false.

1. Suppose you survey 40 random student voters and ask them if they will vote for Selene.
   a. What outcomes would you expect if our assumption is true?

   b. What outcomes would make you believe the assumption is false? Explain your thinking.

   c. Now, imagine that you go out and conduct a random sample of 40 student voters. If you find that 22 of those students support Selene, does that suggest she will win the election? Explain.

   d. Repeat part (c) if 25 of the 40 students surveyed say they will vote for Selene.

   e. Repeat part (c) if 29 of the 40 students surveyed say they will vote for Selene.

Algebra II

Module 10: Investigation 5
Hypothesis Testing

2. Suppose a soda company claims that 3 out of 5 soda drinkers prefer its soda over the best-selling soda brand in a blind taste-test. You work for the best-selling soda brand and are asked to explore whether or not this claim is true.

   a. You will go out and conduct a random sample of soda drinkers. Explain how you will determine ahead of time what results you might expect.

   b. If you randomly sample 40 soda drinkers, what results would make you doubt your competitor's claim?

   c. Suppose 17 of the 40 soda drinkers in your sample prefer your competitor's brand. What does this suggest?

   d. Repeat part (c) if 26 of the 40 soda drinkers in your sample prefer your competitor's brand.

Algebra II

Module 10: Investigation 5
Hypothesis Testing

Exercises #1 and #2 are good examples of how statisticians use the results of samples. We know that a random sample is, for the most part, a proportional representation of the population as a whole, but that there is some amount of variation. (*See the diagram in Investigation #3.*)

When statisticians explore a population, they begin by making an assumption. This could be based on initial research about a population, advice from experts, or simply something that they want to test. Using the assumption, they study what might happen when samples are generated from that population. After conducting a random sample, they compare the sample to the expectations and then must decide on a conclusion.

If the sample results represent a rare occurrence from the assumed population, then they will conclude that the original assumption is probably incorrect. If the sample results are a common result from the assumed population, then they will accept the assumption as possible (and maybe even likely). These conclusions are based on the belief that **a sample should reflect at least a somewhat common result from the population it comes from.**

So the question statisticians try to answer is, "If this sample is at least a somewhat common outcome for the population as a whole, are my population assumptions reasonable?" The formal process for this is called ***hypothesis testing***.

Hypothesis Testing

Hypothesis testing is the process of
1) making an assumption about a population (we call this the *hypothesis*),
2) looking at what this assumption would mean for samples drawn from the population,
3) conducting a real sample, and
4) determining if the sample looks like it supports the hypothesis.

If the sample represents a common result based on our simulation, then we conclude that the hypothesis is reasonable. If we think the sample is pretty unusual under our assumptions, then we usually reason that it's more likely to have come from a population that differs from our assumption.

One useful way of thinking about hypothesis testing is to compare it to the way the U.S. court system works and imagine you are the member of a jury. When a defendant is brought to trial, your initial assumption is supposed to be "the person is innocent" (hence the phrase "Innocent until proven guilty"). It is the prosecutor's job to convince you "beyond a reasonable doubt" that the defendant is guilty.

So the process begins with an assumption ("the defendant is innocent"). You then collect information (the evidence and testimony of the trial) and compare it to the original assumption. How likely do you think it is that the defendant is innocent if the testimony and evidence are true? If you think it's pretty unlikely ("beyond a reasonable doubt"), then you reject the hypothesis and choose a "guilty" verdict. If you don't think the evidence is strong enough, you do not reject the hypothesis and vote "not guilty."

Note: Just like it's not really possible to prove innocence in court, we don't prove the original assumption is true. We just decide that we don't have enough evidence to say the hypothesis is false.

Algebra II

Module 10: Investigation 5
Hypothesis Testing

In Exercises #3-6, you are given a situation and the results of a sample from the population of interest. For each, do the following.
   a) Decide what assumption to test about the population.
   b) Run a simulation using the hypothesis.
   c) Determine if the sample would be fairly common or unusual if the hypothesis is true.
   d) State your conclusion (i.e., decide whether or not to reject the original assumption and explain what this means in your own words).

3. Student Council is planning a school carnival and wants to determine how many people to expect. Last year, 32% of the students at the school attended and they are interested to find out if they should plan for similar numbers this year. They randomly sample 38 students at the school (14 say they will attend).

4. Last month park rangers noticed that a bacterial infection seemed to be impacting the local deer population. Evidence suggested that about 45% of the deer in the park were infected. The rangers want to see if conditions have improved this month, so they randomly select 30 deer from the park and test them for the bacteria. (7 of the 30 deer tested positive for the bacteria).

Algebra II

Module 10: Investigation 5
Hypothesis Testing

5. Two candidates are running for office and you want to find out who is likely to win. You randomly poll 35 voters and find that 24 support Candidate A.

6. A soft drink company wants to test a new soda formula to see if customers like it better than the old formula. They randomly sample 40 customers (16 of the customers say they prefer the new formula).

Usually when statisticians reject the initial assumption the work is just beginning. For example, in Exercise #4 you probably rejected the assumption that 45% of the deer are still infected with the bacteria. But what should this mean to the Park Rangers?

7. Give at least two different reasons that could account for the reduced percentage of infected deer.

Algebra II

Module 10: Investigation 6
Cumulative Probability

In the previous investigation we explored what might happen when samples are taken from populations with different characteristics. Specifically, we explored what would happen if we sampled 40 student voters from a school with different assumptions about the proportion of those students who supported two different candidates.

The graphs that follow show two sets of simulation results based on assumptions of i) a 50%-50% split in student support in the population and ii) a population where 30% of student voters support Selene. In each case we simulate taking samples of 40 students and repeat the process 500 times.

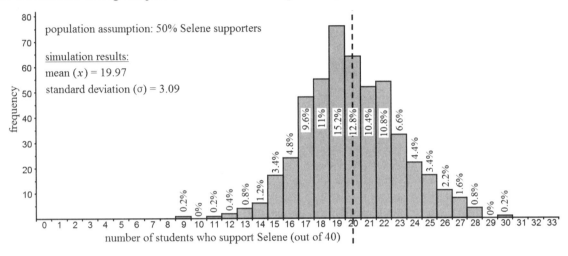

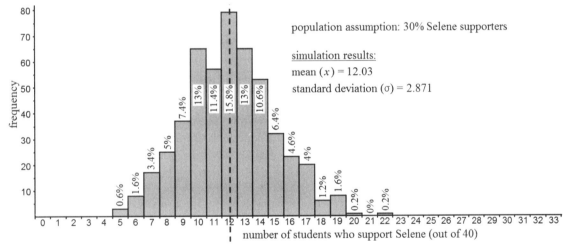

Important! *Remember that it's usually impossible to know the true proportions for an entire population. What these graphs represent are the types of outcomes we can expect **assuming** certain proportions of students who support Selene compared to those who support Tyrese.*

1. Suppose you have a student population evenly split in their support of Selene and Tyrese.
   a. Assume you survey a sample of 40 students and find that 23 of them supported Selene.
      i. Based on our simulation results, about what percent of samples from this population will have 23 or fewer students who support Selene?

      ii. Based on our simulation results, about what percent of samples from this population will have more than 23 students who support Selene?

Algebra II

Module 10: Investigation 6
Cumulative Probability

    b. Find the approximate "middle 60%" of sample results. That is, find an interval <u>centered around the mean</u> that contains about 60% of all sample results. Then explain the meaning of this interval in your own words. (*Note: Round the mean to the nearest whole number first.*)

    c. How far above and below the mean does your interval extend? How far is this when you measure it in *number of standard deviations*?

2. Suppose you have a student population split 30% in favor of Selene and 70% in favor of Tyrese.
    a. Assume you survey a sample of 40 students and find that 13 of them supported Selene.
       i. Based on our simulation results, about what percent of samples from this population will have 13 or fewer students who support Selene?

       ii. Based on our simulation results, about what percent of samples from this population will have more than 13 students who support Selene?

    b. Find the approximate "middle 95%" of sample results. That is, find an interval centered around the mean that contains about 95% of all sample results. Then explain the meaning of this interval in your own words. (*Note: Round the mean to the nearest whole number first.*)

    c. How far above and below the mean does your interval extend? How far is this when you measure it in *number of standard deviations*?

Algebra II

Module 10: Investigation 6
Cumulative Probability

The ideas we explored in Exercises #1 and #2 can help us be more specific in testing hypotheses. We can create an exact cutoff for what we consider unusual outcomes given a certain population assumption. Suppose you are curious whether or not Selene will win the election and you therefore want to test the hypothesis "Selene and Tyrese have equal support among student voters." [*Remember that we test this hypothesis because if it seems reasonable then the election may be too close to call ahead of time. If we reject this hypothesis then we would conclude that one candidate has a significant advantage and will likely win.*]

Before taking a sample and then deciding if the result would be common or unusual if the population were evenly split, let's create a specific rule. **We will reject the hypothesis if the actual sample result is not within the middle 90% of samples from the simulation.** [*Note: This is only one possible rule we could make. Statisticians decide on the rule they want to use in advance of doing their data collection.*]

3. Consider the hypothesis "Selene and Tyrese have equal support among student voters."
   a. Identify the approximate "middle 90%" interval for sample results based on this hypothesis.

   b. If you sample 40 student voters and 26 support Selene, will you accept or reject the hypothesis? What does this mean?

   c. If you sample 40 student voters and 23 support Selene, will you accept or reject the hypothesis? What does this mean?

   d. If you sample 40 student voters and 15 support Selene, will you accept or reject the hypothesis?

Algebra II

Module 10: Investigation 6
Cumulative Probability

Let's return to some contexts we explored in the previous investigation and apply our rule for accepting or rejecting a hypothesis. <u>We will reject the hypothesis if the actual sample result is not within the middle 90% of samples from the simulation.</u>

4. Suppose a soda company claims that 3 out of 5 soda drinkers prefer its soda (Brand A) over the best-selling soda brand (Brand B) in a blind taste-test. You work for Brand B and are asked to explore whether or not this claim is true.

   a. Run a simulation to see what types of sample results to expect if you survey groups of 40 soda drinkers and ask them if they prefer Brand A assuming the claim is true. Record the mean and standard deviation for your simulation results.

   b. Find the approximate "middle 90%" of sample results. (*Round your mean to the nearest whole number first.*)

   c. How far above and below the mean does your interval extend? How far is this when you measure it in *number of standard deviations*?

   d. If you went out and actually sampled 40 soda drinkers and found that 21 prefer Brand A, what should you conclude based on our rule?

5. Student Council is planning a school carnival and wants to determine how many people to expect. Last year, 28% of students at the school attended and they are interested to find out if they should plan for similar numbers this year.

   a. Run a simulation to see what types of sample results to expect if you survey groups of 32 students and ask them if they will attend the carnival this year. Record the mean and standard deviation for your simulation results.

   b. Find the approximate "middle 90%" of sample results. (*Round your mean to the nearest whole number first.*)

   c. If you went out and actually sampled 32 students and found that 14 plan to attend the carnival, what should you conclude based on our rule?

Algebra II

Module 10: Investigation 6
Cumulative Probability

6. Last month park rangers noticed that a bacterial infection seemed to be impacting the local deer population. Evidence suggested that about 47% of the deer in the park were infected. The rangers want to see if conditions have changed this month, so they randomly select 30 deer from the park and test them for the bacteria.
   a. Run a simulation to see what types of sample results to expect if you test groups of 30 deer and see if they are infected. Record the mean and standard deviation for your simulation results.

   b. Find the approximate "middle 90%" of sample results. (*Round your mean to the nearest whole number first.*)

   c. If you went out and actually sampled 30 deer and found that 8 are infected, what should you conclude based on our rule?

7. When we use our rule (*we will reject the hypothesis if the actual sample result is not within the middle 90% of samples from the simulation*), it's possible to make a mistake.
   a. How often will we make the mistake of incorrectly rejecting a hypothesis about our population proportion when we conduct a sample?

   b. Why don't we just change the rule so that the mistake in part (a) won't occur? For example, why don't we use the following rule instead? *We will reject the hypothesis if the actual sample result is outside of 100% of samples in the simulation.*

Algebra II

Module 10: Investigation 7
Normal Distribution

In the last several investigations we have focused on generating histograms based on simulating samples from different populations where every member of the population had one of two possible characteristics. Depending on what we are measuring, however, we may need to draw the distribution differently.

For example, suppose you are interested in studying the heights of all female students at your school. Since height is a continuous measurement, if you measure the shortest girl at your school and the tallest girl, the remaining female students could have *any* height in between. This makes creating categories for the histogram difficult and potentially misleading.

For this reason, relative frequency graphs for continuous data (like measuring heights, weights, time to complete tasks, etc.) look like "smoothed out" versions of the histograms we've been studying. For example, the following graph shows the probability distribution for the weight of grapefruit (in ounces) from trees in a certain orchard.

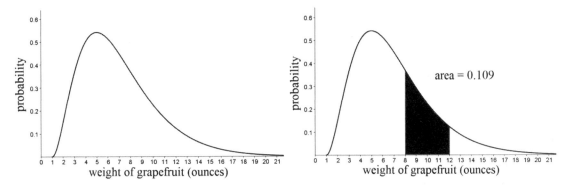

The information isn't broken into natural categories any longer since *any* weight in the domain is possible (2.589 ounces, 10.113 ounces, etc.), so we calculate percentages of data between different values by determining the area of the corresponding region as in the graph on the right. We see that 10.9% of the grapefruit collected from this orchard weigh between 8 and 12 ounces.

The following graph shows a comparison between the histograms we've been working with and the probability distributions for continuous variables.

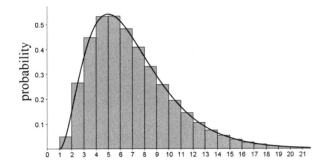

Normal Distribution

Some characteristics of a population have a special kind of distribution called a ***normal distribution***. As you can guess from its name, this type of distribution is fairly common. It's also very useful and very important to the study of statistics.

Algebra II
Module 10: Investigation 7
Normal Distribution

Data is said to have a normal distribution when most of the data (about 68%) falls within one standard deviation above and below the mean, and almost all of the data (about 95%) falls within two standard deviations above and below the mean. Thus only 1 out of every 20 members of the population would have a measurement more extreme than two standard deviations away from the mean.

Normal Distribution

When the measurement of a population characteristic is **_normally distributed_** with a mean of $\bar{x}$ and a standard deviation of σ, it has the following features.

a) The mean is also the median (so 50% of the measurements are larger than the mean and 50% are smaller than the mean).

b) 68% of the measurements are within one standard deviation of the mean (between $\bar{x} - \sigma$ and $\bar{x} + \sigma$).

c) 95% of the measurements are within two standard deviations of the mean (between $\bar{x} - 2\sigma$ and $\bar{x} + 2\sigma$).

d) 99.7% of the measurements are within three standard deviations of the mean (between $\bar{x} - 3\sigma$ and $\bar{x} + 3\sigma$).

e) 0.03% of the measurements are outside of three standard deviations from the mean.

These facts are sometimes referred to as **_the 68-95-99.7 rule_**. These features allow us to completely describe a normal distribution using only the mean and standard deviation.

A visual representation of the normal distribution is shown. Since the distribution is bell-shaped, we often refer to this as the **bell curve**.

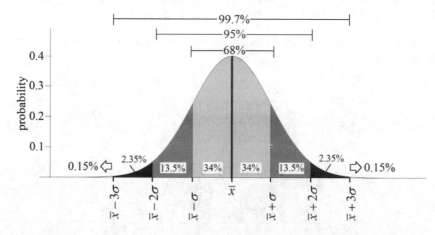

1. The height of adults in a large population generally follows a normal distribution. The average height for an adult man in the U.S. is about 69.5 inches with a standard deviation of about 2.75 inches.
   (*For reference: 60 inches = 5 feet, 72 inches = 6 feet, 84 inches = 7 feet.*)
   a. Between what two heights are the "middle 68%" of adult men in the U.S.?

b. Suppose a man is 75 inches tall (6' 3").
   i) How many standard deviations above the mean is his height?

   ii) About what percent of adult men in the U.S. are taller than him? Shorter?

c. How "rare" is it for an adult man to be shorter than 61.25 inches tall (5' 1.25")? Explain.

d. Fill in the boxes in the graph based on the information for this context.

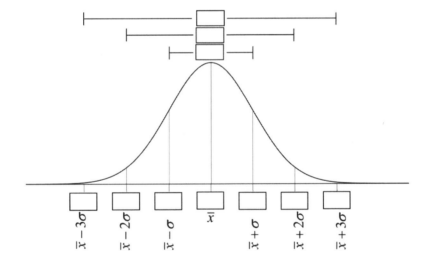

2. Suppose high school seniors from across the country competed in a math competition where they were asked to solve 25 challenging problems. Students could earn between 0 and 8 points per question depending on their answer and explanation for a total score between 0 and 200. The scores were normally distributed with $\bar{x}=112$ and $\sigma=23.5$.

   a. Fill in the boxes in the graph based on the information for this context.

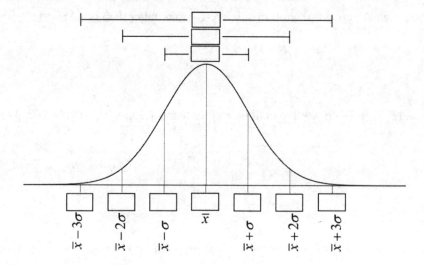

   b. Suppose 15,400 students took the test.
   i) How many students scored between 112 and 135.5?

   ii) How many students scored between 65 and 112?

   iii) How many students scored better than 159?

   iv) How many students scored below 135.5?

Algebra II

Module 10: Investigation 7
Normal Distribution

3. Scores on the IQ test are normalized so that the average score in the population is 100 and the standard deviation is 15. [*This means that they take the "raw" score based on your responses, compare it to other raw scores, and give you an IQ score that shows you rank relative to the rest of the population.*]

   a. Fill in the boxes in the graph based on the information for this context.

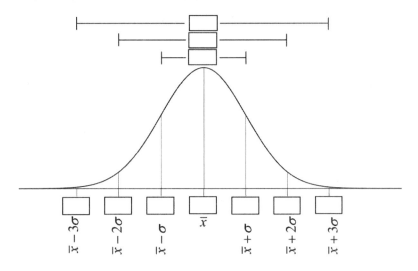

   b. What percent of the population has an IQ between 85 and 130?

   c. What percent of the population has an IQ score below 115?

   d. Mensa is a famous non-profit organization where only people with an IQ score in the top 2% of the population are eligible to apply for membership. Approximately what IQ score is needed to be able to apply for membership in Mensa?

When discussing normally distributed data, we often calculate the **percentile** for a given measurement. In fact, you've already done this in some of the previous exercises.

Algebra II

Module 10: Investigation 7
Normal Distribution

Percentile

The ***percentile*** for a measurement is the percent of the population (or possibly observations in the sample) that are less than that measurement.

Ex1: Trenton and Sherri had a baby daughter. At the child's 6-month checkup the doctor measured her height at 27.5 inches and told the parents that this placed her height in the 90th percentile. [The baby is taller than 90% of 6-month old girls.]

Ex2: Alex took the SAT. His score report said that his score on the mathematics portion was in the 47th percentile. [His score was higher than 47% of people taking the math portion of the SAT.]

---

4. A fitness club was interested in tracking how much time people spend running on treadmills. Over the course of two months they kept a log of how much time each runner spent on a treadmill. The results appeared to be normally distributed with $\bar{x} = 31.4$ minutes and $\sigma = 5.8$ minutes.
   a. What is the percentile for 43 minutes spent on the treadmill? What does this mean?

   b. What is the percentile for 19.8 minutes spent on the treadmill? What does this mean?

   c. How many minutes spent on the treadmill corresponds to the 84th percentile?

Algebra II

Module 10: Investigation 8
Comparing Normal Distributions

In Investigation #1 we learned about the mean and standard deviation for a set of measurements. When the standard deviation is small, the measurements are clustered more closely around the mean and when the standard deviation is large the data is more spread out from the mean. This is true for normal distributions as well.

1. The following graph shows normal distributions for the same measurements taken on three different populations (labeled A, B, and C).

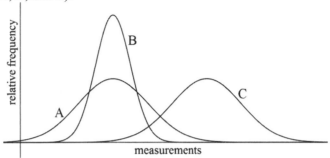

   Compare the means and standard deviations for the three distributions and write a brief explanation of what the similarities/differences tell us about the populations measured.

In Investigation #7 we learned how to determine the percent of population measurements that fall between two values if we know the mean and standard deviation. However, it can be very difficult to find this information since it would require measuring every member of the population. For example, if we want to know the mean and standard deviation for the heights of 16-year-old girls in the U.S., we would need to collect the measurements for every 16-year-old girl in the country.

2. What could we do to estimate the mean and standard deviation for the heights of 16-year-old girls in the U.S. without measuring every member of the population? Discuss the pros and cons of your strategy.

Algebra II

Module 10: Investigation 8
Comparing Normal Distributions

Using a TI-83/84 Calculator to Determine Means and Standard Deviations

Suppose we have the set of measurements 7, 14, 4, 12, 10, 9, 9, and 8 which represent the weight loss in one month (in pounds) for eight people on a certain diet plan.

- Press STAT and select **Edit...** to access the data lists in your calculator.
- Enter the numbers from the data set.
- Press STAT again and select the **CALC** menu. Then choose **1-Var Stats**.
- Press ENTER, and then ENTER again when the command appears on your home screen.

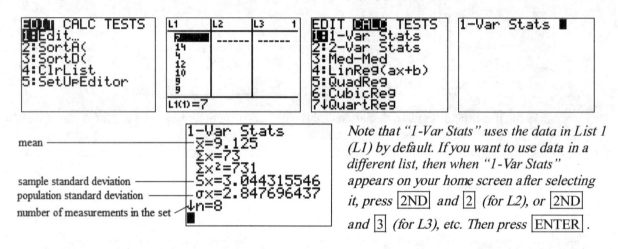

Note that "1-Var Stats" uses the data in List 1 (L1) by default. If you want to use data in a different list, then when "1-Var Stats" appears on your home screen after selecting it, press 2ND and 2 (for L2), or 2ND and 3 (for L3), etc. Then press ENTER.

You'll notice that the calculator gives you two different values for the standard deviation. The population standard deviation is the measure of variability for the data *assuming that you measured the entire population*. So if the eight weights represent the only population you're interested in, you would use this value. The calculations shown in Investigation #1 are for the population standard deviation.

However, if the eight measurements are a *sample* from a larger population of interest then you should use the sample standard deviation. The difference in the calculations is that when you determine the variance (average of the squared deviations from the mean) you divide by $n - 1$ instead of n. The square root of this number is the standard deviation. This modified standard deviation ends up being a better estimate for the standard deviation of the population.

Suppose we work for a company with a new diet program involving special menus. We want to show that our program (*Algebraic Health Diet*) is better than traditional weight loss methods of either i) cutting calories or ii) exercising. We get 90 volunteers and randomly place them in three groups of 30 people.

- Group A: These 30 people will follow the *Algebraic Health* meal plan for 45 days.
- Group B: These 30 people will cut their calorie intake by 500 calories per day for 45 days.
- Group C: These 30 people will exercise for 30 minutes each day for 45 days.

Note: We will assume that the weight loss for each group is normally distributed.

Algebra II

Module 10: Investigation 8
Comparing Normal Distributions

3. After 45 days we record each person's total weight loss. The following results are for people in Group A (*Algebraic Health* diet).

Weight loss after 45 days for people in Group A									
11.2	11.6	10.8	10.2	10.8	9.6	13.2	14.2	13.6	12.2
9.9	14.3	15.1	8.5	10.1	15.6	16.4	12.7	10.0	12.7
13.0	12.5	7.1	14.3	9.3	14.1	12.4	12.6	14.4	14.4

  a. Determine the mean and sample standard deviation for the data. [*These values will be our estimates for the mean and standard deviation for the weight loss of anyone who would use our program.*]

  b. Fill in the following graph showing the estimated distribution of weight loss for people on the *Algebraic Health* diet.

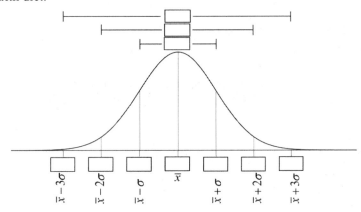

4. The following weight loss totals are for people in Group B (reducing calorie intake).

Weight loss after 45 days for people in Group B									
12.0	12.1	10.0	16.4	12.5	16.1	16.4	15.9	15.0	6.5
9.3	9.1	17.6	14.2	8.0	10.7	12.5	7.1	12.4	10.1
12.3	8.0	11.8	7.7	10.1	17.6	11.5	12.6	15.0	11.0

  a. Determine the mean and sample standard deviation for the data. [*These values will be our estimates for the mean and standard deviation for the weight loss of anyone cutting 500 calories per day.*]

  b. Fill in the following graph showing the estimated distribution of weight loss for people who cut 500 calories per day from their diet.

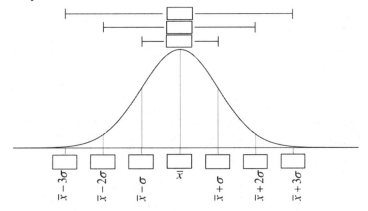

5. Compare your results in Exercises #3 and #4. Do you think it would be right to advertise the *Algebraic Health Diet* as "a significant improvement over dieting alone" for losing weight? Explain.

6. For Group C, the mean weight loss is 15.9 pounds and the sample standard deviation is 4.1 pounds.
   a. Fill in the following graph showing the estimated distribution of weight loss for people exercising 30 minutes per day.

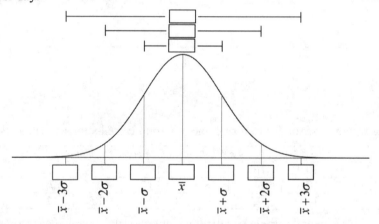

   b. To the best of your ability, sketch all three normal distributions (*Algebraic Health*, dieting alone, and exercise alone) together.

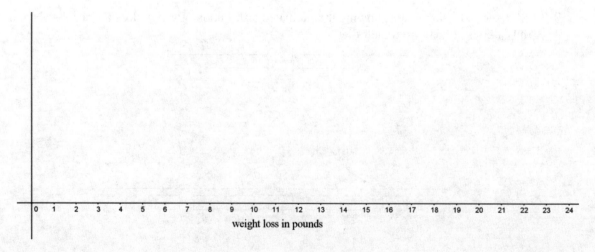

c. Is exercise alone significantly better or worse than the other options? Justify your conclusion.

We conclude this investigation with an extended comment. Normal distributions (or at least approximately normal distributions) are both very common in the real world and a very important foundation for a lot of statistical theory. However, other distribution types do exist so we should be careful not to assume that every set of measurements on a population follow a normal distribution. Some examples of alternative distributions are given below.

Uniform Distribution

In a uniform distribution, if we divide the interval between the smallest and largest values into equal subintervals, each of these will contain the same portion of the total population measurements. The simplest example is distribution we expect if we roll a single six-sided die infinitely many times [1/6 of the outcomes will be 1, 1/6 will be 2, and so on].

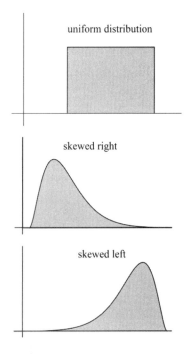

Skewed

If a distribution is skewed it means that it isn't symmetric. **Right skewed** (or positively skewed) means that the tail on the right is longer or thicker than the tail on the left. These distributions tend to happen when we have a restriction on how small a measurement can be but no clear restriction on how large it can be.

For example, if you weighed all U.S. adult men who are between 6' and 6' 2" tall you might see a right skewed distribution because there are many more examples of men who are 50 pounds, 100 pounds, or even 200 pounds heavier than the average weight compared to the number who are 50, 100, or 200 pounds lighter than the average weight. [*Another example that produces a right skewed distribution is to measure household income for U.S. families.*]

Left skewed (or negatively skewed) have a longer or thicker tail on the left side. These are less common than right skewed distributions but can happen if there is a definite restriction on the largest possible measurement but a less clear restriction on the smallest possible measurement.

Algebra II

Module 10: Investigation 9
z-scores

When data follows a normal distribution we learned that we can predict the percentage of measurements that fall within one, two, and three standard deviations of the mean (the 68-95-99.7 rule). However, this alone is not always helpful because most measurements will not fall *exactly* one, two, or three standard deviations from the mean.

1. In Investigation 7 we considered the following context. *Suppose high school seniors from across the country competed in a math competition where they were asked to solve 25 challenging problems. Students could earn between 0 and 8 points per question depending on their answer and explanation for a total score between 0 and 200. The scores were normally distributed with $\bar{x}=112$ and $\sigma=23.5$.*

   a. How many standard deviations from the mean is a score of 127? Round your answer to two decimal places. (*Use a negative number if the measurement is below the mean and a positive number if the measurement is above the mean.*)

   b. How many standard deviations from the mean is a score of 78? Round your answer to two decimal places. (*Use a negative number if the measurement is below the mean and a positive number if the measurement is above the mean.*)

   c. What is the percentile for each of the scores in parts (a) and (b)? If you have difficulty answering this question, then explain why.

Fortunately, normal distributions are so predictable that not only can we predict the percentiles for measurements that fall exactly one, two, or three standard deviations from the mean, but we can accurately predict the percentiles for *any* measurement. The first step is to understand how to **standardize** the measurements in a data set. Standardizing our data involves converting all of our measurements into a number of standard deviations from the mean [as you did in Exercise #1 parts (a) and (b)]. The standardized measurement is called a **z-score**.

2. Determine the z-score for each measurement. Round your answers to two decimal places. (*The z-score is the number of standard deviations from the mean. We use negative numbers if the measurement is below the mean and positive numbers if the measurement is above the mean.*)

   a. $\bar{x}=42$ and $\sigma=9$; find the z-score for a measurement of 45

   b. $\bar{x}=852.8$ and $\sigma=118.2$; find the z-score for a measurement of 700

c. $\bar{x} = 2.965$ and $\sigma = 0.146$; find the z-score for a measurement of 2.911

d. $\bar{x}$ is the mean and σ is the standard deviation; find the z-score for a measurement of x

z-scores

A **z-score**, or **standardized measurement**, is the number of standard deviations from the mean for a given measurement.

If the mean for a population characteristic is $\bar{x}$, the standard deviation is σ, and if we have a measurement of x for a member of the population, the associated z-score is

z-score =

One benefit to standardizing our data is that *every* normal distribution can then be summarized by the same probability distribution graph centered at 0.

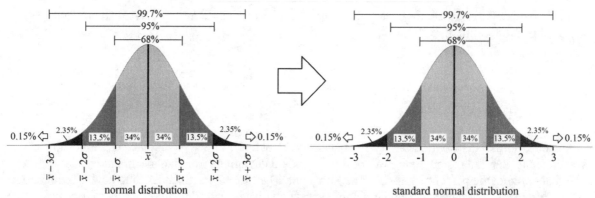

normal distribution   standard normal distribution

The other benefit is that once we know the z-score there are methods for determining the percentile. The traditional method is to use the **z-score table** (also called the **standard normal table**). [*The full table is located at the end of this Module.*]

In Exercise #1 parts (a) and (b) we found that math test scores of 127 and 78 had z-scores of 0.64 and –1.45. To find the associated percentiles, we use the left column and top row of the table to find the entry in the table associated with our z-score. The result is the percentile for that z-score.

z	0.00	0.01	0.02	0.03	0.04	0.05	0.06	0.07	0.08	0.09
...	...	...	...	...	...	...	...	...	...	...
0.5	0.6915	0.6950	0.6985	0.7019	0.7054	0.7088	0.7123	0.7157	0.7190	0.7224
0.6	0.7257	0.7291	0.7324	0.7357	0.7389	0.7422	0.7454	0.7486	0.7517	0.7549
0.7	0.7580	0.7611	0.7642	0.7673	0.7704	0.7734	0.7764	0.7794	0.7823	0.7852
...	...	...	...	...	...	...	...	...	...	...

Algebra II **Module 10: Investigation 9**
z-scores

A score of 127 on the math test (with a z-score of 0.64) has a percentile of 73.89. About 74% of scores on the test were lower than 127. We can also conclude that about 26% of scores were higher.

z	0.00	0.01	0.02	0.03	0.04	0.05	0.06	0.07	0.08	0.09
...	...	...	...	...	...	...	...	...	...	...
−1.5	0.0668	0.0655	0.0643	0.0630	0.0618	0.0606	0.0594	0.0582	0.0571	0.0559
−1.4	0.0808	0.0793	0.0778	0.0764	0.0749	0.0735	0.0721	0.0708	0.0694	0.0681
−1.3	0.0968	0.0951	0.0934	0.0918	0.0901	0.0885	0.0869	0.0853	0.0838	0.0823
...	...	...	...	...	...	...	...	...	...	...

A score of 78 on the math test (with a z-score of −1.45) has a percentile of 7.35. About 7% of scores on the test were lower than 78. We can also conclude that about 93% of scores were higher.

3. Find the percentiles for the measurements in Exercise #2 parts (a) through (c).

We can also use technology to help us determine the percentiles. The following instructions demonstrate how to determine the percentile for a given z-score using a TI-83/84 calculator. [*If you have a different calculator, consult your user's manual for instructions.*]

Press $\boxed{\text{2ND}}$ and then $\boxed{\text{VARS}}$ to get the probability distribution menu. Select the option **normalcdf(**.

normalcdf(requires two pieces of information to get the percentile – a lower boundary and an upper boundary. Since we want all of the values below a certain z-score (such as 0.64), we need a lower boundary of −∞. We can't enter −∞ into our calculator so we will just type in a very large magnitude negative number (like −100,000,000,000), then a comma, and then our z-score. After closing the parentheses, press $\boxed{\text{ENTER}}$.

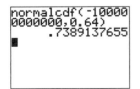

The result matches our answer using the standard normal table (0.7389). About 74% of the students' scores are less than one that is 0.64 standard deviations above the mean.

Algebra II

Module 10: Investigation 9
z-scores

4. Refer to the context in Exercise #1. Find the percentiles for each of the following scores on the math test and interpret your answer.
   a. 60 (out of 200)

   b. 141 (out of 200)

   c. 185 (out of 200)

5. Suppose a student is told that he scored in the 32nd percentile. Determine his score on the test.

6. Repeat Exercise #5 for each of the following percentiles.
   a. 46th percentile
   b. 55th percentile
   c. 93rd percentile

7. What percent of students scored between 100 and 150? If 15,400 students took the test, how many students is this?

Algebra II   **Module 10: Investigation 9**
                                                                                                                                        z-scores

8. Repeat Exercise #7 for each of the following.
   a. between 70 and 140

   b. between 105 and 110

   c. between 130 and 160

9. The height of adults in a large population generally follows a normal distribution. The average height for an adult man in the U.S. is about 69.5 inches with a standard deviation of about 2.75 inches.
   a. Determine the z-score for a man who is 74.1 inches tall. Explain what your answer represents.

   b. Find the percentile for your answer in part (a) and interpret your answer.

   c. What percent of U.S. adult males are between 66 and 68 inches tall?

d. If you randomly select a sample of 250,000 adult men in the U.S., about how many of them will be shorter than 71 inches?

e. If you randomly select a sample of 250,000 adult men in the U.S., about how many of them will be taller than 71 inches?

f. If a man is told that he is in the 60th percentile for height, about how tall is he? What if he is in the 10th percentile?

Module 10: Appendix
z-score (Standard Normal) Table

The following tables show percentiles for z-scores between –3.49 and 3.49. Use the bolded column and row to find the appropriate number of standard deviations from the mean. The entry in the table tells us the percentage (in decimal form) of data in the population less than the given value.

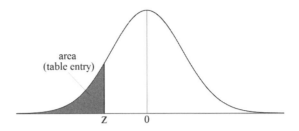

Table 1: negative z-scores

z	0.00	0.01	0.02	0.03	0.04	0.05	0.06	0.07	0.08	0.09
–3.4	0.0003	0.0003	0.0003	0.0003	0.0003	0.0003	0.0003	0.0003	0.0003	0.0002
–3.3	0.0005	0.0005	0.0005	0.0004	0.0004	0.0004	0.0004	0.0004	0.0004	0.0003
–3.2	0.0007	0.0007	0.0006	0.0006	0.0006	0.0006	0.0006	0.0005	0.0005	0.0005
–3.1	0.0010	0.0009	0.0009	0.0009	0.0008	0.0008	0.0008	0.0008	0.0007	0.0007
–3.0	0.0013	0.0013	0.0013	0.0012	0.0012	0.0011	0.0011	0.0011	0.0010	0.0010
–2.9	0.0019	0.0018	0.0018	0.0017	0.0016	0.0016	0.0015	0.0015	0.0014	0.0014
–2.8	0.0026	0.0025	0.0024	0.0023	0.0023	0.0022	0.0021	0.0021	0.0020	0.0019
–2.7	0.0035	0.0034	0.0033	0.0032	0.0031	0.0030	0.0029	0.0028	0.0027	0.0026
–2.6	0.0047	0.0045	0.0044	0.0043	0.0041	0.0040	0.0039	0.0038	0.0037	0.0036
–2.5	0.0062	0.0060	0.0059	0.0057	0.0055	0.0054	0.0052	0.0051	0.0049	0.0048
–2.4	0.0082	0.0080	0.0078	0.0075	0.0073	0.0071	0.0069	0.0068	0.0066	0.0064
–2.3	0.0107	0.0104	0.0102	0.0099	0.0096	0.0094	0.0091	0.0089	0.0087	0.0084
–2.2	0.0139	0.0136	0.0132	0.0129	0.0125	0.0122	0.0119	0.0116	0.0113	0.0110
–2.1	0.0179	0.0174	0.0170	0.0166	0.0162	0.0158	0.0154	0.0150	0.0146	0.0143
–2.0	0.0228	0.0222	0.0217	0.0212	0.0207	0.0202	0.0197	0.0192	0.0188	0.0183
–1.9	0.0287	0.0281	0.0274	0.0268	0.0262	0.0256	0.0250	0.0244	0.0239	0.0233
–1.8	0.0359	0.0351	0.0344	0.0336	0.0329	0.0322	0.0314	0.0307	0.0301	0.0294
–1.7	0.0446	0.0436	0.0427	0.0418	0.0409	0.0401	0.0392	0.0384	0.0375	0.0367
–1.6	0.0548	0.0537	0.0526	0.0516	0.0505	0.0495	0.0485	0.0475	0.0465	0.0455
–1.5	0.0668	0.0655	0.0643	0.0630	0.0618	0.0606	0.0594	0.0582	0.0571	0.0559
–1.4	0.0808	0.0793	0.0778	0.0764	0.0749	0.0735	0.0721	0.0708	0.0694	0.0681
–1.3	0.0968	0.0951	0.0934	0.0918	0.0901	0.0885	0.0869	0.0853	0.0838	0.0823
–1.2	0.1151	0.1131	0.1112	0.1093	0.1075	0.1056	0.1038	0.1020	0.1003	0.0985
–1.1	0.1357	0.1335	0.1314	0.1292	0.1271	0.1251	0.1230	0.1210	0.1190	0.1170
–1.0	0.1587	0.1562	0.1539	0.1515	0.1492	0.1469	0.1446	0.1423	0.1401	0.1379
–0.9	0.1841	0.1814	0.1788	0.1762	0.1736	0.1711	0.1685	0.1660	0.1635	0.1611
–0.8	0.2119	0.2090	0.2061	0.2033	0.2005	0.1977	0.1949	0.1922	0.1894	0.1867
–0.7	0.2420	0.2389	0.2358	0.2327	0.2296	0.2266	0.2236	0.2206	0.2177	0.2148
–0.6	0.2743	0.2709	0.2676	0.2643	0.2611	0.2578	0.2546	0.2514	0.2483	0.2451
–0.5	0.3085	0.3050	0.3015	0.2981	0.2946	0.2912	0.2877	0.2843	0.2810	0.2776
–0.4	0.3446	0.3409	0.3372	0.3336	0.3300	0.3264	0.3228	0.3192	0.3156	0.3121
–0.3	0.3821	0.3783	0.3745	0.3707	0.3669	0.3632	0.3594	0.3557	0.3520	0.3483
–0.2	0.4207	0.4168	0.4129	0.4090	0.4052	0.4013	0.3974	0.3936	0.3897	0.3859
–0.1	0.4602	0.4562	0.4522	0.4483	0.4443	0.4404	0.4364	0.4325	0.4286	0.4247
0.0	0.5000	0.4960	0.4920	0.4880	0.4840	0.4801	0.4761	0.4721	0.4681	0.4641

Module 10: Appendix
z-score (Standard Normal) Table

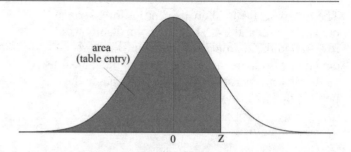

area (table entry)

Table 2: positive z-scores

z	0.00	0.01	0.02	0.03	0.04	0.05	0.06	0.07	0.08	0.09
0.0	0.5000	0.5040	0.5080	0.5120	0.5160	0.5199	0.5239	0.5279	0.5319	0.5359
0.1	0.5398	0.5438	0.5478	0.5517	0.5557	0.5596	0.5636	0.5675	0.5714	0.5753
0.2	0.5793	0.5832	0.5871	0.5910	0.5948	0.5987	0.6026	0.6064	0.6103	0.6141
0.3	0.6179	0.6217	0.6255	0.6293	0.6331	0.6368	0.6406	0.6443	0.6480	0.6517
0.4	0.6554	0.6591	0.6628	0.6664	0.6700	0.6736	0.6772	0.6808	0.6844	0.6879
0.5	0.6915	0.6950	0.6985	0.7019	0.7054	0.7088	0.7123	0.7157	0.7190	0.7224
0.6	0.7257	0.7291	0.7324	0.7357	0.7389	0.7422	0.7454	0.7486	0.7517	0.7549
0.7	0.7580	0.7611	0.7642	0.7673	0.7704	0.7734	0.7764	0.7794	0.7823	0.7852
0.8	0.7881	0.7910	0.7939	0.7967	0.7995	0.8023	0.8051	0.8078	0.8106	0.8133
0.9	0.8159	0.8186	0.8212	0.8238	0.8264	0.8289	0.8315	0.8340	0.8365	0.8389
1.0	0.8413	0.8438	0.8461	0.8485	0.8508	0.8531	0.8554	0.8577	0.8599	0.8621
1.1	0.8643	0.8665	0.8686	0.8708	0.8729	0.8749	0.8770	0.8790	0.8810	0.8830
1.2	0.8849	0.8869	0.8888	0.8907	0.8925	0.8944	0.8962	0.8980	0.8997	0.9015
1.3	0.9032	0.9049	0.9066	0.9082	0.9099	0.9115	0.9131	0.9147	0.9162	0.9177
1.4	0.9192	0.9207	0.9222	0.9236	0.9251	0.9265	0.9279	0.9292	0.9306	0.9319
1.5	0.9332	0.9345	0.9357	0.9370	0.9382	0.9394	0.9406	0.9418	0.9429	0.9441
1.6	0.9452	0.9463	0.9474	0.9484	0.9495	0.9505	0.9515	0.9525	0.9535	0.9545
1.7	0.9554	0.9564	0.9573	0.9582	0.9591	0.9599	0.9608	0.9616	0.9625	0.9633
1.8	0.9641	0.9649	0.9656	0.9664	0.9671	0.9678	0.9686	0.9693	0.9699	0.9706
1.9	0.9713	0.9719	0.9726	0.9732	0.9738	0.9744	0.9750	0.9756	0.9761	0.9767
2.0	0.9772	0.9778	0.9783	0.9788	0.9793	0.9798	0.9803	0.9808	0.9812	0.9817
2.1	0.9821	0.9826	0.9830	0.9834	0.9838	0.9842	0.9846	0.9850	0.9854	0.9857
2.2	0.9861	0.9864	0.9868	0.9871	0.9875	0.9878	0.9881	0.9884	0.9887	0.9890
2.3	0.9893	0.9896	0.9898	0.9901	0.9904	0.9906	0.9909	0.9911	0.9913	0.9916
2.4	0.9918	0.9920	0.9922	0.9925	0.9927	0.9929	0.9931	0.9932	0.9934	0.9936
2.5	0.9938	0.9940	0.9941	0.9943	0.9945	0.9946	0.9948	0.9949	0.9951	0.9952
2.6	0.9953	0.9955	0.9956	0.9957	0.9959	0.9960	0.9961	0.9962	0.9963	0.9964
2.7	0.9965	0.9966	0.9967	0.9968	0.9969	0.9970	0.9971	0.9972	0.9973	0.9974
2.8	0.9974	0.9975	0.9976	0.9977	0.9977	0.9978	0.9979	0.9979	0.9980	0.9981
2.9	0.9981	0.9982	0.9982	0.9983	0.9984	0.9984	0.9985	0.9985	0.9986	0.9986
3.0	0.9987	0.9987	0.9987	0.9988	0.9988	0.9989	0.9989	0.9989	0.9990	0.9990
3.1	0.9990	0.9991	0.9991	0.9991	0.9992	0.9992	0.9992	0.9992	0.9993	0.9993
3.2	0.9993	0.9993	0.9994	0.9994	0.9994	0.9994	0.9994	0.9995	0.9995	0.9995
3.3	0.9995	0.9995	0.9995	0.9996	0.9996	0.9996	0.9996	0.9996	0.9996	0.9997
3.4	0.9997	0.9997	0.9997	0.9997	0.9997	0.9997	0.9997	0.9997	0.9997	0.9998

Algebra II

Module 10: Homework
Univariate Statistics

INVESTIGATION 1: MEAN AND STANDARD DEVIATION

In Exercises #1-4 find the mean of each set of numbers and explain how we can interpret the mean.

1. 5, 9, 0, 1, 21, 15, 2, 13

2. 102, 97, 118, 119, 95, 101

3. −7, −1, 3, −5, −11, −10, 5, 0, −2, 4

4. −13, −14, −15, −10, −20, −5, −8, −1

5. Julian scored 83, 88, 85, and 90 on his last four math quizzes. Calculate and interpret the mean of these scores.

6. Sabina scored 96, 92, 82, 89 and 100 on her last five tests in Physics. Calculate and interpret the mean of these scores.

7. In the 2014 NBA Finals Lebron James played 32.9 minutes in one game, 37.6 minutes in the next game, and 39.8, 37.7, and 41.3 minutes in the final three games. Find and interpret the mean of these values.

8. In the five games of the 2014 NBA Finals Tony Parker scored 19 points, 21 points, 15 points, 19 points, and 16 points. Find and interpret the mean of these values.

9. In a 2012 report on www.businessinsider.com, Henry Blodget wrote that the "average American" consumes 100 pounds of sugar per year. How should we interpret this number?

10. (See Exercise #9.) In the same report Blodget says that the "average American" in 1822 consumed 6.4 pounds pf sugar per year. How should we interpret this number?

11. Why is it useful to calculate a measure of variability (such as the *standard deviation*) for a set of numbers in addition to calculating the mean?

12. Suppose two sets of numbers have the same mean but different standard deviations. What do we know about the numbers in the set with the larger standard deviation?

In Exercises #13-16, calculate the mean of each set of numbers and then follow the given process to find the standard deviation. (*See Investigation 1 for an example of using a table to organize the process.*)
  i. Take each number in the set and subtract the mean. (These values are called the *deviations* or *deviations from the mean*).
  ii. Square each of the deviations. (These are called the *squared deviations.*)
  iii. Find the mean of the squared deviations. (This value is called the *variance* for the data set.)
  iv. Take the square root of the variance. (This value is called the *standard deviation.*)

13. 9, 11, 2, 8, 6, 6

14. 16, 8, 22, 21, 19, 10

15. 5, 9, 0, 1, 21, 15, 2, 13

16. 102, 97, 118, 119, 95, 101

17. You are given four histograms representing sets of 20 numbers each with the same mean (12) but with different standard deviations. Match each standard deviation to the correct histogram.

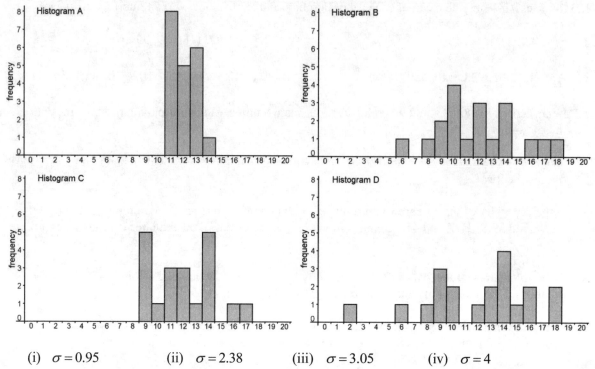

(i) $\sigma = 0.95$ (ii) $\sigma = 2.38$ (iii) $\sigma = 3.05$ (iv) $\sigma = 4$

18. Two quarterbacks (Devon and Travis) from different football teams are comparing their statistics from last season. Devon averaged 218 yards per game passing with a standard deviation of 15 yards. Travis averaged 205 yards per game passing with a standard deviation of 27 yards. Compare the performance of these players.

19. Three students (Alicia, Gina, and Matt) took a series of timed multiplication tests. You are analyzing their scores and find the following. Alicia had an average of 40 correct with a standard deviation of 6, Gina had an average of 38 correct with a standard deviation of 3, and Matt had an average of 38 correct with a standard deviation of 6. Compare the performance of these students.

20. Jackson has scored 74, 80, 78, and 83 on his first four tests. What must he score on the next test so that his average score is at least 80?

21. A set of numbers includes 15, 19, 10, 26, 32, 25, 28 and one additional number that is unknown. If the mean of the set is 21.75, what is the missing number?

Algebra II

Module 10: Homework
Univariate Statistics

INVESTIGATION 2: MEAN AND EXPECTED VALUE

22. A scratcher lottery game from Wisconsin was called *Pharaoh's Gold* with tickets costing $2 each. The probability of winning different amounts in the game is shown in the following table.

Pharaoh's Gold Payoffs and Probabilities

payoff	probability	payoff	probability
$2	1/10	$30	1/201
$3	1/15	$50	1/3,301
$4	1/44	$90	1/2,292
$6	1/48	$100	1/40,001
$9	1/102	$900	1/120,001
$12	1/152	$1,000	1/600,001

http://www.wilottery.com/scratchgames/historical.aspx

   a. We did not include an entry for $0 (not a winner). Is this value needed to calculate the expected payoff per ticket? If so, find the probability that a ticket doesn't win any money. If not, explain why not.
   b. Determine the expected payoff for a single ticket and explain how to interpret this value.
   c. Estimate the total payouts on 500,000 tickets.
   d. Estimate the amount of money kept by the state from 500,000 tickets.

23. A scratcher lottery game from Wisconsin was called *Gigantic Crossword* with tickets costing $5 each. The probability of winning different amounts in the game is shown in the following table.

Gigantic Crossword's Payoffs and Probabilities

payoff	probability	payoff	probability
$5	1/8	$50	1/91
$10	1/18	$100	1/1,501
$15	1/62	$500	1/44,446
$20	1/52	$5,000	1/600,001
$25	1/62	$50,000	1/600,001
$35	1/302		

http://www.wilottery.com/scratchgames/historical.aspx

   a. We did not include an entry for $0 (not a winner). Is this value needed to calculate the expected payoff per ticket? If so, find the probability that a ticket doesn't win any money. If not, explain why not.
   b. Determine the expected payoff for a single ticket and explain how to interpret this value.
   c. Estimate the total payouts on 750,000 tickets.
   d. Estimate the amount of money kept by the state from 750,000 tickets.

24. Reflect on your work calculating expected payoffs for scratcher tickets (both in the investigation and/or in the homework). Are scratcher tickets a good investment for players?

Algebra II

Module 10: Homework
Univariate Statistics

25. Insurance companies frequently make use of expected value calculations when deciding how much to charge a customer for insurance. Suppose that, based on past experience and industry guidelines, they believe that they will have to pay out insurance claims with the following frequencies for customers each year.

approximate payout amount for insurance claims in one year for a customer	probability
$500	0.061
$1,000	0.038
$5,000	0.043
$10,000	0.032
$25,000	0.021
$50,000	0.009
$100,000	0.007
$1,000,000	0.0006

Note: In this exercise we are ignoring the idea that different customers might have different risk levels. In real life insurance companies adapt the calculations to better model their risks, but the results are used in similar ways to what we demonstrate here.

a. Pick a row in the table and explain how to interpret the values.

b. What is the expected cost for insuring each customer?

c. Suppose that the company has 3,400,000 customers, and they want to price their policies so that they earn $1,000,000,000 per year after payouts (this money will go to pay salaries, business expenses, and so on, with the remaining money as profit). How much should they charge a person for insurance for one year? (*Assume every person will pay the same rate.*)

d. Assume that you are a customer of this insurance company. They allow you to either pay your total annual bill in one lump sum or you can split it into 12 monthly payments. If you choose the monthly payment option you are also charged a 5% convenience fee. Suppose you pay monthly. What is your monthly bill?

26. (See Exercise #25) Suppose that increasing costs have lead the company to revise their table as follows.

approximate payout amount for insurance claims in one year for a customer	probability
$750	0.062
$1,500	0.036
$10,000	0.039
$18,000	0.032
$30,000	0.021
$60,000	0.008
$115,000	0.004
$1,250,000	0.0003

Repeat Exercise #25 using the new table.

Algebra II

Module 10: Homework
Univariate Statistics

In Exercises #27-30, find the expected value on one spin of each spinner. *Estimate probabilities if necessary.*

27.

28.
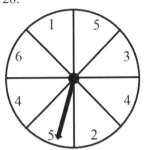

29.

30.

31. Find the expected value when rolling two standard six-sided dice.

32. Find the expected value for two spins of the spinner shown in Exercise #27.

INVESTIGATION 3: SAMPLING FROM A POPULATION

33. Janice opened a 1000-piece puzzle box and dumped out the pieces. She mixed them up and randomly drew 40 pieces and 3 of them had straight edges. She then states that there must be 75 edge pieces in the box. Do you agree with her conclusion? Explain.

34. Quentin was curious how many students at his school (with 2,460 students) were planning on going to the Homecoming dance. He asked the 30 students in his first period class and 8 said they were planning to go. Trevor uses this result to estimate that between 600 and 700 students are planning to attend the dance. Do you agree with his conclusion? Explain?

35. A teacher had a bag of marbles (red and blue) and his 30 students each randomly drew 20 marbles and counted the number of red marbles. Their results are given in the table below.

number of red marbles drawn									
15	19	13	11	15	16	18	17	13	16
16	12	15	15	17	16	9	13	11	15
15	13	11	17	13	12	16	15	14	19

   a. Draw a histogram to represent the class sampling results.

(continues...)

b. Suppose you sample **50** marbles from the container and that you get each of the following results. Give a ranking from 1 (very normal) to 4 (very unusual) for how unusual you would find the result.

Result	(very common)	(somewhat common)	(a little unusual)	(very unusual)
20 marbles are red	1	2	3	4
16 marbles are red	1	2	3	4
30 marbles are red	1	2	3	4
42 marbles are red	1	2	3	4
27 marbles are red	1	2	3	4
38 marbles are red	1	2	3	4

c. Suppose there are 160 total marbles in the container. Provide a range of possible values that you are confident contains the true number of red marbles.

d. Let's call the container from earlier in this exercise *Container 1*. Imagine that we have another container with red and blue marbles and suppose you sample 20 marbles at random and 9 of them are red. Which of the following conclusions do you think is more likely? Defend your answer.
   (i) The new container has a similar proportion of red marbles compared to Container 1.
   (ii) The new container has a very different proportion of red marbles compared to the Container 1.

36. A teacher had a bag of marbles (green and white) and her 30 students each randomly drew 20 marbles and counted the number of green marbles. Their results are given in the table below.

number of green marbles drawn									
7	5	11	6	3	8	7	5	10	8
10	8	7	5	5	11	9	10	7	8
11	6	9	8	12	7	8	8	11	5

a. Draw a histogram to represent the class sampling results.

b. Suppose you sample **40** objects from the container and that you get each of the following results. Give a ranking from 1 (very normal) to 4 (very unusual) for how unusual you would find the result.

Result	(very common)	(somewhat common)	(a little unusual)	(very unusual)
26 marbles are green	1	2	3	4
15 marbles are green	1	2	3	4
7 marbles are green	1	2	3	4
17 marbles are green	1	2	3	4
3 marbles are green	1	2	3	4
22 marbles are green	1	2	3	4

c. Suppose there are 115 total marbles in the container. Provide a range of possible values that you are confident contains the true number of red marbles.

d. Let's call the container from earlier in this exercise *Container 1*. Imagine that we have another container with green and white marbles and suppose you sample 20 marbles at random and 9 of them are green. Which of the following conclusions do you think is more likely? Defend your answer.
   (i) The new container has a similar proportion of red marbles compared to Container 1.
   (ii) The new container has a very different proportion of red marbles compared to the Container 1.

Algebra II

Module 10: Homework
Univariate Statistics

In Exercises #37-38 you are given a scenario. For each scenario do the following.
  (i) describe how a random sample could be generated and
  (ii) describe a way to generate a biased sample.

37. A university created a math placement test to give to incoming freshmen in order to decide which math class they should register for in their first semester. They are interested to see how high school students who have completed precalculus will do on the test.

38. The owner of an online retail store is curious about how often his employees forget to sign the "inspected by _____" receipt when shipping out orders.

INVESTIGATION 4: PERFORMING SIMULATIONS

In Exercises #39-40 you are given the results of a simulation where 40 people are asked whether they prefer Soda Brand A or Soda Brand B under different assumptions (with the process repeated 300 times). For each, study the simulation results and then answer the questions that follow.

39. Assume that 44% of the population prefers Soda Brand A over Soda Brand B. Using this assumption we ran a simulation of random sampling.

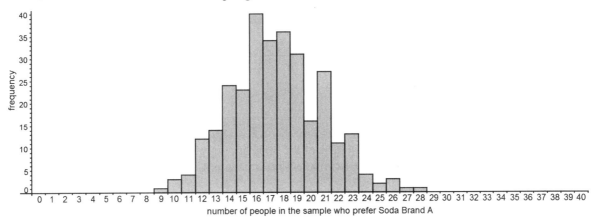

  a. In your own words, explain what the histogram represents.

  b. Suppose you randomly sample 40 real people assuming that 44% of the population prefers Soda Brand A. Give a ranking of 1 (very common), 2 (somewhat common), 3 (a little unusual), or 4 (very unusual) for how unusual you would find each of the following results.

Result	(very common)	(somewhat common)	(a little unusual)	(very unusual)
8 people prefer Brand A	1	2	3	4
26 people prefer Brand A	1	2	3	4
15 people prefer Brand A	1	2	3	4
30 people prefer Brand A	1	2	3	4
18 people prefer Brand A	1	2	3	4
20 people prefer Brand A	1	2	3	4

  c. Our simulation results assumed that 44% of the population prefers Soda Brand A over Soda Brand B. If you go out and randomly sample 40 real people, what results might make you believe this assumption is wrong?

40. Assume that 62% of the population prefers Soda Brand A over Soda Brand B. Using this assumption we ran a simulation of random sampling.

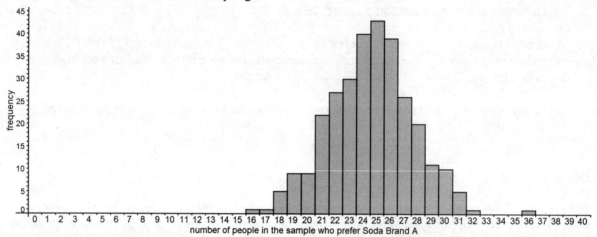
number of people in the sample who prefer Soda Brand A

a. Suppose you randomly sample 40 real people assuming that 62% of the population prefers Soda Brand A. Give a ranking of 1 (very common), 2 (somewhat common), 3 (a little unusual), or 4 (very unusual) for how unusual you would find each of the following results.

Result	(very common)	(somewhat common)	(a little unusual)	(very unusual)
8 people prefer Brand A	1	2	3	4
26 people prefer Brand A	1	2	3	4
15 people prefer Brand A	1	2	3	4
30 people prefer Brand A	1	2	3	4
18 people prefer Brand A	1	2	3	4
20 people prefer Brand A	1	2	3	4

b. Our simulation results assumed that 62% of the population prefers Soda Brand A over Soda Brand B. If you go out and randomly sample 40 real people, what results might make you believe this assumption is wrong?

Algebra II

Module 10: Homework
Univariate Statistics

INVESTIGATION 5: HYPOTHESIS TESTING

In Exercises #41-42 you are given histograms that show simulated results of polling 40 soda drinkers and asking them if they prefer Soda Brand A or Soda Brand B, and then repeating the process 600 times.

41. population assumption: 35% of the population prefers Soda Brand A

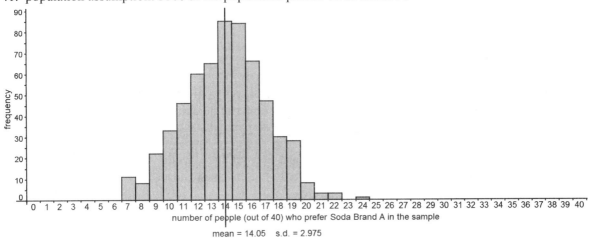

mean = 14.05   s.d. = 2.975

a. Suppose you survey 40 random soda drinkers and ask them if they prefer Soda Brand A or Soda Brand B. What outcomes would you expect if our assumption is true?
b. If you performed an actual survey of 40 randomly chosen soda drinkers, what outcomes would make you believe the assumption is false? Explain your thinking.

42. population assumption: 58% of the population prefers Soda Brand A

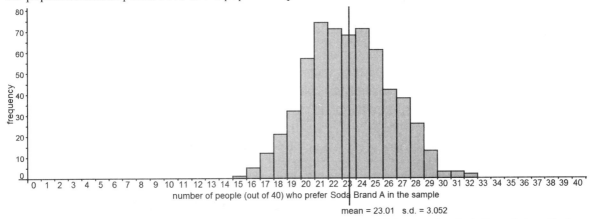

mean = 23.01   s.d. = 3.052

a. Suppose you survey 40 random soda drinkers and ask them if they prefer Soda Brand A or Soda Brand B. What outcomes would you expect if our assumption is true?
b. If you performed an actual survey of 40 randomly chosen soda drinkers, what outcomes would make you believe the assumption is false? Explain your thinking.

43. Suppose a soda company wants to change the formula for one of its drinks. They make samples of the new version and conduct a random sampling of consumers to see if they prefer the new version or the original version. Out of 80 people sampled, 43 say they prefer the new formula. What is a reasonable conclusion?

Algebra II

Module 10: Homework
Univariate Statistics

44. Repeat Exercise #43 if
    a) the results showed that 58 out of the 80 people surveyed prefer the new formula and
    b) the results showed that 25 out of the 80 people surveyed prefer the new formula.

In Exercises #45-48, do the following for each situation.
   a) Decide what assumption to test about the population.
   b) Suppose that you run a simulation using the assumption in part (a). Explain how you would use the simulation results along with an actual sample from the population to test the assumption.

45. Suppose you work for a state highway safety organization and a tire manufacturer tells you that his company estimates that 25% of cars on the road have underinflated tires. This concerns you because underinflated tires decrease gas mileage and increase stopping distances for cars.

46. Suppose you work for the national park service and find out that a beetle infestation is killing trees in a certain national park. A researcher in the field tells you that she believes about 30% of trees are currently affected.

47. Joseph and Farrah are running against each other for student body president of your high school.

48. Suppose that a certain ballot measure in your state proposes a change to the state constitution and requires more than 2/3 of voters to approve of the change for it to take effect.

INVESTIGATION 6: CUMULATIVE PROBABILITY

49. The given histogram shows simulated results of polling 40 soda drinkers and asking them if they prefer Soda Brand A or Soda Brand B, and then repeating the process 300 times with the assumption that 65% of soda drinkers prefer Brand A.

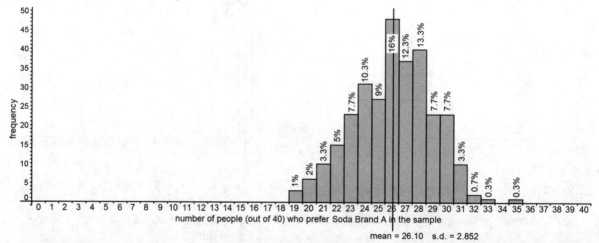

   a. Assume you survey a sample of 40 soda drinkers and find that 21 of them prefer Soda Brand A.
      i. Based on our simulation results, about what percent of samples from this population will have 21 or fewer people with the same preference?
      ii. Based on our simulation results, about what percent of samples from this population will have more than 21 people with the same preference?

(*continues...*)

Algebra II

Module 10: Homework
Univariate Statistics

b. Find the approximate "middle 60%" of sample results. That is, find an interval <u>centered around the mean</u> that contains about 60% of all sample results. Then explain the meaning of this interval in your own words. (*Note: Round the mean to the nearest whole number first.*)

c. How far above and below the mean does your interval extend? How far is this when you measure it in *number of standard deviations*?

50. The given histogram shows simulated results of polling 40 soda drinkers and asking them if they prefer Soda Brand A or Soda Brand B, and then repeating the process 300 times with the assumption that 44% of soda drinkers prefer Brand A.

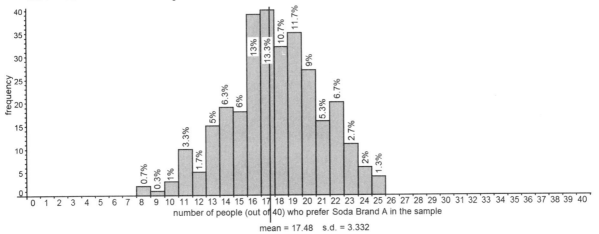

a. Assume you survey a sample of 40 soda drinkers and find that 19 of them prefer Soda Brand A.
   i. Based on our simulation results, about what percent of samples from this population will have 19 or fewer people with the same preference?
   ii. Based on our simulation results, about what percent of samples from this population will have more than 19 people with the same preference?

b. Find the approximate "middle 80%" of sample results. That is, find an interval <u>centered around the mean</u> that contains about 60% of all sample results. Then explain the meaning of this interval in your own words. (*Note: Round the mean to the nearest whole number first.*)

c. How far above and below the mean does your interval extend? How far is this when you measure it in *number of standard deviations*?

51. To formally test an assumption we decided to create the following rule: We will reject the hypothesis if the actual sample result is not within the middle 90% of samples from the simulation. [*Note: This is only one possible rule we could use.*]

a. Return to the context in Exercise #49. Identify the approximate "middle 90%" interval for sample results based on the assumption that 65% of soda drinkers prefer Soda Brand A.

b. If you sample 40 soda drinkers and 23 say they prefer Soda Brand A, will you accept or reject the assumption? What does this mean?

c. If you sample 40 soda drinkers and 32 say they prefer Soda Brand A, will you accept or reject the assumption? What does this mean?

52. To formally test an assumption we decided to create the following rule: We will reject the hypothesis if the actual sample result is not within the middle 90% of samples from the simulation. [*Note: This is only one possible rule we could use.*]
    a. Return to the context in Exercise #50. Identify the approximate "middle 90%" interval for sample results based on the assumption that 44% of soda drinkers prefer Soda Brand A.
    b. If you sample 40 soda drinkers and 23 say they prefer Soda Brand A, will you accept or reject the assumption? What does this mean?
    c. If you sample 40 soda drinkers and 13 say they prefer Soda Brand A, will you accept or reject the assumption? What does this mean?

INVESTIGATION 7: NORMAL DISTRIBUTION

53. The height of adults in a large population generally follows a normal distribution. Suppose the average height for an adult woman in the U.S. is about 65 inches with a standard deviation of about 2.25 inches. (*For reference: 60 inches = 5 feet and 72 inches = 6 feet*)
    a. Between what two heights are the "middle 68%" of adult women in the U.S.?
    b. Between what two heights are the "middle 95%" of adult women in the U.S.?
    c. Fill in the boxes in the graph based on the information for this context.

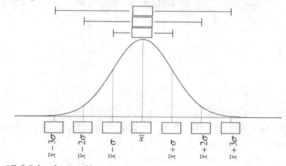

    d. Suppose a woman is 67.25 inches tall.
        i) How many standard deviations from the mean is her height?
        ii) About what percent of adult women in the U.S. are taller than her? Shorter?
    e. Suppose a woman is 58.25 inches tall.
        i) How many standard deviations from the mean is her height?
        ii) About what percent of adult women in the U.S. are taller than her? Shorter?

54. Suppose the lengths of dogs of a certain breed are normally distributed with a mean of 28 inches and a standard deviation of 1.8 inches.
    a. Between what two lengths are the "middle 68%" of dogs from this breed?
    b. Between what two lengths are the "middle 99.7%" of dogs from this breed?
    c. Fill in the boxes in the graph based on the information for this context.

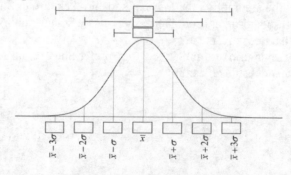

(*continues…*)

d. Suppose a dog from this breed is 31.6 inches long.
   i) How many standard deviations from the mean is its length?
   ii) About what percent of the breed have a greater length? Shorter length?
e. Suppose a dog from this breed is 26.2 inches long.
   i) How many standard deviations from the mean is its length?
   ii) About what percent of the breed have a greater length? Shorter length?

55. Suppose students graduating from high school in Germany were given a mathematics assessment (with a maximum score of 100) and scored an average of 78 with a standard deviation of 5.
    a. Fill in the boxes in the graph based on the information for this context.

    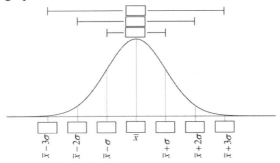

    b. What percent of students taking the test…
       i) had a score less than 73?
       ii) had a score greater than 68?
       iii) had a score less than 93?
       iv) had a score greater than 88?
    c. Repeat part (b) to find the *number* of students in each range if 24,600 students took the test.

56. Suppose the same students (see Exercise #57) took a science reasoning assessment (with a maximum score of 100) and scored an average of 82 with a standard deviation of 4.
    a. Fill in the boxes in the graph based on the information for this context.

    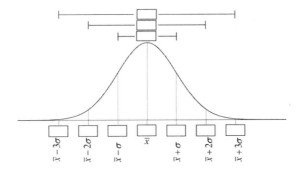

    b. What percent of students taking the test…
       i) had a score less than 82?
       ii) had a score greater than 90?
       iii) had a score less than 70?
       iv) had a score greater than 78?
    c. Repeat part (b) to find the *number* of students in each range if 24,600 students took the test.

Algebra II
Module 10: Homework
Univariate Statistics

57. A machine designed to fill and seal jars of peanut butter labeled as containing 454 grams has some variation in how much peanut butter ends up in the final product. Suppose that the average peanut butter in jar is 454 grams with a standard deviation of 2.3 grams.
    a. What is the percentile for a jar with 458.6 grams of peanut butter? What does this mean?
    b. How much peanut butter in a jar corresponds with the 2.5th percentile?

58. (See Exercise #57) Suppose a different machine fills and seals jars of peanut butter labeled as containing 751 grams. The average peanut butter in the final product is 751 grams with a standard deviation of 3.1 grams.
    a. What is the percentile for a jar with 744.8 grams of peanut butter? What does this mean?
    b. How much peanut butter in a jar corresponds with the 99.85th percentile?

INVESTIGATION 8: COMPARING NORMAL DISTRIBUTIONS

59. The following graph shows normal distributions for the same measurements taken on three different populations (labeled A, B, and C).

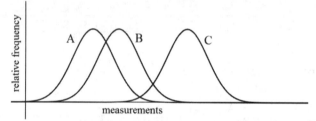

Compare the means and standard deviations for the three distributions and write a brief explanation of what the similarities/differences tell us about the populations measured.

60. The following graph shows normal distributions for the same measurements taken on three different populations (labeled A, B, and C).

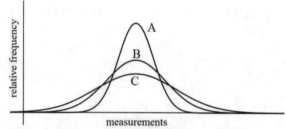

Compare the means and standard deviations for the three distributions and write a brief explanation of what the similarities/differences tell us about the populations measured.

Algebra II

Module 10: Homework
Univariate Statistics

For Exercises #61-64 use the following information. A new clinical trial is designed to test the effectiveness of different methods for reducing a patient's LDL cholesterol level. Many people signed up to participate in the study. Participants were randomly assigned to one of the following four groups.
- A. a control group (they were told not to change anything about their diet or lifestyle)
- B. a "diet only" group (they were given a menu of low-fat and low-cholesterol foods and were asked to follow the diet plan)
- C. a "medication only" group (they were told to change nothing about their diet or lifestyle but to take a new medication)
- D. a "diet and medication" group (they were asked to take the medication and follow the low-fat and low-cholesterol diet)

61. The following graph shows the distribution for changes in LDL levels for Group (A the control group) after 6 months. Explain what information we can gather from the graph.

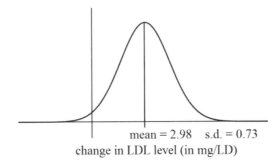

mean = 2.98   s.d. = 0.73
change in LDL level (in mg/LD)

62. The following results represent changes in LDL levels for Group B (the "diet only" group) after 6 months.

Change in LDL levels (in mg/DL)										
−10.4	−13.2	−7.8	−7.5	−9.8	−8.1	−8.3	−8.4	−8.3	−7.3
−9	−4.7	−8.3	−9.7	−5.8	−9	−8.2	−8.3	−10.1	−7
−8.4	−11.8	−9.6	−6	−9.9	−6.7	−13.7	−10.2	−2.8	−10.5

a. Determine the mean and sample standard deviation for the data.

b. Fill in the following graph showing the estimated distribution of changes in LDL levels for the population represented by Group B.

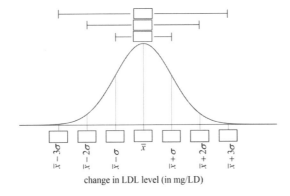

change in LDL level (in mg/LD)

c. Compare the results for Groups A and B.

Algebra II

Module 10: Homework
Univariate Statistics

63. The following results represent changes in LDL levels for Group C (the "medication only" group) after 6 months.

Change in LDL levels (in mg/DL)									
−8	−8	−10.4	−8.7	−7.8	−8.1	−7.2	−6.6	−9.3	−9.9
−5.3	−7.1	−8.5	−5.6	−8.3	−7	−5.8	−6.4	−7.8	−4.9
−11.2	−7.5	−9.1	−8.2	−7.2	−8.2	−9.8	−6.3	−10.3	−10.2

a. Determine the mean and sample standard deviation for the data.
b. Fill in the following graph showing the estimated distribution of changes in LDL levels for the population represented by Group C.

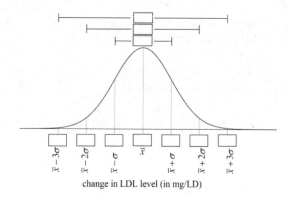

change in LDL level (in mg/LD)

c. Compare the results for Group C to the results for Groups A and B.

64. The following results represent changes in LDL levels for Group D (the "diet and medication" group) after 6 months.

Change in LDL levels (in mg/DL)									
−8.1	−11.2	−12.4	−9.5	−12.3	−9.7	−6.9	−15.8	−14.3	−11.9
−6.4	−10.3	−13.1	−8.1	−12.7	−13.1	−13	−15.9	−10	−9.8
−18.8	−13.1	−11.4	−18.9	−13.2	−9.1	−14.9	−14	−11.4	−10.8

a. Determine the mean and sample standard deviation for the data.
b. Fill in the following graph showing the estimated distribution of changes in LDL levels for the population represented by Group D.

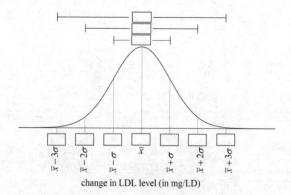

change in LDL level (in mg/LD)

c. Compare the results for Group D to the results for the other three groups.

Algebra II

Module 10: Homework
Univariate Statistics

INVESTIGATION 9: Z-SCORES

Use the following information for Exercises #65-72. The Intelligence Quotient (IQ) test is standardized so that the mean score is 100 with a standard deviation of 15.

65. Suppose a person takes the IQ test and scores 113.
    a. How many standard deviations from the mean is a score of 113? Round your answer to two decimal places. (*Use a negative number if the measurement is below the mean and a positive number if the measurement is above the mean.*)

    b. What is the percentile associated with this score? What does this number mean?

66. Suppose a person takes the IQ test and scores 143.
    a. How many standard deviations from the mean is a score of 143? Round your answer to two decimal places. (*Use a negative number if the measurement is below the mean and a positive number if the measurement is above the mean.*)

    b. What is the percentile associated with this score? What does this number mean?

67. Suppose a person takes the IQ test and scores 95.
    a. How many standard deviations from the mean is a score of 95? Round your answer to two decimal places. (*Use a negative number if the measurement is below the mean and a positive number if the measurement is above the mean.*)

    b. If 75,000 people take the test, about how many do you expect to have an IQ score higher than 95?

68. Suppose a person takes the IQ test and scores 77.
    a. How many standard deviations from the mean is a score of 77? Round your answer to two decimal places. (*Use a negative number if the measurement is below the mean and a positive number if the measurement is above the mean.*)

    b. If 75,000 people take the test, about how many do you expect to have an IQ score higher than 77?

69. Suppose a person takes the IQ test and is told his or her score is in the 48th percentile. What is the person's score?

70. Suppose a person takes the IQ test and is told his or her score is in the 3rd percentile. What is the person's score?

71. Suppose a person takes the IQ test and is told his or her score is in the 59th percentile. What is the person's score?

72. Suppose a person takes the IQ test and is told his or her score is in the 97th percentile. What is the person's score?

Algebra II

Module 10: Homework
Univariate Statistics

In Exercises #73-76 you are given a mean and standard deviation along with a measurement. For each, do the following.
    a. Calculate the z-score for the given measurement (round to two decimal places).
    b. Determine the percentile for the measurement.

73. $\bar{x} = 63$ and $\sigma = 5.2$; the measurement is 55

74. $\bar{x} = 119$ and $\sigma = 14.1$; the measurement is 137

75. $\bar{x} = 17.4$ and $\sigma = 0.7$; the measurement is 18

76. $\bar{x} = 1.08$ and $\sigma = 0.13$; the measurement is 0.99

In Exercises #77-80 you are given a mean and standard deviation along with a percentile. Find the measurement associated with each percentile.

77. $\bar{x} = 63$ and $\sigma = 5.2$; percentile is 1.07

78. $\bar{x} = 119$ and $\sigma = 14.1$; percentile is 11.31

79. $\bar{x} = 17.4$ and $\sigma = 0.7$; percentile is 71.57

80. $\bar{x} = 1.08$ and $\sigma = 0.13$; percentile is 99.4

Algebra II **Module 11: Investigation 1**
(Linear) Correlation

In the previous module we studied the distributions when measuring one characteristic of a population. In this module we will look at more than one measurement and determine if a relationship between the measurements exists.

When a member of the population is chosen and two measurements are made for that member (such as a height and weight measurement for a male high school senior at a school), we plot a point showing the corresponding measurements using a coordinate plane. We call this graph a *scatterplot*.

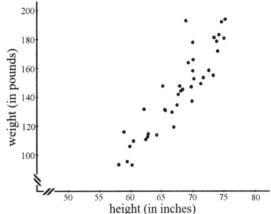

1. Suppose we sampled 40 male high school seniors at a certain school and recorded their heights (in inches) and weights (in pounds) and plotted this information to create the given scatterplot.

   a. Explain what a point on this graph represents.

   b. What general statement(s) can we make about the relationship between the heights and weights of male high school seniors at this school?

2. A common phrase in sports is "Defense wins championships." So how does a team's defense correspond to its ability to win games? The following scatterplot shows data for the last three 82-game seasons in the National Hockey League (NHL) comparing the total number of season wins to the number of goals the team allowed that season. *Data collected from www.nhl.com.*

   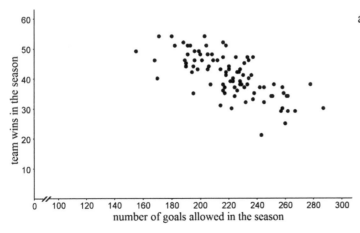

   a. Explain what a point on this graph represents.

Module 11: Investigation 1   © 2014 Carlson, O'Bryan & Joyner   Page 541

b. What general statement(s) can we make about the relationship between the number of wins in a season and the number of goals the team allowed in that season?

3. Suppose a test is designed to measure student learning on key mathematical ideas taught in high school. The test writers randomly selected 50 high school seniors from one school to take the test. They recorded each student's score on the test (out of 100) and compared it to that student's GPA.

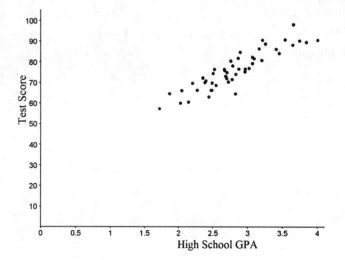

What general statement(s) can we make about the relationship between a student's score on the test and that student's GPA?

4. The following scatterplot compares the population of a state (in millions) in the United States to its land area (in thousands of square miles). *Data collected from www.ipl.org. Note that 3 states were excluded because their data falls outside of the ranges chosen for each axis.*

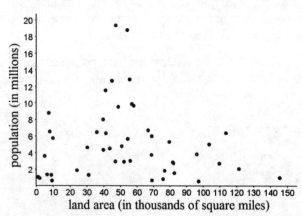

What general statement(s) can we make about the relationship between a state's population and its physical size (as measured by its land area)?

Algebra II

Module 11: Investigation 1
(Linear) Correlation

Positive Correlation, Negative Correlation, and No Correlation

Suppose we make a scatterplot showing the relationship between two measurements. If the points tend to form a linear "band" then we say the two measurements are (*linearly*) *correlated*. If a non-linear trend exists, or if no obvious pattern is visible, then the measurements are not (linearly) correlated.

If a (linear) correlation exists, and if the points rise from left to right, we say that there is a *positive correlation* between the measurements. *Note that in a positive correlation small values of one measurement correspond with small values of the other measurement and large values of one measurement correspond with large values of the other measurement.*

If a (linear) correlation exists, and if the points fall from left to right, we say that there is a *negative correlation* between the measurements. *Note that in a negative correlation large values of one measurement correspond with small values of the other measurement and vice versa.*

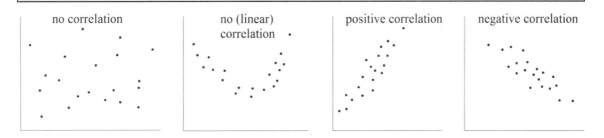

no correlation   no (linear) correlation   positive correlation   negative correlation

5. Categorize each of the relationships in Exercises #1-4 as having either positive correlation, negative correlation, or no correlation.

Strong vs. Weak (Linear) Correlation

If the points of a scatterplot lie relatively tightly along a line, then we say that the two measurements have a *strong correlation*. If the points lie in a linear band, but are not tightly focused along a single line, then we say that the two measurements have a *weak correlation*.

6. For any relationships in Exercise #1-4 that you believe are correlated, rank them from strongest correlation to weakest correlation.

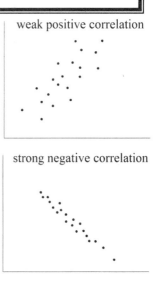

weak positive correlation

strong negative correlation

6. For any relationships in Exercise #1-4 that you believe are correlated, rank them from strongest correlation to weakest correlation.

In Investigation 2 we will begin to add trend lines to our plots. The notion of *strong correlation* vs. *weak correlation* is an indication of how well the trend line can be used to predict the values for additional data points.

Correlation is an extremely important concept that has wide applications in the real world. For example, if you've every shopped with an online retailer like Amazon.com you've probably noticed that they recommend products to you based on your purchase history and your recent searches. The products they suggest are based on noticing correlations where people who purchase one product (or group of products) tend to purchase other products. Similarly, companies like Netflix use correlation studies to predict how you will rate specific movies and TV shows based on your viewing and rating history.

It's important to note that, while we generally only use positive and negative correlation to refer to linear trends we observe in the data, a lack of linear correlation does not mean that no pattern or relationship exists.

7. The following scatterplot shows the total number of drowning deaths in Arizona from 1992 to 2010 that occurred during specific months of the year. *Data collected by the Arizona Department of Health Services, www.azdhs.gov.*

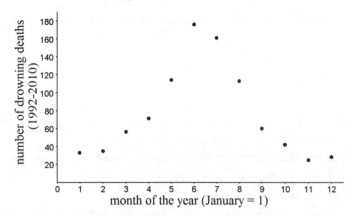

a. Does the scatterplot demonstrate a positive or negative correlation?

b. Does a predictable relationship exist between drowning deaths and the month of the year? Explain.

We conclude this investigation with an important point about correlation.

Correlation vs. Causation

When we find that a correlation between two measurements exists, it means that they tend to be related in a predictable way. It does *not* necessarily mean that changes in one measurement *cause* the corresponding changes in the other measurement.

Note: Experiments can be designed to attempt to determine causal links, but that topic is beyond the scope of this course.

Algebra II **Module 11: Investigation 1**
(Linear) Correlation

8. Consider the information in Exercise #3 (test score vs. GPA). Discuss why it might be incorrect to claim that a higher GPA *causes* higher tests scores (or that a lower GPA *causes* lower test scores).

9. Suppose you collect data on the percent of monthly household energy used for heating in one state and compare this to the average number of drowning deaths in the corresponding month for the same state.
    a. Imagine that, after plotting the points, you notice a negative correlation. Explain what this tells us about energy spending and drowning deaths in the state.

    b. Explain why we would not say that there is a causal relationship between energy spending and drowning deaths. (*Hint: What other factor(s) might account for the relationship?*)

Algebra II

Module 11: Investigation 2
Exploring Lines of Best Fit

The following distributions show the lengths (in cm) of baby girls at 4 months (mean = 63, s.d. = 2.25), 8 months (mean = 69, s.d. = 2.25), 12 months (mean = 74, s.d. = 2.325), and 16 months (mean = 78, s.d. = 2.5). *Data adapted from World Health Organization Child Growth Standards, ©2006.*

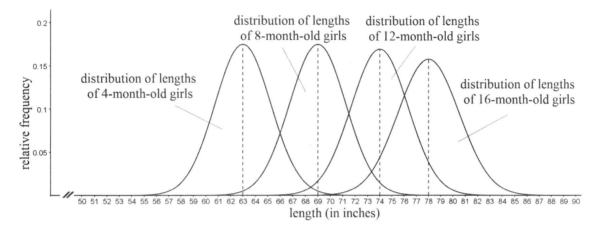

A graph like this can certainly give us a lot of information about expected lengths of infant girls at specific ages, but oftentimes we are more interested in studying trends so that we can make predictions about the expected lengths of infant girls of *any* age.

To examine the trend, let's plot the mean lengths with the corresponding ages for infant girls (**Figure 1**) and then draw a curve that passes through these points (**Figure 2**).

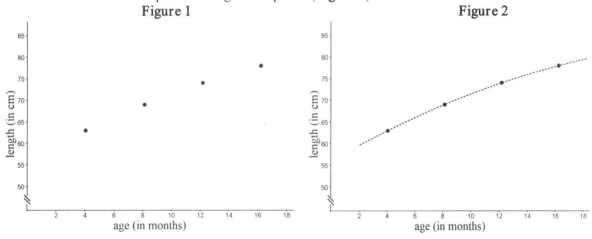

The curve represents the trend and shows how the mean lengths change for infant girls as they age. We could use this trend to make predictions about the mean lengths of infant girls at, say, 5 months, 11 months, and so on. The important point to understand, however, is that this line ***shows the expected (mean) lengths only*** – it is not a perfect predictor for the length of a randomly chosen infant girl.

1. Discuss the difference between predicting the mean length of infant girls at specific ages and predicting the weight of a randomly chosen infant girl of that age.

Algebra II

Module 11: Investigation 2
Exploring Lines of Best Fit

We know that each age has a distribution of lengths centered at the mean, and if we assume normal distributions we know that most (68%) of the infant girls at a given age will be within one standard deviation of the mean and that 95% of them will be within two standard deviations of the mean (see **Figure 3**).

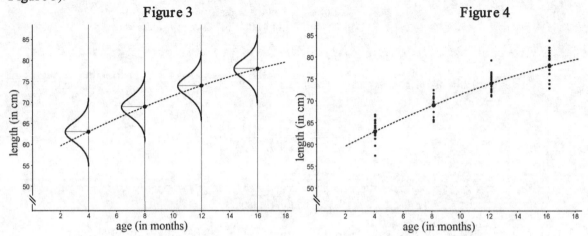

So if we randomly sample infant girls of ages 4, 8, 12, and 16 months and plot their lengths, we expect to see something similar to **Figure 4**, and if we expanded our sampling to include other ages we would expect to see something similar to **Figure 5**.

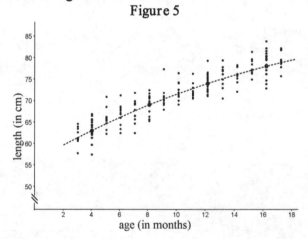

2. In your own words, explain your understanding of what Figure 5 is telling us.

In the previous example we were *given* the means and standard deviations for the lengths of infant girls at various ages and we made sense of how to predict the trend and how to interpret the trend curve. As we discussed in Module 9, however, this isn't common. Typically we are sampling a population in order to estimate the mean and standard deviation for the population. Therefore, it's much more common to sample from a population and attempt to use this sample to predict trends.

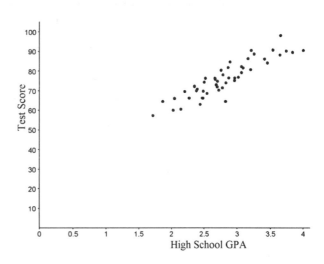

In investigation #1 we examined how scores on a test designed to measure student learning on key mathematical ideas compared to the GPA of test-takers.

Our assumption is that for each GPA there is an associated mean score on the test for students with that GPA, but also that there is a distribution of scores for students with that GPA that extends above and below the mean.

A typical goal in statistics for this contest is to determine the trend line that shows us how the **mean** test scores compare to GPA. The basic process is to draw a line that passes as close as possible to all of the data points. *Note that for the rest of this investigation we will focusing exclusively on trend **lines** – we will not consider whether or not the trend line should be curved.*

3. a. Using a ruler or other straightedge, draw (to the best of your ability) the line that you think passes as close as possible to all of the points.

    b. Estimate the slope and vertical intercept.

    c. Interpret the slope and vertical intercept of your line in this context.

The line you drew in Exercise #3 is an attempt to create **a line of best fit** to predict the mean values of one measurement (in this case *score on the test*) for values of another measurement (in this case *student high school GPA*). There is a very good reason why the line that passes most closely to our given data points is a good estimate for these mean values. The reasoning is as follows. *Note that we assume distributions are normal.*

  i.   Measurements in a data distribution tend to be close to the mean. It is rare for a measurement to be far from the mean.
  ii. Therefore for a given x-value on the scatterplot most of the y-values should be close to the mean y-value at that x-value (and conversely the mean should be close to them).
  iii. If we draw a line that passes as close as possible to all of the points then the line should pass through (or at least very close to) the mean y-values for each x-value.

Algebra II

Module 11: Investigation 2
Exploring Lines of Best Fit

Line of Best Fit

A *line of best fit* shows the estimated mean *y*-value in a data set for each *x*-value. The line is not intended to perfectly represent how two measurements vary together. It is intended to show the trend in how the mean values of one measurement co-vary with the values of another measurement.

In Investigation 1 we examined the scatterplot shown in **Figure 6** (the heights and weights for a sample of 40 male high school seniors at a certain school). In **Figure 7** we've sketched an approximate line of best fit.

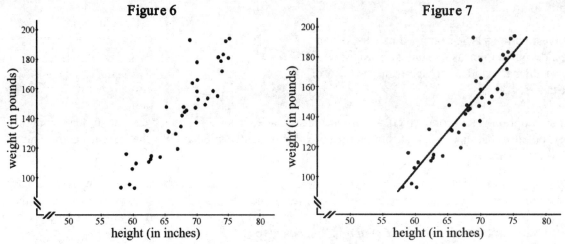

4. a. Explain what we are trying to estimate using the line of best fit.

   b. Estimate the slope of the best fit line and explain what this represents in context.

   c. Explain why the vertical intercept is not a very useful piece of information in this context.

Algebra II

d. Write a formula to define the line of best fit in this context. (*Using estimated values are okay.*)

5. In Investigation 1 we examined the following scatterplot comparing the total number of season wins for NHL teams over the last three full 82-game seasons to the number of goals the team allowed that season.

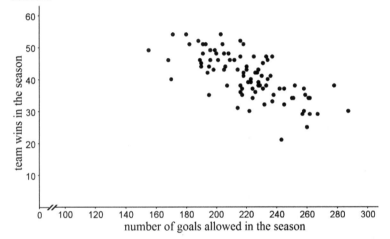

a. Use a ruler or other straightedge to draw (to the best of your ability) the line that you think passes as close as possible to all of the points.

b. Explain what the line of best fit is supposed to represent in this context.

c. Estimate the slope and vertical intercept (if applicable) and interpret their meaning in the given context.

Algebra II

Module 11: Investigation 2
Exploring Lines of Best Fit

Lines of best fit are more useful when the correlation between two measurements seems strong. If the correlation is weak then even if our line accurately predicts the mean values, those mean values aren't necessarily good representations of the data.

6. The following scatter plot shows sale prices (in thousands of dollars) for a sample of houses sold in one city last year compared to the age of the home.

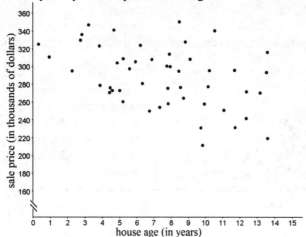

   a. Describe any correlation you observe.

   b. Sketch (to the best of your ability) a line of best fit for the scatterplot.

   c. Discuss the usefulness of the best fit line in this context.

Algebra II

Module 11: Investigation 3
Residuals

In the previous investigation you attempted to draw a best fit line by making it pass as close as possible to all points on the scatterplot. In this investigation we will focus on how we specifically measure the distance between a point and a best fit line in statistics.

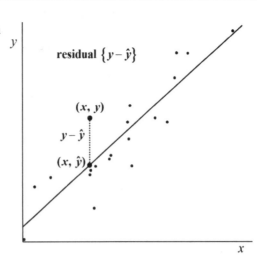

When we draw a best fit line to try to capture a trend, then we can compare each point (x, y) on the scatterplot to the point on the best fit line with the same *x*-value. *Note: We use the symbol ^ above y to represent the y-values on the best fit line. We call $\hat{y}$ "y hat" and will use this notation for the rest of the module.*

For a given data point, the difference between the *y*-value for that point and the $\hat{y}$-value at the same *x*-value is called the ***residual*** (i.e., $y - \hat{y}$).

"*y*hat"

We use $\hat{y}$ ("*y*hat") to represent the *y*-values on the best fit line. The $\hat{y}$-values are sometimes called the ***fitted values*** or ***predicted values***.

From now on when we write the formula that defines the best fit line we will use $\hat{y}$ to represent the output values, such as $\hat{y} = 1.27x + 16.48$.

Residual

A ***residual*** is the difference between the *y*-value for a point and the $\hat{y}$-value at the same *x*-value (i.e., $y - \hat{y}$).

---

1. Calculate the six residuals using the given scatterplot and best fit line ($\hat{y} = 2.89x - 3.05$).

x-value	*y*-value	$\hat{y}$-value	residual ($y - \hat{y}$)
2.1	1.12		
2.3	6.84		
3.8	4.27		
5.0	15.13		
8.2	16.19		
9.4	27.06		

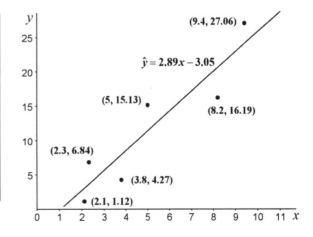

Algebra II

Module 11: Investigation 3
Residuals

Suppose that three of your classmates tried to draw a best fit line for the data shown in the scatterplot to the right (their attempts are shown in Graphs A, B, and C). Since there can only be one line that passes as close as possible to all of the points, at most one of these lines can be correct (*and it's possible that none of them drew the perfect line of best fit*).

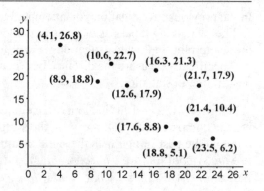

The way we compare potential lines of best fit is to compare their residuals. However, since some residuals are positive and some are negative, we square all of the residuals to make them positive. We then add up these squared residuals and look for the smallest sum.

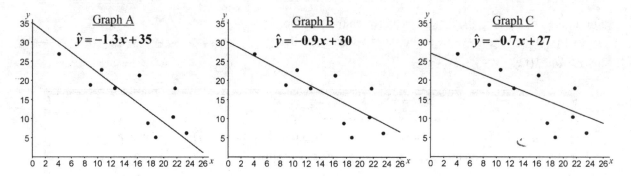

2. Find the sum of the squared residuals for the line in Graph A ($\hat{y} = -1.3x + 35$).

x-value	y-value	$\hat{y}$-value	residual ($y - \hat{y}$)	squared residual $(y - \hat{y})^2$
4.1	26.8			
8.9	18.8			
10.6	22.7			
12.6	17.9			
16.3	21.3			
17.6	8.8			
18.8	5.1			
21.4	10.4			
21.7	17.9			
23.5	6.2			

sum of squared residuals

Algebra II

Module 11: Investigation 3
Residuals

3. Find the sum of the squared residuals for the line in Graph B ($\hat{y} = -0.9x + 30$).

x-value	y-value	$\hat{y}$-value	residual $(y - \hat{y})$	squared residual $(y - \hat{y})^2$
4.1	26.8			
8.9	18.8			
10.6	22.7			
12.6	17.9			
16.3	21.3			
17.6	8.8			
18.8	5.1			
21.4	10.4			
21.7	17.9			
23.5	6.2			

sum of squared residuals

4. Find the sum of the squared residuals for the line in Graph C ($\hat{y} = -0.7x + 27$).

x-value	y-value	$\hat{y}$-value	residual $(y - \hat{y})$	squared residual $(y - \hat{y})^2$
4.1	26.8			
8.9	18.8			
10.6	22.7			
12.6	17.9			
16.3	21.3			
17.6	8.8			
18.8	5.1			
21.4	10.4			
21.7	17.9			
23.5	6.2			

sum of squared residuals

5. Which of the three lines fits the data the best?

Algebra II

Module 11: Investigation 3
Residuals

It turns out that the true line of best fit is $\hat{y} = -0.9085x + 29.7171$ because it has the minimum possible sum of squared residuals [204.66]. This means that none of your classmates' predictions were ***the best*** solution, although the prediction in Graph B was very close. We will learn how to use technology to help us find the true line of best fit in Investigation 4.

We end this investigation by making a connection between residuals and strong vs. weak correlation.

6. Compare the scatterplots below and their lines of best fit. Which line of best fit will have the smallest sum of squared residuals? The largest? Why?

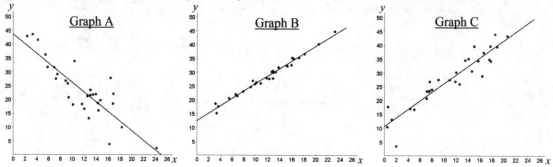

7. Do you think there is a connection is between strong vs. weak correlation and residuals? Explain.

Algebra II

Module 11: Investigation 4
Linear Regression

Note: See the module appendix for a tutorial on using a TI-83/84 calculator to complete the exercises in this investigation. If you have a different graphing calculator or want to use programs such as Microsoft Excel you should consult your owner's manual or online help documents and tutorials.

1. In Investigation 3 we examined the following two scatterplots. Use technology to determine the formula for the best fit line. [We call this the **regression function**.] In addition, your calculator will report an r^2 and r value. Record these values for use later in the investigation.

   a.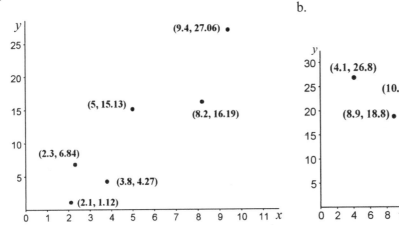
   b.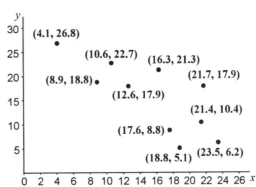

2. a. What is the criterion for choosing the best fit (regression) line? (*Review Investigation 3 if necessary.*)

   b. What are the points on the best fit (regression) line intended to represent? (*Review Investigation 2 if necessary.*)

Algebra II

Module 11: Investigation 4
Linear Regression

3. The following table shows the number of completed passes and the number of total passing yards for NFL quarterbacks during the 2013 season. *Data collected from www.espn.com. Note that only quarterbacks with at least 2,000 passing yards were included in the data set.*

Completed passes	Passing yards	Completed passes	Passing yards	Completed passes	Passing yards
450	5,477	362	3,912	247	3,046
446	5,162	342	3,828	203	2,891
371	4,650	343	3,822	224	2,621
439	4,515	317	3,818	247	2,608
378	4,478	292	3,379	193	2,536
380	4,343	257	3,357	217	2,454
363	4,293	308	3,313	219	2,310
362	4,274	305	3,241	180	2,015
375	4,261	274	3,203		
355	3,913	243	3,197		

a. Use technology to plot the points, calculate the linear regression function for passing yards vs. completed passes, and graph the line of best fit. In the space below write down the linear regression function and the r^2 and r values.

b. Do you think a linear function is a good model for this data? Explain.

c. Explain how we can interpret the slope of the regression function.

d. Use the regression function to predict
   i) how many passing yards would correlate with 260 completed passes during this season.

   ii) how many completed passes would correlate with 4,000 passing yards during this season.

Algebra II

Module 11: Investigation 4
Linear Regression

4. The following data shows the total annual precipitation and average annual temperature for Syracuse, New York from 1990 to 2012. *Data collected from www.weather.com.*

Year	Average temp (°F)	Total precipitation (inches)	Year	Average temp (°F)	Total precipitation (inches)
2012	52.5	35.11	2000	47.2	37.01
2011	50.6	48.04	1999	49.5	30.88
2010	49.7	41.47	1998	50.8	37.06
2009	47.6	35.36	1997	47.2	32.62
2008	48.2	41.91	1996	47.1	39.39
2007	48.1	41.62	1995	48	31.34
2006	50.3	47.2	1994	46.7	36.63
2005	49.1	40.1	1993	46.8	43.38
2004	47.5	43.14	1992	46.6	43.61
2003	47.1	37.6	1991	50.2	37.05
2002	50.4	40.26	1990	50.2	49.47
2001	50	34.3			

a. Use technology to plot the points, calculate the linear regression function for total precipitation (in inches) vs. average temperature (in °F), and graph the line of best fit. In the space below write down the linear regression function and the r^2 and r values.

b. Do you think a linear function is a good model for this data? What does this mean for the regression function you generated in part (a)?

Throughout this investigation we've asked you to record the r^2 and r values that your calculator reports when determine a line of best fit. Let's explore what these values mean by returning to the scatterplot in Exercise #1a.

Let's begin by finding the mean for the set of *y*-values (or $\bar{y}$) in the scatterplot and mark this value on our graph as a horizontal line (**Figure 1**).

$$\bar{y} = \frac{1.12 + 6.84 + 4.27 + 15.13 + 16.19 + 27.06}{6} = \frac{70.61}{6} \approx 11.77$$

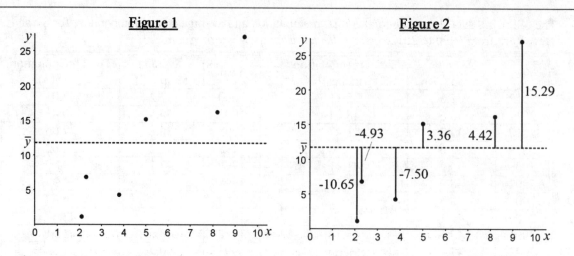

We then determine the difference between each *y*-value and the overall mean *y*-value (**Figure 2**), square these differences, and add up the results. [*Therefore we are summing up values of* $(y-\bar{y})^2$. *We call this sum* **the total sum of squares**, *or SSTO.*] For this example,

$$\text{SSTO} \approx (-10.65)^2 + (-4.93)^2 + (-7.50)^2 + (3.36)^2 + (4.42)^2 + (15.29)^2 \approx 458.5875$$

Our goal in creating a regression function is to explain as much of the variation away from $\bar{y}$ as possible. When we draw in the best-fit line determined by the calculator ($\hat{y} = 2.89x - 3.05$), we can see that the total difference between a given *y*-value and $\bar{y}$ is broken into two parts (**Figure 3**).

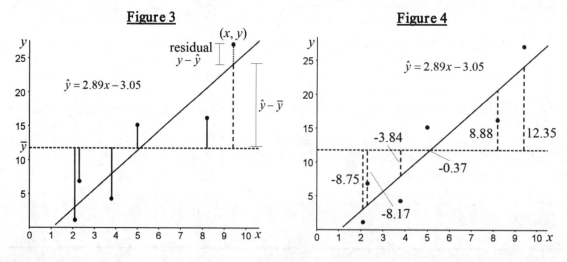

The first part is the difference between the $\hat{y}$-value given by the regression line and the overall mean *y*-value $\bar{y}$ (so $\hat{y} - \bar{y}$). The second part is the residual (which we know is $y - \hat{y}$). If we calculate each difference $\hat{y} - \bar{y}$, square them, and add them up we get a sum called **the regression sum of squares**, or SSR (**Figure 4**).

$$\text{SSR} \approx (-8.75)^2 + (-8.17)^2 + (-3.84)^2 + (-0.37)^2 + (8.88)^2 + (12.35)^2 \approx 389.5708$$

The value r^2 is called *the coefficient of determination* and is the ratio of the regression sum of squares and the total sum of squares.

$$r^2 = \frac{SSR}{SSTO} \approx \frac{389.5708}{458.5875} \approx 0.85$$

5. Compare this value to the r^2 value reported by your calculator in Exercise #1a.

The Coefficient of Determination r^2 for a Linear Regression Function

The value r^2 is called *the coefficient of determination* and is the ratio of the regression sum of squares and the total sum of squares.

$$r^2 = \frac{SSR}{SSTO}$$

For a linear regression function we interpret this number as the portion of the variability in the data explained by the regression function.

Values of r^2 will fall between 0 and 1. The closer r^2 is to 1 the greater the linear association between the measurements. If r^2 is close to 0 then the linear regression function does not do a good job of explaining the variability.

Ex: For the example presented, $r^2 \approx 0.85$, meaning that the linear regression function accounts for about 85% of the variability in the data. The remaining variability could be due to normal variations in data distributions or they may be additional factors we aren't accounting for.

---

6. In a previous investigation we examined the relationship between the high school GPA of a student and that student's score on a new math test. For the linear regression function $\hat{y} = 16.45x + 29.88$ we have $r^2 \approx 0.81$. Interpret this value.

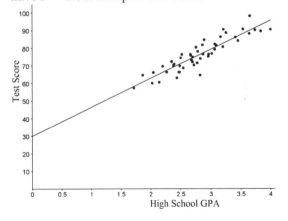

Algebra II

Module 11: Investigation 4
Linear Regression

7. Suppose we want to know how much consistency exists in Major League Baseball teams' wins from one year to the next. The given scatterplot shows team wins in the 2012 season on the horizontal axis and team wins in the 2013 season on the vertical axis. *Data collected from www.mlb.com.*

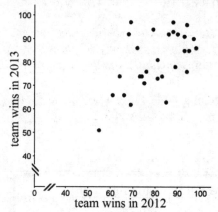

   a. Are team wins in 2013 correlated with team wins in 2012? Explain.

   b. The best-fit line is about $\hat{y} = 0.6x + 32.9$ with $r^2 \approx 0.335$. Sketch this line on the graph and interpret the r^2 value.

8. Compare the r^2 values in Exercises #7 and #8. Explain what their difference indicates about the data sets.

In some cases the **correlation coefficient r** is used to measure linear association where $r = \pm\sqrt{r^2}$. [*We choose + or − depending on whether the data shows positive or negative correlation.*] Since $0 \leq r^2 \leq 1$, then $-1 \leq r \leq 1$. When $|r|$ is closer to 1 it indicates a stronger linear association. When $|r|$ is close to 0 it indicates little or no linear association.

Algebra II

Module 11: Investigation 5
Interpolation and Extrapolation

In a previous investigation we examined the relationship between high school GPA and scores on a new math test. The linear regression function is $\hat{y} = 16.45x + 29.88$ with $r^2 \approx 0.81$ where x is a student's GPA and $\hat{y}$ is the estimated mean test score for students with that GPA.

Since we believe that a linear function is a good model for this data set, we can use the regression function to predict a student's score on the test using his/her GPA. *We must acknowledge, however, that there will be variability and that the value returned by the regression function is simply the average score for students with that GPA.*

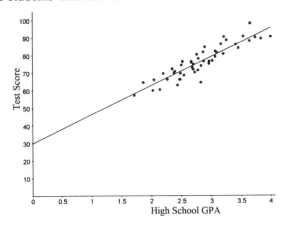

1. Predict a student's score on the test if that student has a GPA of 3.2.

2. Predict a student's score on the test if that student has a GPA of 0.8.

3. Consider your answers to Exercises #1 and #2. Are you concerned about the accuracy of your estimate for either student? Discuss.

In Investigation #4 we looked at the relationship between total passing yards and completed passes for NFL quarterbacks in 2013. The linear regression function is $\hat{y} = 10.74x + 270.40$ with $r^2 \approx 0.92$ where x is the number of completed passes and $\hat{y}$ is the estimated mean number of total passing yards for quarterbacks with that many completions.

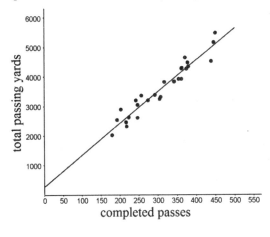

4. Predict a quarterback's total passing yards if he had 425 completed passes.

5. Predict a quarterback's total passing yards if he had 50 completed passes.

6. Consider your answers to Exercises #4 and #5. Are you concerned about the accuracy of either of your estimates? Discuss.

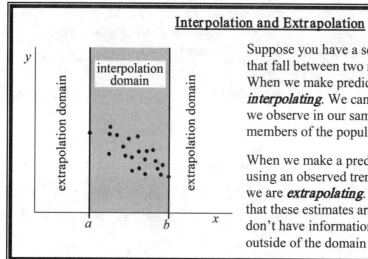

Interpolation and Extrapolation

Suppose you have a set of data with x-values that fall between two real numbers a and b. When we make predictions with $a \leq x \leq b$ we are *interpolating*. We can be confident that trends we observe in our sample data will apply to most members of the population in this domain.

When we make a prediction for $x < a$ or $x > b$ using an observed trend or regression function we are *extrapolating*. We are less confident that these estimates are accurate because we don't have information from our sample outside of the domain $a \leq x \leq b$.

To understand the potential dangers of extrapolation we can look at two examples from this module.

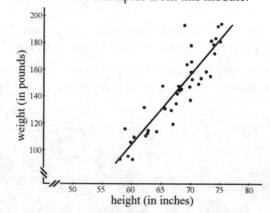

Recall the data set showing weight x (in pounds) compared to height y (in inches) for male high school seniors at one school. The regression function is $\hat{y} = 5.21x - 208.81$, and this is based on a sample with males between about 57 inches and 76 inches tall. If we use this trend to predict the weight for someone who is 39 inches tall (extrapolation) we get −5.62 pounds (an answer that makes no sense in this context). Our model doesn't account for trends beyond the scope of our sample data.

Similarly, our data about total passing yards compared to completions for NFL quarterbacks (see Exercises #4-6) only shows information for quarterbacks with between about 175 and 450 completions. If we use the regression function $\hat{y} = 10.74x + 270.40$ to predict the total number of passing yards for a quarterback with 2 completions (extrapolation), we get about 292 total passing yards. This number is impossible.

Algebra II

Module 11: Investigation 5
Interpolation and Extrapolation

7. The following data shows the closing stock price (in dollars per share) for Facebook, Inc. for 16 of the first 18 trading days of December 2013. *Data collected from www.thestockmarketwatch.com.* **Be careful entering the data.**

Trading day	Closing stock price (dollars)
1	47.06
2	46.73
4	48.34
5	47.94
6	48.84
7	50.24

Trading day	Closing stock price (dollars)
8	49.38
9	51.83
10	53.32
11	53.81
12	54.86
13	55.57

Trading day	Closing stock price (dollars)
15	55.12
16	57.77
17	57.96
18	57.96

a. Calculate the linear regression function and verify that a linear model is a reasonable choice given the data.

b. Use your linear regression function to estimate the value of the Facebook, Inc. stock on trading days 3 and 14.

c. An investor, seeing this trend, purchases 1,000 shares of Facebook, Inc. stock on trading day 18 (for $57.96 per share). He plans to hold on to the stock for 19 trading days and then sell it. Use extrapolation to predict
   i. the closing stock price 19 days after he purchases the stock (trading day 37) and

   ii. his profit from selling off his shares at this price.

d. The real closing stock price 19 trading days later was $53.55 per share. Compare this to your answers in part (c).

e. Discuss how this relates to our discussion of the potential dangers of extrapolation.

8. Summarize the important ideas in this investigation.

Algebra II

Module 11: Appendix
Linear Regression with a TI-83/84 Calculator

The following instructions will help you perform linear regression with a TI-83/84 calculator.

Suppose you want to plot the following points and generate the best-fit line.

x	y
6.4	10.7
10.3	15.7
9.8	19.6
16.7	26.5
9.9	16.4
14.2	25.4
4.4	8.9
10.2	10.7
7.7	11.1
3.6	4.4

Press STAT to bring up the Statistics menu and select **Edit...** Type your *x*-values in **L1** and the corresponding *y*-values in **L2**.

If you want to view the scatterplot, press 2ND and Y= to bring up the Stat Plots menu. Select one of the plots (such as **Plot 1...**). Select **On** to graph the points and make sure that the first type of graph is selected and that the **Xlist** is set to **L1** and the **Ylist** is set to **L2**. Press GRAPH to see the scatterplot.

  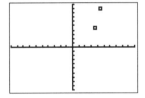

In many cases you won't be able to see all of the points. Press ZOOM and scroll until you see **ZoomStat** and select that option.

 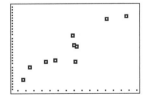

Algebra II

Module 11: Appendix
Linear Regression with a TI-83/84 Calculator

To calculate the linear regression function/line of best fit, press $\boxed{\text{STAT}}$ and then use the right arrow to select the **CALC** menu. Select the option **LinReg(ax+b)**. The command will appear on your home screen. Press $\boxed{\text{ENTER}}$. Your calculator will give you the parameters for the regression function and the values for r^2 and r.

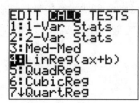 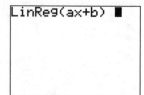 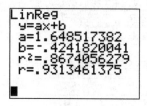

Based on the results, the line that best fits the data is about $\hat{y} = 1.6485x - 0.4242$, and $r^2 \approx 0.8674$ and $r \approx 0.9313$.

If you want to graph the line of best fit with your scatterplot you have two options. The first option is to write down the regression function formula, then select $\boxed{\text{Y=}}$ and type the formula into $Y_1=$. Then press $\boxed{\text{GRAPH}}$ and (assuming you also followed the instructions to graph the scatterplot) you will see your best fit line along with the scatterplot of the actual data.

 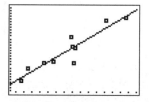

The second option is to have the calculator automatically enter the regression function formula into $Y_1=$ for you. After selecting **LinReg(ax+b)** and having it appear on your home screen, do *not* press $\boxed{\text{ENTER}}$. Instead, press $\boxed{\text{VARS}}$ and use the right arrow to get the **Y-VARS** menu. Select **Function...** and then Y_1.

Your home screen will now show **LinReg(ax+b) Y_1**. Press $\boxed{\text{ENTER}}$. You get the same information as before, but if you now press $\boxed{\text{Y=}}$ you will see that the regression function formula has been copied into $Y_1=$. Press $\boxed{\text{GRAPH}}$ to see the graph.

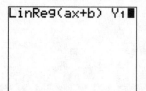 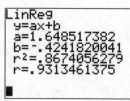  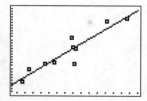

Algebra II

Module 11: Homework
Bivariate Statistics

INVESTIGATION 1: (LINEAR) CORRELATION

In Exercises #1-4 you are given a scatterplot showing the relationship between two quantities. For each scatterplot answer the following questions.
  a) What does a point on the graph represent?
  b) What general statement(s) can we make about the relationship between the quantities?
  c) Is there a (linear) correlation between the quantities? If so, is it positive or negative? Strong or weak?

1. (source: www.fueleconomy.gov)

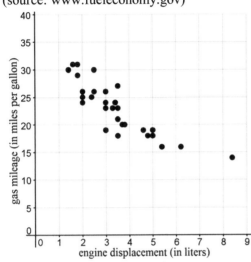

2. (source: www.infoplease.com)

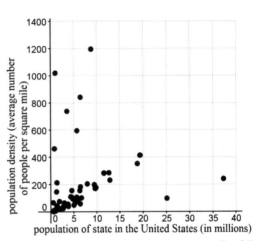

Note: Engine displacement is a measurement of engine size. It represents the total volume all of the pistons in an engine move through in one cycle.

3.

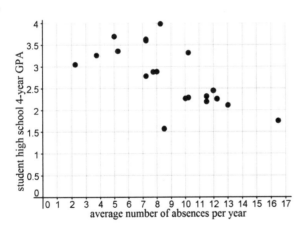

4.

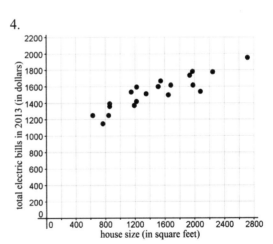

Module 11: Homework

© 2014 Carlson, O'Bryan & Joyner

Page 569

Algebra II

Module 11: Homework
Bivariate Statistics

In 2013, professional baseball player Justin Upton was traded from the Arizona Diamondbacks to the Atlanta Braves. His first season with the Braves started very well and a selection of his statistics for the first 23 games are given here. Use this information for Exercises #6-9.

Games Played	Number of At Bats	Total Runs Scored	Total Hits	Total Homeruns	Total Runs Batted Inn	Total Strike Outs	Season Batting Average
1	4	1	1	1	1	2	0.25
2	7	3	2	2	3	4	0.286
4	13	4	4	3	5	5	0.308
5	18	7	7	5	7	7	0.389
7	26	9	11	6	8	11	0.423
12	46	12	16	7	11	13	0.348
13	50	13	17	8	12	14	0.34
14	54	13	18	8	12	16	0.333
15	58	15	19	9	13	17	0.328
19	71	16	21	10	14	22	0.296
20	75	18	23	11	16	22	0.307
23	86	20	26	12	18	25	0.302

(*Source: www.mlb.com*)

6. Without making a graph, explain why comparing any two quantities from the table (except Season Batting Average) can never show a negative correlation relationship.

7. a. Plot Justin Upton's Total Homeruns (vertical axis) vs. Number of At Bats (horizontal axis) through his first 23 games.
   b. What general statement(s) can we make about the relationship between the quantities?
   c. Is there a (linear) correlation between the quantities? If so, is it positive or negative? Strong or weak?

8. a. Plot Justin Upton's Season Batting Average (vertical axis) vs. Games Played (horizontal axis) through his first 23 games.
   b. What general statement(s) can we make about the relationship between the quantities?
   c. Is there a (linear) correlation between the quantities? If so, is it positive or negative? Strong or weak?

9. If we compare Total Hits and Total Strikeouts for Justin Upton through his first 23 games we find a strong positive correlation. Explain why there is no "cause and effect" relationship. (*In other words, explain why it's correct to say the quantities are correlated but not to say there is a causal relationship.*)

10. Suppose you collect data on the number of snow shovels sold in various cities and the number of snow tires sold in the same cities.
    a. Imagine that, after plotting the points, you notice a positive correlation. Explain what this tells us about snow shovel sales and snow tire sales.
    b. Explain why we would not say that there is a causal relationship between snow shovel sales and snow tire sales. (*Hint: What other factor(s) might account for the relationship?*)

INVESTIGATION 2: EXPLORING LINES OF BEST FIT

11. Explain the purpose of drawing a line (or curve) of best on a scatterplot.

Algebra II
Module 11: Homework
Bivariate Statistics

12. Use the scatterplot in Exercise #1 from the homework.
    a. Sketch an approximate line of best fit.
    b. Estimate the slope of the line you drew and explain how to interpret this slope in context.
    c. Write the (approximate) formula for your line of best fit.

13. Use the scatterplot in Exercise #3 from the homework.
    a. Sketch an approximate line of best fit.
    b. Estimate the slope of the line you drew and explain how to interpret this slope in context.
    c. Write the (approximate) formula for your line of best fit.

14. Use the scatterplot in Exercise #4 from the homework.
    a. Sketch an approximate line of best fit.
    b. Estimate the slope of the line you drew and explain how to interpret this slope in context.
    c. Write the (approximate) formula for your line of best fit.

For Exercises #15-16, use the Justin Upton 2013 statistics for his first 23 games found in Homework Exercises #6-9.

15. a. Create a scatterplot showing Justin's Total Strike Outs (vertical axis) vs. Total Homeruns (horizontal axis) and sketch an approximate line of best fit.
    b. Estimate the slope of the line of best fit and explain what it represents in this context.

16. a. Create a scatterplot showing Justin's Total Runs Batted In (vertical axis) vs. Total Hits (horizontal axis) and sketch an approximate line of best fit.
    b. Estimate the slope of the line of best fit and explain what it represents in this context.

INVESTIGATION 3: RESIDUALS

Exercises #17-19 show three attempts to sketch a line of best fit for the same scatterplot. For each attempt, complete the table showing the predicted values based on the given formula, the residuals, the squared residuals, and the sum of the squared residuals.

17.

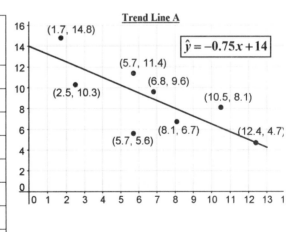

x-value	y-value	$\hat{y}$-value	residual $(y-\hat{y})$	squared residual $(y-\hat{y})^2$
1.7	14.8			
2.5	10.3			
5.7	5.6			
5.7	11.4			
6.8	9.6			
8.1	6.7			
10.5	8.1			
12.4	4.7			
			sum	

18.

x-value	y-value	$\hat{y}$-value	residual $(y-\hat{y})$	squared residual $(y-\hat{y})^2$
1.7	14.8			
2.5	10.3			
5.7	5.6			
5.7	11.4			
6.8	9.6			
8.1	6.7			
10.5	8.1			
12.4	4.7			
			sum	

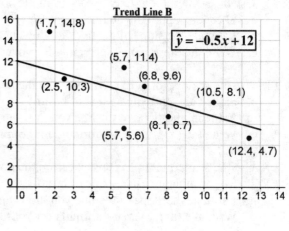

Trend Line B

$\hat{y} = -0.5x + 12$

19.

x-value	y-value	$\hat{y}$-value	residual $(y-\hat{y})$	squared residual $(y-\hat{y})^2$
1.7	14.8			
2.5	10.3			
5.7	5.6			
5.7	11.4			
6.8	9.6			
8.1	6.7			
10.5	8.1			
12.4	4.7			
			sum	

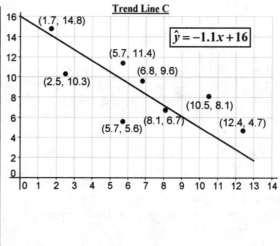

Trend Line C

$\hat{y} = -1.1x + 16$

20. Which of the attempts in Exercises #17-19 is the better fit? How do we know?

For Exercises #21-22, use the Justin Upton 2013 statistics for his first 23 games found in Homework Exercises #6-9.

21. The line of best fit relating Total Runs Batted In (y) and Number of At Bats (x) is $\hat{y} = 0.18x + 0.23$. Calculate the sum of the squared residuals.

22. The line of best fit relating Total Homeruns (y) and Total Hits (x) is $\hat{y} = 0.41x + 1.17$. Calculate the sum of the squared residuals.

Algebra II

Module 11: Homework
Bivariate Statistics

23. We are given two trend lines for the same scatterplot. How do we know that using Trend Line A will produce a smaller sum of squared residuals than using Trend Line B?

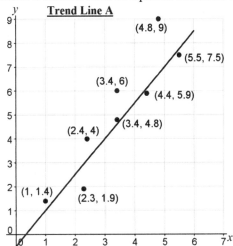
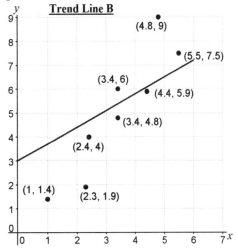

24. Compare the two given scatterplots and their lines of best. Which relationship will have the smallest sum of squared residuals? How do we know?

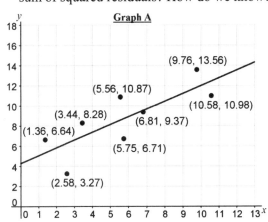
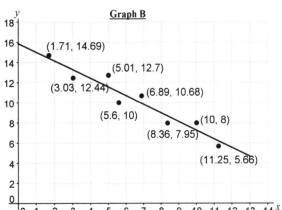

INVESTIGATION 4: LINEAR REGRESSION

25. Find the true line of best fit for the relationship given in Homework Exercises #17-19. In addition, record the r^2 and r values associated with the line of best fit.

26. Find the true line of best fit for the relationship given in Homework Exercise #23. In addition, record the r^2 and r values associated with the line of best fit.

27. Find the actual line of best fit for both relationships given in Homework Exercise #24. In addition, record the r^2 and r values associated with the lines of best fit.

Algebra II   **Module 11: Homework**
                                                                             Bivariate Statistics

28. The data set associated with Homework Exercise #1 is given. Find the line of best fit and record the r^2 and r values associated with the line of best fit.

engine displacement (liters)	gas mileage (mpg)	engine displacement (liters)	gas mileage (mpg)	engine displacement (liters)	gas mileage (mpg)	engine displacement (liters)	gas mileage (mpg)
1.4	30	2.4	25	3.4	24	4.8	18
1.6	31	2.5	32	3.5	30	5.0	19
1.8	29	2.5	26	3.5	23	5.0	18
1.8	31	3.0	24	3.5	18	5.4	16
2.0	25	3.0	26	3.5	21	6.2	16
2.0	24	3.0	19	3.5	18	6.2	16
2.0	26	3.0	23	3.7	20	6.2	16
2.0	25	3.0	19	3.8	20	8.4	15
2.4	25	3.3	23	4.6	19		

29. The data set associated with Homework Exercise #3 is given. Find the line of best fit and record the r^2 and r values associated with the line of best fit.

average absences per year	GPA	average absences per year	GPA	average absences per year	GPA	average absences per year	GPA
7.25	3.6	3.75	3.26	10	2.27	12.25	2.26
2.25	3.05	10.25	2.29	7.25	3.63	5	3.7
8	2.89	8.5	1.57	8.25	3.98	12	2.45
7.25	2.78	5.25	3.35	11.5	2.32	13	2.12
16.5	1.75	11.5	2.19	10.25	3.32	7.75	2.88

30. For any of the exercises you completed from #25-29 explain what the r^2 value tells us about the line of best fit.

31. Are all lines of best fit a useful tool for explaining the relationship between two quantities? Explain.

For Exercises #32-33, use the Justin Upton 2013 statistics for his first 23 games found in Homework Exercises #6-9.

32. a. Plot Justin Upton's Total Runs Batted In (vertical axis) vs. Total Homeruns (horizontal axis) through his first 23 games.
    b. Determine the regression (best fit) line for the data. Explain the meaning of the slope of the regression line.
    c. Record the r^2 value and explain its meaning.

33. a. Plot Justin Upton's Total Strikeouts (vertical axis) vs. Number of At Bats (horizontal axis) through his first 23 games.
    b. Determine the regression (best fit) line for the data. Explain the meaning of the slope of the regression line.
    c. Record the r^2 value and explain its meaning.

Algebra II

Module 11: Homework
Bivariate Statistics

INVESTIGATION 5: INTERPOLATION AND EXTRAPOLATION

34. In Homework Exercise #28 you calculated the regression function for the relationship shown in Homework Exercise #1. (If not, complete Homework Exercise #28 now.)
    a. Use your regression function to predict the gas mileage for a car if you know its engine displacement is 6 liters.
    b. Repeat part (a) for another new car with an engine displacement of 4.2 liters.
    c. Explain why your answers in parts (a) and (b) are considered *interpolations* and should be reasonable predictions.
    d. Explain why we should be careful when performing *extrapolations* using this regression function.

35. The best-fit line for the relationship in Homework Exercise #4 is $\hat{y} = 0.32x + 1054.63$ with $r^2 \approx 0.806$. [*Note that y values represent total electric bills in 2013 in dollars and x values represent house sizes in square feet.*]
    a. Use the regression function to predict the total electric bills in 2013 for a 2,250 square foot house.
    b. Repeat part (a) for a 910 square foot house.
    c. Explain why your answers in parts (a) and (b) are considered *interpolations* and should be reasonable predictions.
    d. Explain why we should be careful when performing *extrapolations* using this regression function.

For Exercises #36-37, use the Justin Upton 2013 statistics for his first 23 games found in Homework Exercises #6-9.

36. a. Calculate the regression function relating Justin's Total Homeruns (y) and Games Played (x).
    b. Justin played in 149 regular season games in 2013. Use your regression function from part (a) to predict how many homeruns he hit during the season.
    c. Calculate the regression function relating Justin's Total Homeruns (y) and Number of At Bats (x).
    d. Justin had 558 total at bats during the 2013 regular season. Use your regression function from part (c) to predict how many homeruns he hit during the season.
    e. Compare your answers in parts (b) and (d) and comment on the similarity or difference. In addition, compare these predictions to the Major League Baseball record for most homeruns hit in one season (73 by Barry Bonds).
    f. In the 2013 regular season Justin had a total of 27 homeruns. Comment on the dangers of extrapolating from a regression function.

37. a. Calculate the regression function relating Justin's Total Hits (y) and Games Played (x).
    b. Justin played in 149 regular season games in 2013. Use your regression function from part (a) to predict how many hits he had during the season.
    c. Calculate the regression function relating Justin's Total Hits (y) and Number of At Bats (x).
    d. Justin had 558 total at bats during the 2013 regular season. Use your regression function from part (c) to predict how many hits he had during the season.
    e. Compare your answers in parts (b) and (d) and comment on the similarity/difference.
    f. In the 2013 regular season Justin had a total of 147 hits. Comment on the dangers of extrapolating from a regression function.